ELECTRIC POWER PRINCIPLES

ELECTRIC POWER PRINCIPLES

Sources, Conversion, Distribution and Use

James L. Kirtley

Massachusetts Institute of Technology, USA

A John Wiley and Sons, Ltd., Publication

This edition first published 2010

Registered office
John Wiley & Sons Ltd, The Atrium, Southern Gate, Chichester, West Sussex, PO19 8SQ, United Kingdom

For details of our global editorial offices, for customer services and for information about how to apply for permission to reuse the copyright material in this book please see our website at www.wiley.com.

Library of Congress Cataloging-in-Publication Data

Kirtley, James L.
Electric power principles: sources, conversion, distribution and use / James L. Kirtley.
p. cm.
Includes index.
ISBN 978-0-470-68636-2 (cloth)
1. Electric power production. I. Title.
TK1001.K544 2010
621.3–dc22 2010010755

A catalogue record for this book is available from the British Library.

ISBN: 978-0-470-68636-2

Typeset in 10/12pt Times by Aptara Inc., New Delhi, India
Printed in Singapore by Markono Print Media Pte Ltd

2 2011

Contents

Preface

At the time of this writing (autumn of 2009), there appears to be a heightened awareness of the importance of energy to our social welfare. There are, in my opinion, two reasons for this. First, there is an awareness of the finite supply of fossil fuels stored beneath the surface of our planet, and that one day we will have to make do with sustainable sources of energy. The other is the fact that use of these fossil fuels releases carbon back into the atmosphere, leading to possible changes in heat transfer from the surface of the earth to space with attendant climate change. For both of these reasons the traditional methods for producing electric power may have to change, and this will mean the need for well educated, innovative engineers to build the power systems of the future.

In addition to the need for engineering of the electric power system itself is the plain fact that electric power, in the broad sense, is being used for a wider range of applications as time goes on. Virtually all rail transportation employs electric propulsion; hybrid electric automobiles have become important items of commerce and promise to become part of our energy future. Reduction of the need for energy (that is conservation) requires enhanced efficiency and effectiveness of the use of energy, and very often that involves the use of electricity.

The implications for education are clear: we in the academic world must educate engineers to be the leaders in designing, building and operating new types of electric power systems. Perhaps even more important, we must also educate a broader class of students who will become leaders in the industrial and political realms to understand at least the rudiments and implications of energy, including electric power.

This book is the descendant of sets of lecture notes that I have used in two subjects at the Massachusetts Institute of Technology: *6.061, Introduction to Electric Power Systems* and *6.685, Electric Machines*. These notes have existed in various forms for about three decades, and they have benefited from the experience of being used by multiple generations of MIT undergraduate and graduate students.

It is my hope that this book be used by students who want to gain a broad understanding of how electric power is generated, transmitted, distributed and used. Thus there is material here beyond the traditional electric power system and electric machinery disciplines. That said, this book does have chapters that discuss some of the traditional material of electric power systems: per-unit normalizations, symmetrical components and iterative load flow calculations. In keeping with my feeling that fundamental understanding is important, I have included chapters on the principles of electromechanical energy conversion and on magnetic circuits. To round out the power systems story is a fairly extensive chapter on synchronous machines, which are still the most important generators of electric power. There are also short discussions of the different types of power plants, including both traditional plants and those used for extracting sustainable energy from wind and sun, and topics important to the power system: protection and DC transmission. On the usage side there is a chapter on power electronics and chapters on the major classes of electric motors: induction and direct current. MATLAB is included, and each of the chapters is accompanied by some problems with a fairly wide range of difficulty, from quite easy to fairly challenging.

The material in this book should be accessible to an undergraduate electrical engineering student at the third year level. I have assumed the reader of this book will have a basic but solid background in the fundamentals of electrical engineering, including an understanding of multivariable calculus and basic differential equations, a basic understanding of electric circuit theory and an understanding of Maxwell's equations.

This book could be used for subjects with a number of different areas of emphasis. A 'first course' in electric power systems might use Chapters 1 through 4, 6, 7, 10 and 11. Chapter 7 has an appendix on transmission line inductance parameters that can probably be safely skipped in an introductory subject.

Chapter 9 is about synchronous machines and instructors of many power systems subjects would want to address this subject. Chapter 12 is an introduction to power electronics and this, too, might be considered for a course in power systems.

A 'first course' that deals primarily with electric machines could be taught from Chapters 4, 5, 8, 9 and 12 to 15. Tutors can find solutions for end-of-chapter problems at www.wiley.com/go/kirtley_electric.

The number of students who have influenced, hopefully for the better, the subject material in this book is so large that there would be no hope in calling them all out. However, I must acknowledge a few of the people who have taught me this material and influenced my professional career. These include Herb Woodson, Jim Melcher, Gerald Wilson, Alex Kusko, Joe Smith, Charles Kingsley, Steve Umans and Steve Leeb.

I would also like to thank Steve Sprague of the Electric Motor Education and Research Foundation for the electrical sheet steel data graphics.

1

Electric Power Systems

There are many different types of power systems, such as the propulsion systems in automobiles and trucks and the hydraulic systems used in some industrial robots and for actuating scoops and blades in digging equipment. All power systems have certain fundamental elements. There is some sort of prime mover (such as a gasoline engine), a means of transport of the power produced (such as the drive shaft, transmission, differential and axles), and a means of using that power (wheels on the road). The focus of this book is on *electric* power systems, in which the means of transporting energy is the flow of electrical current against an electric potential (voltage). There are many different types of electric power systems as well, including the electrical systems in cars and trucks, propulsion systems in electric trains and cruise ships. The primary focus in this book will be the kinds of electric power systems incorporated in public utilities, but it must be kept in mind that all electric power systems have many features in common. Thus the lessons learned here will have applicability well beyond the utility system.

It has become all too easy to take for granted the electric utility service that is ubiquitous in the developed countries. Electric utilities are wired to nearly every business and residence, and standardized levels of voltage and frequency permit a wide range of appliances to be simply 'plugged in' and operated. Consumers don't have to give any thought to whether or not an appliance such as a television set, a computer or an egg beater will work. Not only will these appliances work when plugged in, but the electric power to make them work is quite reliable and cheap. In fact, the absence of useful electric power is quite rare in the developed countries in the world. Widespread failure to deliver electric power has become known as a 'blackout', and such events are rare enough to make the nightly news across the country. Even substantial distribution system failures due to weather are newsworthy events and very often cause substantial hardship, because we have all come to depend on electric power to not only keep the lights on, but also to control heating, cooling, cooking and refrigeration systems in our homes and businesses.

At the time of this writing, electric power systems in the United States and most of the developing world use as their primary sources of energy fossil fuels (coal and natural gas), falling water (hydroelectric power), and heat from nuclear fission. There are small but rapidly growing amounts of electric power generated from wind and solar sources and some

Electric Power Principles: Sources, Conversion, Distribution and Use James L. Kirtley

electric power is generated from volcanic heat (geothermal energy). These 'renewables' are expected to grow in importance in the future, as the environmental impacts of the use of fossil fuels become more noticeable and as the fossil fuels themselves are exhausted. There are some differences between technologies involved in the older, existing power generation sources and the newer, sustainable technologies, and so in this book we will discuss not only how existing utility systems work but also how the emerging technologies are expected to function.

1.1 Electric Utility Systems

A very 'cartoon-ish' drawing of a simple power system is shown in Figure 1.1.

Electric power originates in 'power plants'. It is transmitted by 'transmission lines' from the power plants to the loads. Along the way the voltage is first stepped up by transformers, generally within the power plants, from a level that is practical for the generators to a level that provides adequate efficiency for long-distance transmission. Then, near the loads the electric power is stepped down, also by transformers, to a voltage useable by the customer. This picture is actually quite simplified. In modern utility systems there are thousands of power plants connected together through networks, and many more connections to loads than are indicated in Figure 1.1. The connections to actual loads is usually a bit more like what is shown in Figure 1.2. At the distribution level the connection is 'radial', in that there is one connection from the source of electric power (the 'grid'), and that is broken down into many load connections. Usually the distribution primary line is at a voltage level intermediate between the transmission level and the voltage that is actually used by customers.

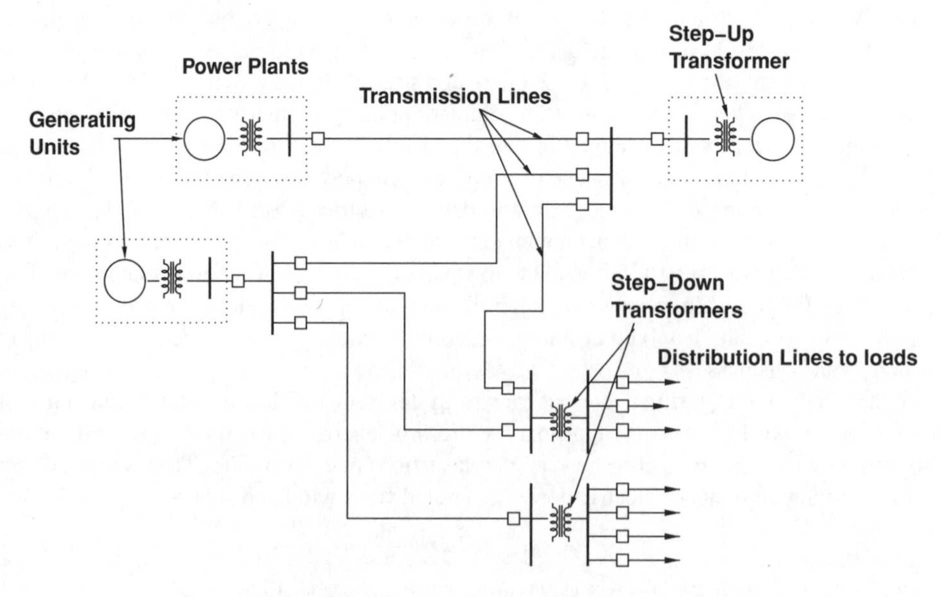

Figure 1.1 Cartoon of a simple power system

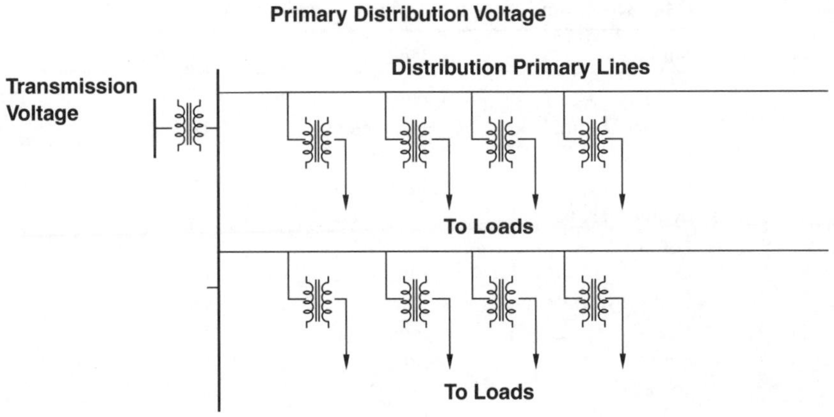

Figure 1.2 Distribution circuits

1.2 Energy and Power

1.2.1 Basics and Units

Before starting to talk about electric power systems it is important to understand some of the basics of energy and power. In the international system of units (SI), there are two basic units of energy. One is the joule (J), which is the energy expended by pushing a newton (N), a unit of force, over one meter. So a joule is a newton-meter. (A kilogram 'weighs' about 9.8 newtons at the surface of the earth). The other unit of energy is related to heat, and it is the Calorie. This story is complicated by the fact that there are actually two definitions of the Calorie. One is the heat (amount of energy) required to heat 1 gram of water 1 degree Celsius. This amounts to about 4.184 joules. The second definition is often called the 'kilogram Calorie', the amount of energy required to heat 1 kilogram of water 1 degree Celsius. This is obviously just 1,000 of the 'gram Calories', or 4,184 joules.

The basic unit of power is the watt, which is one joule/second. As it our predecessors crafted it, 1 watt is also 1 volt $\times 1$ ampere. The volt is a unit of electrical potential and the ampere is a unit of current flow. Power is expressed in watts, kilowatts, etc., and a basic unit of energy is the kilowatt-hour (kWh), (3.6×10^6 J). Electricity is sold at retail by the kilowatt-hour and, usually, at wholesale by the megawatt-hour.

Another unit of heat that is commonly used in discussing power plants is the British Thermal Unit (BTU), which is the amount of heat required to raise 1 pound of water 1 degree Fahrenheit. This is about 0.252 kilogram calories or 1054 joules. In the United States, fuels are often sold based on their energy content as measured in BTUs, or more commonly in millions of BTU's (MBTU). See Tables 1.1 and 1.2.

1.3 Sources of Electric Power

There are two basic ways in which electric power is produced: by generators turned by some sort of 'prime mover' or by direct conversion from a primary source such as sunlight, or conversion of chemical energy in fuel cells. The prime movers that turn generators can be heat

Table 1.1 Some of the unit symbols used in this book

Unit	Unit of	Symbol
Ampere	current	A
British Thermal Unit	heat energy	BTU
Coulomb	charge	C
Calorie	heat energy	Cal
degree Celsius	temperature	°C
Farad	capacitance	F
Gauss	flux density	G
Hertz (cycles/second)	frequency	Hz
Henry	inductance	H
hour	time	h
Joule	energy	J
Kelvin	temperature	K
kilogram	mass	kg
meter	length	m
Newton	force	N
volt	electric potential	V
volt-ampere	apparent power	VA
watt	power	W
Weber	flux	Wb

engines such as steam turbines, gas turbines, internal combustion engines burning diesel fuel, natural gas or (rarely) gasoline, or turbines that convert power directly from falling water or wind. Geothermal heat is sometimes used to power heat engines in places where that heat is accessible (this is the major source of electric power in Iceland). Even sunlight has been used as the power input to heat engines.

1.3.1 Heat Engines

Most power plant 'prime movers' are heat engines that burn a primary fuel such as coal or natural gas and that use the energy released by combustion to produce mechanical power (generally turning a shaft) that is used to drive a generator to produce electrical power. We

Table 1.2 Multiplying prefixes used in this book

Prefix	Symbol	Multiple
tera	T	10^{12}
giga	G	10^{9}
mega	M	10^{6}
kilo	k	10^{3}
centi	c	10^{-2}
milli	m	10^{-3}
micro	μ	10^{-6}
nano	n	10^{-9}

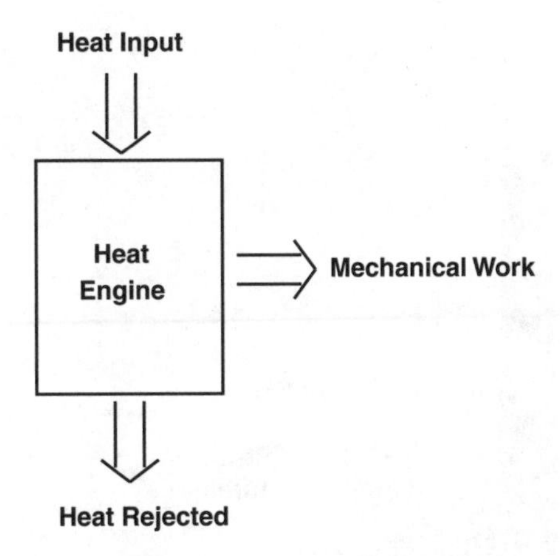

Figure 1.3 Energy balance

will, in later chapters of this book describe how generators work. Heat engines can convert only some of the heat energy that is input to the engines into mechanical work. The details of this are beyond our scope here, but as is shown in Figure 1.3, there will always be waste heat associated with a heat engine. Heat engines take energy at a high temperature and reject heat energy at a lower temperature. The difference between the heat input and rejected heat energy is what is converted to mechanical power, and efficiency is the ratio of mechanical power output to heat power input.

There is a well known bound on efficiency of a heat engine, called the 'Carnot efficiency', and that is associated with the temperature of the input heat and the temperature of the rejected heat. This is:

$$W_{\mathrm{m}} < Q_{\mathrm{h}} \frac{T_{\mathrm{h}} - T_{\ell}}{T_{\mathrm{h}}}$$

where Q_{h} is the input energy. Mechanical work is the difference between heat input and heat rejected, and the efficiency depends on the heat input temperature T_{h} and heat rejection temperature T_{ℓ}. Practical heat engines do not approach this Carnot limit very closely, but this expression is a guide to heat engine efficiency: generally higher heat input temperatures and lower heat rejection temperatures lead to more efficient heat engines.

In discussing power plant efficiency, we often note that one kilowatt-hour is 3.6 MJ or 3,414 BTU. The fuel energy input to a power plant to produce one kilowatt hour is referred to as its 'heat rate', and this is inversely related to its thermal efficiency. A power plant that has a heat rate of, say, 10,000 BTU/kWh would have a net thermal efficiency of $\eta = \frac{3414}{10000} \approx 0.3414$.

1.3.2 Power Plants

Figure 1.4 shows a cartoon of a power plant that burns fossil fuels. The heat engine in this case is a steam turbine. Water is first compressed and pumped into a 'boiler', where a fire heats it into steam. The steam is expanded through a turbine which turns a generator. The turbine exhaust is then fed to a 'condenser' where the waste heat is rejected. There are several

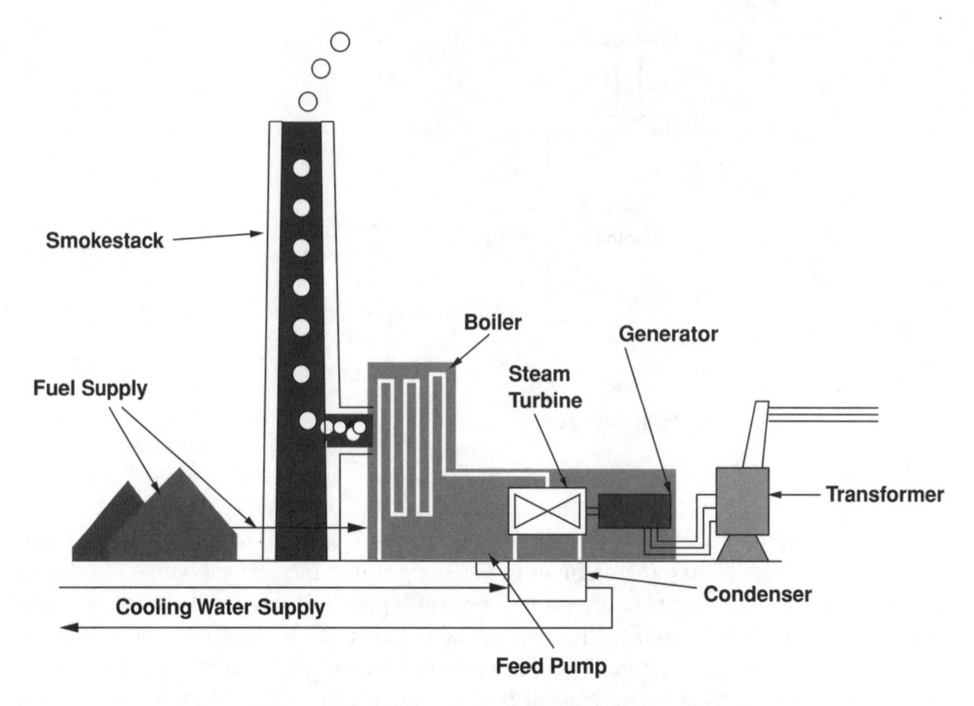

Figure 1.4 Cartoon of a fossil fired power plant

different recipients of the rejected heat, depending on the situation: rivers, lakes, the ocean, purpose built cooling ponds or cooling towers are all used. Generated electricity is generated at 'medium' voltage ('medium voltage' is generally taken to be between 1 kV and 100 kV, but power plant generators are generally limited to about 30 kV) and is usually stepped up to 'high' (100 to 230 kV) or 'extra high' (230 to 800 kV) voltage for transmission.

While Figure 1.4 shows a coal-fired power plant, similar steam turbine-based power plants can burn any of the fossil fuels, wood or even municipal garbage, and often such plants are built in such a way that they can burn different fuels, based on which fuel is cheapest at a given time.

There are also power plants that employ gas turbines, as opposed to steam turbines, or even some power plants that have gas turbine engines on the same shaft as steam turbine engines. The 'simple cycle' gas turbine engines are based on the same technology as jet engines that power aircraft ('aero derivative'). 'Binary cycle' power plants use a gas turbine engine with the exhaust gas rejecting heat to a steam cycle and can achieve higher efficiency than simple cycle gas- or steam- turbine engines, but with a higher level of complexity.

1.3.2.1 Environmental Impact of Burning Fossil Fuels

Fuels such as coal often have contaminants such as sulfur or mercury that have adverse environmental effects, and there has been, in recent years, substantial development of methods to mitigate those effects.

Table 1.3 Carbon analysis of hypothetical power plants

Fossil fuel carbon analysis	Coal	No. 6 Fuel Oil	Natural Gas
Fraction carbon	0.807	0.857	0.750
Fraction hydrogen	0.045	0.105	0.250
HHV (BTU/kg)	30 870	40 263	50 780
HHV (kJ/kg)	32 573	42 438	53 522
kg CO_2/kg fuel	2.959	3.142	2.750
kg CO_2/MBTU	95.9	78.0	54.2
kg CO_2/kWh (at 10 000 BTU/kWh)	0.959	0.780	0.542
kg CO_2/hour (at 1000 MW)	323 939	248 365	196 928
kg CO_2/hour	958 536	780 446	541 552

Mercury, for example, is toxic in surprisingly small quantities. When coal containing trace amounts of mercury is burned, the mercury is released in the effluent gas and/or 'fly ash' (solids in the effluent gas) and then gets into some food chains such as fish. As big fish eat small fish the mercury is concentrated. As fish are generally fish eaters this process is repeated until fish near the top of the food chain are caught by the carnivores at the top of the food chain (people). Sometimes toxic levels of mercury are present in those big fish.

Sulfur oxides and nitrogen oxides, the result of oxidation of nitrogen in the air, are the stuff of 'acid rain'. There are different oxidation states of both nitrogen and sulfur, so that this type of pollution is often referred to as 'SO_X and NO_X'. Not only do these chemicals produce acid rain, but they can (and do) react with hydrocarbons present in the air to form a visible haze that is often referred to as 'smog'. Methods of mitigating these pollutants have been developed but are beyond our scope here.

Fossil fuels generally contain carbon and hydrogen (which is why they are called 'hydrocarbons', and the chief effluents of power plants are water vapor and carbon dioxide. The latter is a 'greenhouse' gas, and while it appears naturally in the atmosphere of the Earth, there are indications that man-made injections of carbon dioxide are raising the levels of CO_2, with possible impacts on the earth's heat balance ('global warming'). For this reason, it seems important to understand the carbon content of fuels.

Table 1.3 shows a simple analysis of carbon effluent for a 1,000 MW power plant assuming a 'heat rate' of 10,000 BTU/kWh. It should be noted that this heat rate, while it is within the range of numbers actually encountered, is not necessarily typical for any particular plant. The fuels assumed in Table 1.3 are bituminous coal, heavy fuel oil (# 6 is what comes out near the bottom of the refinery distillation column) and natural gas. It should also be noted that these numbers are roughly correct, but that all of these fuels come with ranges of the various quantities. For example, the energy content of bituminous coal varies between about 23 000 and about 31,000 BTU/kg. Natural gas is primarily methane, which is 75% carbon and 25% hydrogen, but most sources of natural gas have some heavier components (ethane, propane, butane, etc.). Note that coal, which also can have varying fractions of carbon and hydrogen, has some non-combustible components (water, inorganic solids) as does fuel oil. The 'higher heating value' (HHV) for these fuels assumes that all of the heat released when the fuel is burned can be used, including the heat of vaporization of water that is produced when the hydrogen is combined with oxygen. This is often not the case, and the 'lower heating value' is somewhat less.

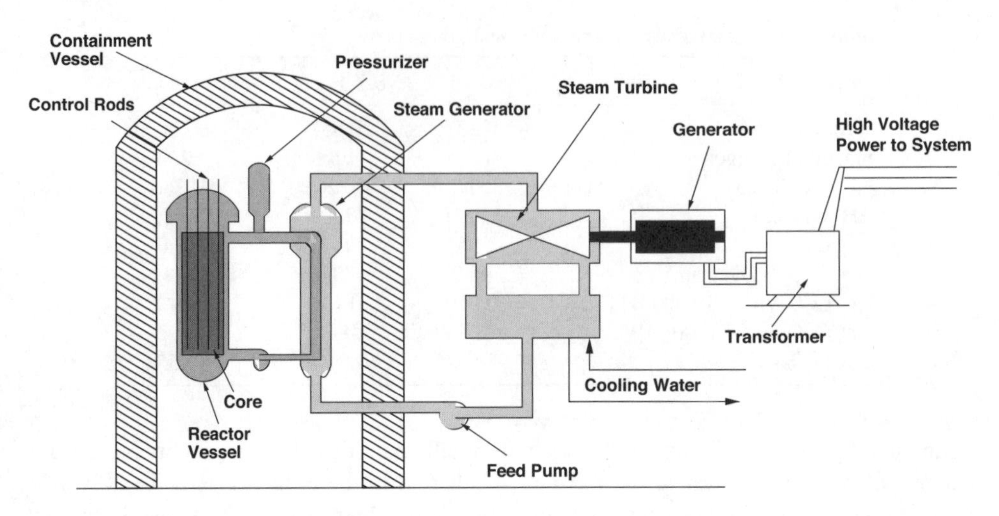

Figure 1.5 Cartoon of a nuclear power plant

Note that the amount of carbon dioxide produced when burning natural gas is substantially smaller, per unit of energy produced, than when coal or fuel oil is burned, and for that reason natural gas is sometimes thought of as a 'cleaner' fuel.

1.3.3 Nuclear Power Plants

Nuclear power plants employ the same thermodynamic cycle as most fossil fueled plants. Because of the relatively difficult environment for the materials that carry high-pressure water (it is radioactive in there), the high end temperature of a nuclear power plant cannot be as high as it can in fossil-fueled plants and so thermal efficiency tends to be a bit lower.

The reactor in a nuclear power plant generates heat through fission of heavy atoms into two (or more) lighter atoms. When an atom of uranium (U^{235}), the isotope of uranium that is capable of fission, splits, about 1/5 of an atomic mass unit (AMU) is converted to energy. Since the mass fraction of U^{235} in natural uranium is about 0.7%, were all of the fissile isotope to be converted to energy, a fraction amounting to about $\frac{0.2}{235} \times 0.007 \approx 6.1 \times 10^{-6}$ of the natural uranium would be converted to energy. That turns out to be quite a lot of energy, however, because $E = MC^2 = M \times 9 \times 10^{16}$ J/kg, or one kilogram of natural uranium would yield about $6.1 \times 10^{-6} \times 9 \times 10^{16} \approx 5.5 \times 10^{11}\text{J} \approx 1.53 \times 10^5\text{kWh}$. If the plant operates with a thermal efficiency of 33%, That would mean about 51,000 kWh/kg of natural uranium. This compares with perhaps 3 or 4 kWh/kg for coal.

Virtually all commercial nuclear power plants are 'light water' moderated (LWR) and are either of the 'pressurized water' or 'boiling water' type. Figure 1.5 is a cartoon sketch of a pressurized water reactor type power plant. Moderation here means reducing the energy of the neutrons that are emitted from fissioning nuclei to the level that is best for initiating fissioning of other nuclei. When a nucleus of uranium splits, it emits, among other things, a few 'fast' neutrons. These fast neutrons, while they can convert U^{238} to plutonium, are not very effective at inducing fission in U^{235}. Passing through the water that surrounds the fuel rods, the neutrons are slowed down, giving up energy and becoming 'thermal neutrons' (about 0.025 eV), to the point where they are effective in inducing fission. In fact, since slower neutrons are more effective in inducing fission, there is a negative reactivity coefficient with temperature that tends to stabilize the chain reaction. Further control is afforded by the 'control rods' that absorb neutrons. Dropping the rods fully into the reactor stops the chain reaction. The plutonium produced by fast neutrons interacting with U^{238} includes a fissile isotope that is subsequently fissioned and this contributes more to the energy produced.

There is no carbon emitted by nuclear power plants in normal operation. The byproducts of the nuclear reaction, however, are really nasty stuff: lethally radioactive, hot and poisonous. Fortunately there is not a great deal of spent fuel produced and so it can be (and is) simply contained. There is still much public debate about what to do with spent fuel and development of techniques for processing it or for stabilizing it so that it can be stored safely. Of particular interest is the fact that the plutonium present in spent fuel can be used to make nuclear explosives. The plutonium can be separated chemically, whereas fissile U^{235} cannot. This is both good and bad news: good because plutonium is a useful fuel that can be mixed in with uranium and burned in reactors; bad because it facilitates fabrication of nuclear explosives, making securing spent fuel from potential terrorists or failed states very important, an added expense of the nuclear fuel cycle.

At the time of this writing (2009), there were 104 nuclear power plants in the United States, producing about 20% of the electric energy used in the country.

1.3.4 Hydroelectric Power

Hydroelectric power plants take advantage of falling water: under the influence of gravity, water descending through a pipe exerts force on a turbine wheel which, in turn, causes a generator to rotate. Figure 1.6 shows a cartoon style cutaway of a hydroelectric unit (or 'waterwheel'). For hydrodynamic reasons these units tend to turn relatively slowly (several tens to a few hundred r.p.m.), and can be physically quite large.

Power generated by a waterwheel unit is:

$$P = \rho_{\text{water}} g h \dot{v} \eta_t$$

where ρ_{water} is mass density of water (1000kg/m^3, $\boldsymbol{g}$ is acceleration due to gravity (about 9.812m/s^2), h is the 'head' or height the water falls, $\dot{v}$ is volume flow of water and η_t is efficiency of the turbine system.

Hydroelectric power plants, even though they produce a relatively small fraction of total generation, are very important because their reservoirs provide energy storage and their generation can be modulated to supply power for variations in load over time. In fact, some 'pumped hydro' plants have been built solely for storage. Two reservoirs are established at different elevations. The hydroelectric generators are built so they can serve not only as generators but also as pumps. When electric power is in surplus (or cheap), water is pumped 'uphill' into the upper reservoir. Then, when electric power is in short supply (expensive), water is allowed to flow out of the upper reservoir to provide for extra generation. The lower reservoir is often a river and the upper reservoir might be formed by hollowing out the top of a mountain.

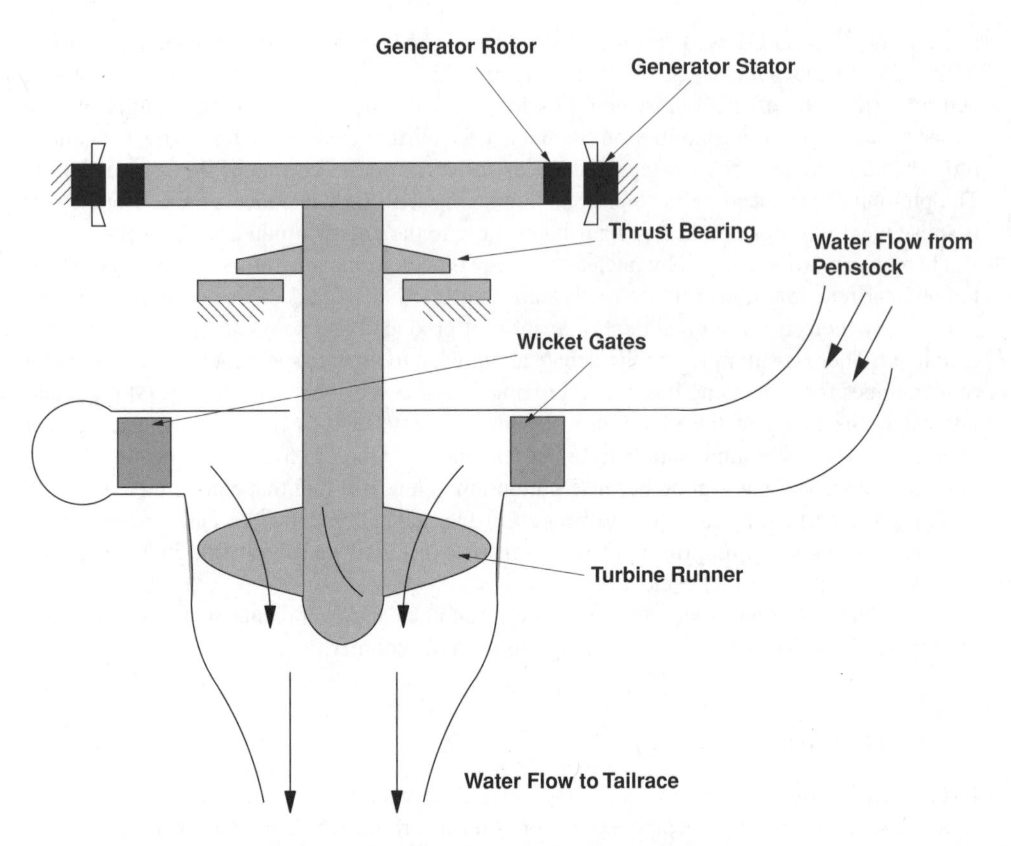

Figure 1.6 Cartoon of a hydroelectric generating unit

Hydroelectric power generation is the oldest and largest source of sustainable electric power, but other renewable sources are emerging.

1.3.5 Wind Turbines

Among the emerging 'sustainable' sources of electric power, wind is both the largest and fastest growing. Figure 1.7 shows a view of a wind farm, with a number of 1.5 MW wind turbines. These units have a nearly horizontal axis with a blade disk diameter of about 77 m and 'hub height' of 65 to 100 m, depending on site details.

Power generated by a wind turbine is, approximately:

$$P = \frac{1}{2} C_{\mathrm{p}} \rho_{\mathrm{air}} A u^3$$

where ρ is air density (about $1.2\mathrm{kg/m^3}$) and u is air velocity, so that $\frac{1}{2}\rho u^2$ is kinetic energy density of wind entering the disk of area A and C_{p} is the 'power coefficient', a characteristic of wind speed, rotor angular velocity and blade pitch angle. It has a theoretical maximum value of about 59% but as a practical matter usually does not exceed about 50%. Because

Figure 1.7 Turbine top view of the Klondike wind farm in Oregon, USA. Photo by Author

this coefficient is a function of wind and rotor tip speed (actually of the advance angle), wind turbines work best if the rotational speed of the rotor is allowed to vary with wind speed. More will be said about this in the discussion the kinds of machines used for generators, but the variable speed, constant frequency (VSCF) machines used for wind generators are among the most sophisticated of electric power generators. They start generating with wind speeds of about 3 m/s, generate power with a roughly cubic characteristic with respect to wind speed until they reach maximum generating capacity at 11–13 m/s, depending on details of the wind turbine itself, and then, using pitch control, maintain constant rotational speed and generated power constant until the wind becomes too strong, at which point the turbine must be shut down. This 'cut out' speed may be on the order of 30 m/s.

A cartoon showing the major elements of a wind turbine is shown in Figure 1.8. Turbine blades are mounted to a nose cone that contains pitch adjusters to control speed. The relatively low turbine speeds are increased by a factor of perhaps 80 by a gear box, usually made up of one planetary and two bull gear and pinion gear stages. The generator is often a doubly fed induction generator: a wound rotor induction machine with a cascade of power electronics to couple the rotor windings with the stator windings and local power system and to provide constant frequency, variable speed capabilities.

The wind turbine is mounted on a tower that is usually implemented as simple steel tube, on the order of 65 to 100 m in height. The nacelle is mounted on a yaw mechanism to point the turbine at the wind. Both the yaw mechanism and the main turbine blades have braking mechanisms (not shown in the figure). In some wind turbines there is a transformer in the nacelle to couple the low voltage of the generator to the medium voltage used to carry electric power from the wind turbines to the point of common contact with the utility system (POCC).

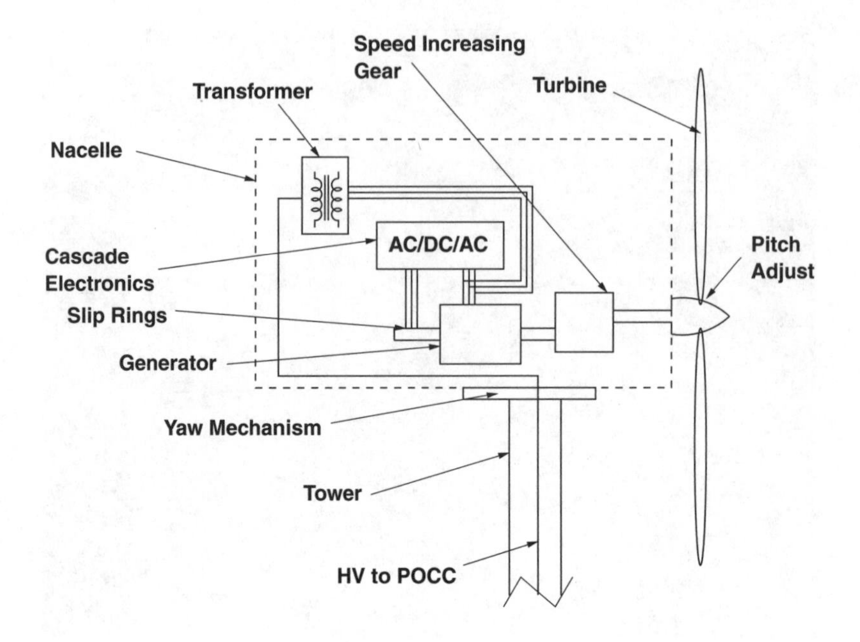

Figure 1.8 Wind turbine components

1.3.6 Solar Power Generation

Generation of electric power is another source of energy that is very small but growing rapidly. Radiation from the sun, in the visible and infrared, amounts to about 1 kW per square meter in the vicinity of the earth. Were it possible to economically capture all of this energy we would not be considering any of the other means of power generation. It is, of course, not possible for a variety of reasons:

1. The atmosphere captures and scatters some of the solar radiation, which is why the sky is blue and sunsets are red. This effect is stronger in latitudes away from the equator.
2. Because the earth turns, half the time the sun is not visible at all. And for much of the day the sun is near or not very far from the horizon. Solar arrays that track the sun are expensive and complex, with moving parts that must be maintained. Solar arrays that do not track the sun absorb less energy than is available.
3. Clouds interfere with solar radiation in most parts of the earth, and surfaces of solar arrays can be fouled by dust and other crud that falls from the sky.
4. Existing technologies for conversion of solar radiation into electricity are, currently, expensive relative to other sources.

There are two principal means of solar generation of electricity. One employs heat engines similar to fossil fuel or nuclear generation, using sunlight to heat the top end of the heat engine cycle. Often this is in the form of a 'solar tower', with the element to be heated at the focus of a lot of mirrors. The operational issues with this sort of a system are chiefly associated with

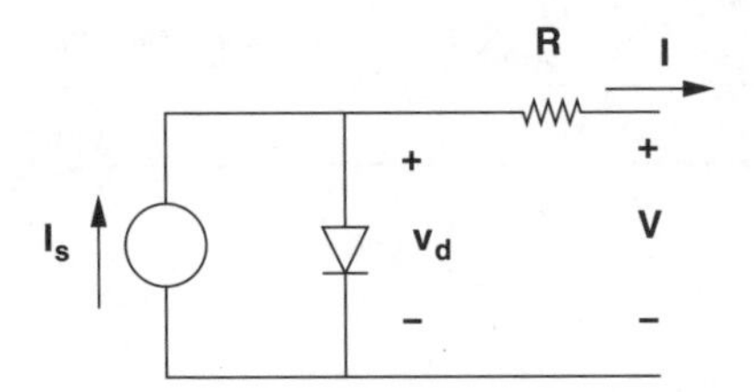

Figure 1.9 Equivalent circuit model of a solar cell

tracking and focusing the sunlight on the top element. The method of operation of the power plant is similar to that of any other heat engine.

The second means of generating electricity from sunlight employs photovoltaic cells. These are large area junction diodes that, when sunlight shines on them and splits electron/hole pairs, produce a current. The cost and efficiency of these cells are not very favorable at the present time, although for certain applications such as powering space stations (where solar energy is more abundant than it is on the surface and where other fuels are very expensive) or powering remote, low power services would otherwise be very expensive, they are the power source of choice. There has been and continues to be substantial development of solar cells and it is to be anticipated that cost and performance will continue to improve.

An equivalent circuit model of a solar array is shown in Figure 1.9. The source current I_s is the result of absorption of photons in sunlight that cause separation of valence electrons from their atoms. The resulting hole/electron pairs fall across the high field gradients present at the diode junction. Because any voltage resulting from this current tends to forward bias the actual junction, the voltage available is limited. One can readily deduce that the cell current is:

$$I = I_s - I_0 \left(e^{\frac{v_d}{\frac{kT}{q}}} - 1 \right)$$

Here, I_s depends on the strength of solar radiation actually reaching the junction and on how strongly it is absorbed and on the junction area. I_0 also depends on junction area and on how the cell was constructed. The voltage $\frac{kT}{q}$ is about 25 mV at room temperature. A representative curve of output current vs. voltage is shown in Figure 1.10.

One aspect of solar cell generation requires some discussion. The characteristic curve of current vs. voltage depends on both solar irradiance and temperature. When shorted, the panel produces a certain current. When open it produces a certain voltage. At both extremes the panel produces no power. The output is maximum somewhere in the middle. The trouble is that the maximum power point as shown in Figure 1.11 is a function of both temperature and radiation, so there is no simple way of loading the cells to get the maximum amount of power from them. The problem of maximum power point tracking (MPPT) has become an item of competitive art among manufacturers of solar cells and the electronics that go with them.

Large solar photovoltaic systems intended for connection with the utility network must employ electronic systems that absorb electric power from the cells, implement maximum power point tracking, and then convert the resulting DC power into utility frequency AC, single-phase (for small systems) or polyphase (for larger systems). The inverter systems involved will be covered later in this text.

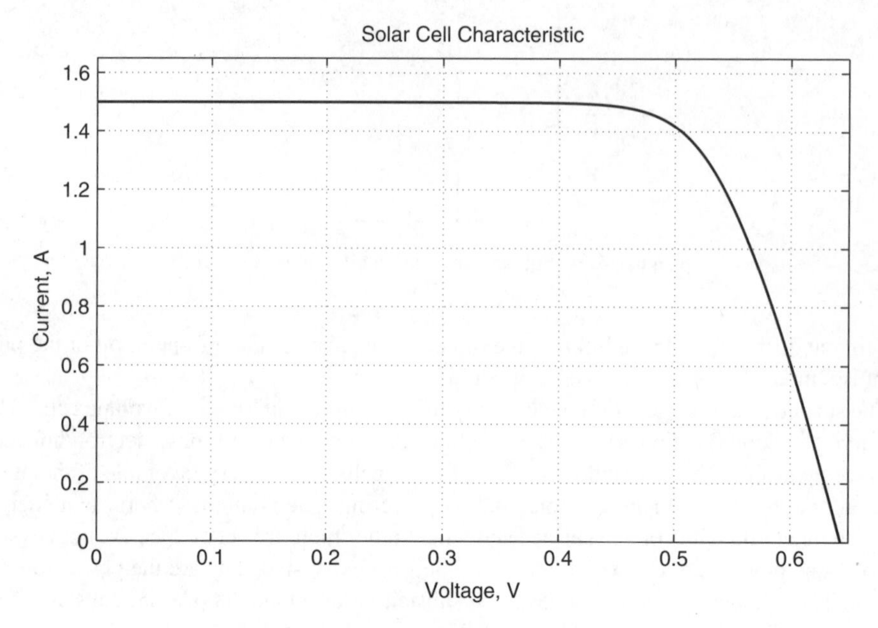

Figure 1.10 Output current vs. voltage

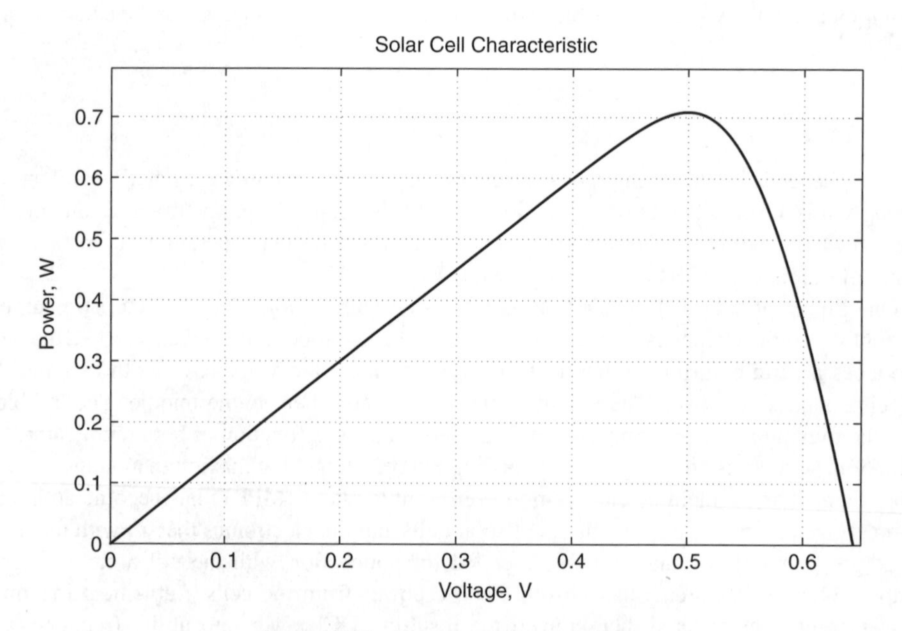

Figure 1.11 Output power vs. voltage

Table 1.4 Fraction of capacity and energy produced

	Generating capacity	Supplied energy
Coal	30.5%	48.5%
Natural gas	40.9%	21.6%
Nuclear	9.9%	19.4%
Conventional hydroelectric	7.5%	6.0%
Petroleum	5.9%	1.6 %
Wood	0.7%	0.9%
Wind	1.6%	0.8%
Other biomass	0.4%	0.4 %
Geothermal	0.2 %	0.4 %
Other gases	0.2 %	0.3 %
Solar	0.04%	0.01%

Source: United States Energy Information Administration, Electric Power Annual 2007.

1.4 Electric Power Plants and Generation

According to the United States Energy Information Administration, at the end of 2007 the country had 17,342 generating facilities with a total capacity of between 995,000 and 1,032,000 MW, depending on season. (Note: heat engines have higher capacity in cold weather.) In 2007 those plants produced a total of 4,156,745 GWh of electrical energy. Table 1.4 shows a breakdown of the fraction of generating capacity and generated electric energy represented by each source technology in the United States in that year.

The differences in fractions of capacity and energy generated are related to economics: nuclear and coal plants are expensive to build but cheap to run; natural gas plants are just the opposite. The share of renewables is very small, but it is growing fast.

Electric power is a big business that has come to have a profound impact on the lives of everyone living in the industrialized world. In the following chapters we will describe generation, transmission, distribution, handling and, to some extent, use of electric power.

1.5 Problems

1. Your household electrical system has a circuit that is single phase and employs a voltage of 240 V, RMS. What can a circuit with a 50 A breaker handle?
   - In Watts?
   - A heater, rated in British Thermal Units/hour.
2. What is the 'heat rate' (BTU/kWh) of a power plant with a net thermal efficiency of 50%?
3. Using the data of Table 1.3, what is the amount of coal required for a power plant with a heat rate of 11,000 BTU/kWh to produce 1000 MW for a year?
4. What is the carbon dioxide emission rate of a coal fired power plant with a heat rate of 9,500 BTU/kWh:
   - Per hour if the rating of the plant is 600 MW?

- Per kWh?

Use the data contained in Table 1.3.

5. What is the carbon dioxide emission rate of a natural gas fired power plant with a thermal efficiency of 53%?
   - Per hour if the rating of the plant is 600 MW?
   - Per kWh?

   Use the data contained in Table 1.3.
6. A nuclear power plant 'burns' Uranium enriched to about 4% U^{235}, the fissile isotope. If this plant achieves a 'burnup' of 50% (that is, it converts half of the fissile component of the fuel), how much enriched uranium is required for the plant to make 1000 MW for a year? Assume a heat rate of 12,000 BTU/kWh.
7. Assume the density of air to be $1.2 \mathrm{kg/m^3}$. What diameter wind turbine is required to capture 1.5 MW at a wind speed of 10 m/s if the turbine coefficient of performance is 40%?
8. What is the water volume flow rate for a 100 MW water turbine operating with a 'head' of 20 meters, assuming an efficiency of the turbine and generator of 80%? (Water has a mass density of $1000 \mathrm{kg/m^3}$).

2

AC Voltage, Current and Power

The basic quantities in electric power systems are voltage and current. Voltage is also called, suggestively, 'electromotive force'. It is the pressure that forces electrons to move. Current is, of course, that flow of electrons. As with other descriptions of other types of power, electric power is that force (voltage) pushing on the flow (current). In order to understand electric power, one must first solve the circuit problems associated with flow of current in response to voltage.

Most electric power systems, including all electric utility systems, employ alternating current. Voltages and currents very closely approximate to sine waves. Thus to understand the circuit issues it is necessary to prepare to analyze systems with sinusoidal voltages and currents. Robust and relatively easy to use methods for handling sinusoidal quantities have been developed, and this is the subject material for this chapter.

In this chapter we first review sinusoidal steady state notation for voltage and current and real and reactive power in single phase systems.

2.1 Sources and Power

2.1.1 Voltage and Current Sources

Consider the interconnection of two sources shown in Figure 2.1. On the right is the symbol for a *voltage source*. This is a circuit element that maintains a voltage at its terminals, conceptually *no matter what* the current. On the left is the symbol for a *current source*. This is the complementary element: it maintains current *no matter what* the voltage. Quite obviously, these are idealizations, but they do serve as good proxies for reality. For example, the power system connection to a customer's site is a good approximation to a voltage source. And some types of generators interconnections to the power system, such as those from solar photovoltaic power plants and some types of wind turbines approximate current sources. So that the situation shown in Figure 2.1 is an approximation of the interconnection of a solar plant to the power system. If the voltage and current are both sine waves at the same frequency, perhaps with a phase shift, they could be represented as:

$$v = V \cos \omega t$$
$$i = I \cos (\omega t - \psi)$$

Electric Power Principles: Sources, Conversion, Distribution and Use James L. Kirtley

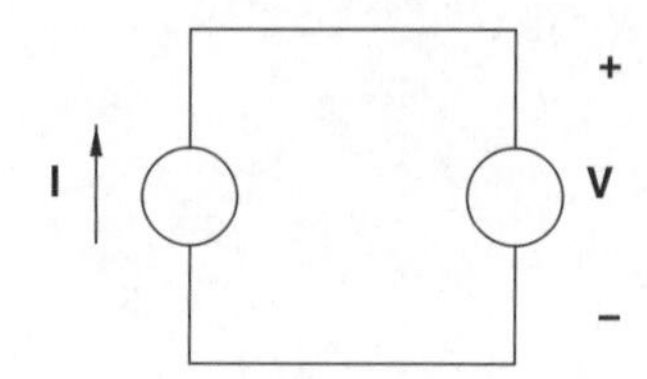

Figure 2.1 Current and voltage sources connected together

2.1.2 *Power*

Note that Kirchhoff's Voltage and Current Laws (KVL and KCL), when applied here, specify that the voltage across the current source and the current through the voltage source are each specified by the other source. So power *out* of the current source is also power *in* to the voltage source. If the angle ψ is small so that the voltage and current are close to being in phase, the direction of positive power flow will be from the current source to the voltage source. That power flow will be:

$$p = vi = VI\cos\omega t\cos(\omega t - \psi)$$

Since $\cos a \cos b = \frac{1}{2}cos(a-b) + \frac{1}{2}\cos(a+b)$, that power flow is:

$$p = \frac{VI}{2}[\cos\psi + \cos(2\omega t + \psi)]$$

As will become clear shortly, it is possible to handle sine waves in a very straightforward way by using complex notation.

2.1.3 *Sinusoidal Steady State*

The key to understanding systems in the sinusoidal steady state is Euler's relation:

$$\mathrm{e}^{jx} = \cos x + j\sin x$$

where e is the base for common logarithms (about 2.718).

From this comes:

$$\mathrm{e}^{j\omega t} = \cos\omega t + j\sin\omega t$$

and:

$$\mathrm{Re}\left\{\mathrm{e}^{j\omega t}\right\} = \cos\omega t$$

If a voltage is a pure sine wave (that is, with no DC components or transients, it can be described as:

$$v(t) = V\cos(\omega t + \phi) = \mathrm{Re}\left\{V\mathrm{e}^{j\phi}\mathrm{e}^{j\omega t}\right\}$$

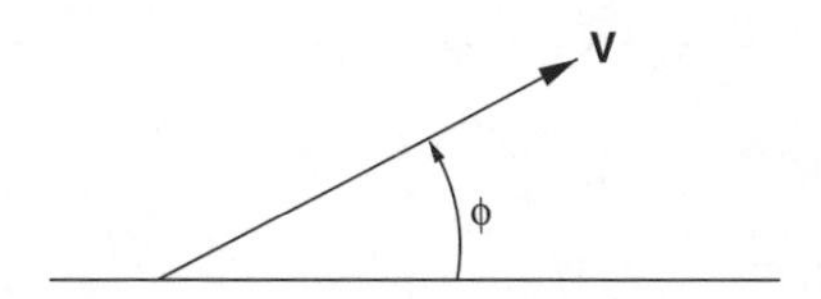

Figure 2.2 Voltage phasor

This is the real part of a complex exponential with continuously increasing phase angle. In this case the complex amplitude of the voltage is as shown in Figure 2.2:

$$\mathbf{V} = V e^{j\phi}$$

2.1.4 *Phasor Notation*

This sinusoidal voltage may be represented graphically as is shown in Figure 2.2. The magnitude of this 'phasor', V, is the length of the vector, while the phase angle ϕ is represented by a rotation of the vector about its origin. The instantaneous value of the voltage is equal to the projection onto the horizontal axis of the tip of a vector that is rotating with angular velocity ω that is at the position of the voltage phasor at time $t = 0$.

2.1.5 *Real and Reactive Power*

In a circuit such as that of Figure 2.1 in which there exists both voltage and current, as is represented by two phasors in Figure 2.3, voltage and current are:

$$v = \text{Re}\left\{\text{V}\, e^{j(\omega t + \phi)}\right\}$$
$$i = \text{Re}\left\{\text{I}\, e^{j(\omega t + \phi - \psi)}\right\}$$

Taking advantage of the fact that the real part of a complex number is simply one half of the sum of that number and its complex conjugate: $\text{Re}\{X\} = \frac{1}{2}(X + X^*)$, one can see that instantaneous power is:

$$p = \frac{1}{2} VI \cos\psi + \frac{1}{2} VI \cos(2(\omega t + \phi) - \psi) \tag{2.1}$$

$$= \frac{1}{2} VI (\cos\psi\,(1 + \cos 2(\omega t + \phi)) + \sin\psi \sin 2(\omega t + \phi)) \tag{2.2}$$

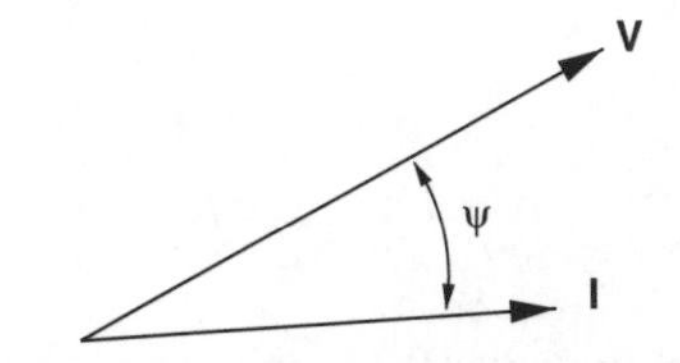

Figure 2.3 Voltage and current phasors

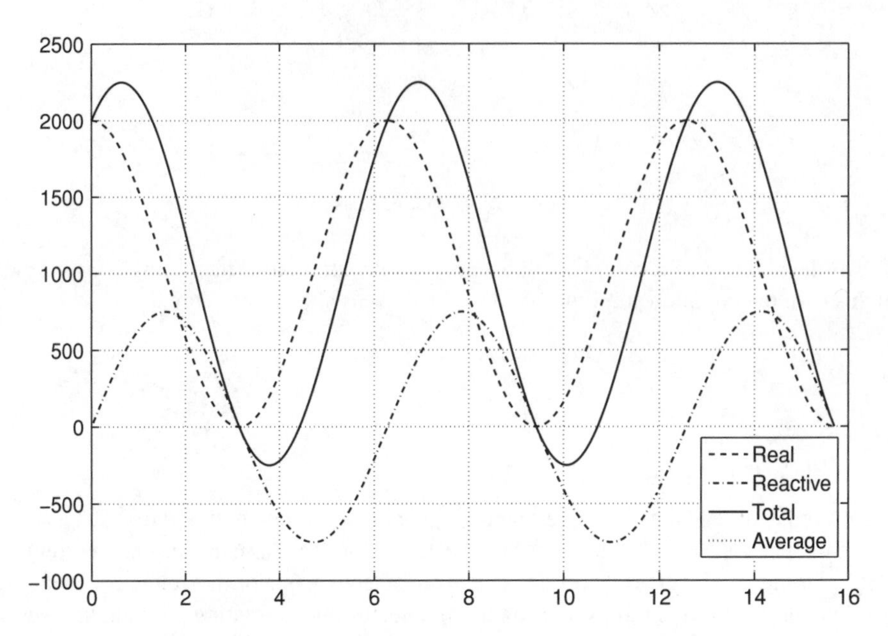

Figure 2.4 Instantaneous power with power factor of 0.8

This is illustrated in Figure 2.4. Note there are two principal terms here. One is for the *real*, or time-average power:

$$P = \frac{1}{2} V I \cos \psi$$

The other term has no time average, but represents power that is seemingly just 'sloshing around', or being exchanged between source and sink at twice the electrical frequency. This is *reactive* power:

$$Q = \frac{1}{2} V I \sin \psi$$

and it plays a very important role in controlling voltage in electric power systems. Note that the *apparent* power is just the magnitude of real plus reactive power, assuming reactive is 'imaginary' in the real/imaginary number plane:

$$|P + jQ| = \frac{1}{2}|V||I|$$

and

$$P + jQ = \frac{1}{2} V I^*$$

Here, *real* power is measured in *watts* (W), *apparent* power is measured in *volt-amperes* (VA) and *reactive* power is measured in *volt-amperes reactive* (VARs).

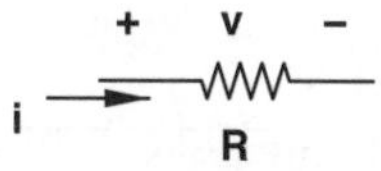

+ V -
i R

Figure 2.5 Resistor

2.1.5.1 Root Mean Square Amplitude

It is common to refer to voltages and currents by their root mean square (RMS) amplitudes, rather than peak. For sine waves the RMS amplitude is $1/\sqrt{2}$ of the peak amplitude. Thus, if the RMS amplitudes are V_{RMS} and I_{RMS} respectively,

$$P + jQ = V_{\text{RMS}} I_{\text{RMS}}^*$$

2.2 Resistors, Inductors and Capacitors

These three linear, passive elements can be used to understand much of what happens in electric power systems.

The resistor, whose symbol is shown in Figure 2.5 has the simple voltage–current relationship:

$$i = \frac{v}{R}$$

Since voltage and current are quite obviously *in phase*, when driven by a voltage source,

$$I_{\text{RMS}} = \frac{V_{\text{RMS}}}{R}$$

and complex power into the resistor is:

$$P + jQ = \frac{V_{\text{RMS}}^2}{R}$$

That is, the resistor draws only real power and reactive power at its terminals is zero.

The inductor, whose symbol is shown in Figure 2.6, has the voltage–current relationship:

$$v = L\frac{\mathrm{d}i}{\mathrm{d}t}$$

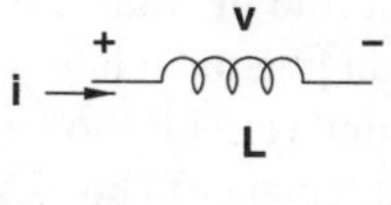

V
+ -
i
L

Figure 2.6 Inductor

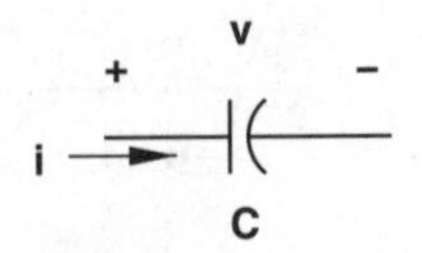

Figure 2.7 Capacitor

In sinusoidal steady state, this statement can be made using complex notation:

$$\mathrm{V} = j\omega L \mathrm{I}$$

This means that real plus reactive power is:

$$P + jQ = V_{\mathrm{RMS}} I^*_{\mathrm{RMS}} = j\frac{|V_{\mathrm{RMS}}|^2}{\omega L}$$

The inductor, then, draws reactive power (that is, the sign of reactive power into the inductor is positive), and for the ideal inductor the real power is zero. Of course real inductors have some resistance, so the real power into an inductor will not be exactly zero, but in most cases it will be small compared with the reactive power drawn. The inductor has reactance $X_L = \omega L$.

The capacitor, whose symbol is shown in Figure 2.7 has a voltage–current relationship given by:

$$i = C\frac{\mathrm{d}v}{\mathrm{d}t}$$

Or, in complex notation:

$$\mathrm{I} = \mathrm{V} j\omega C$$

This means that complex (real plus reactive) power drawn by the capacitor is:

$$P + jQ = V_{\mathrm{RMS}} I^*_{\mathrm{RMS}} = -j|V_{\mathrm{RMS}}|^2 \omega C$$

The capacitor is a source of reactive power. As with the inductor, the capacitor draws little real power. The idealized capacitor sources reactive power and draws zero real power. The capacitor has reactance $X_C = -\frac{1}{\omega C}$.

2.2.1 *Reactive Power and Voltage*

Reactive power plays a very important role in voltage profiles on electric power systems. For that reason, it is useful to start understanding the relationship between reactive power and voltage from the very start. Consider the simple circuit shown in Figure 2.8. A voltage source is connected to a resistive load through an inductance. This is somewhat like a power system, where transmission and distribution lines are largely inductive. The voltage across the resistor is:

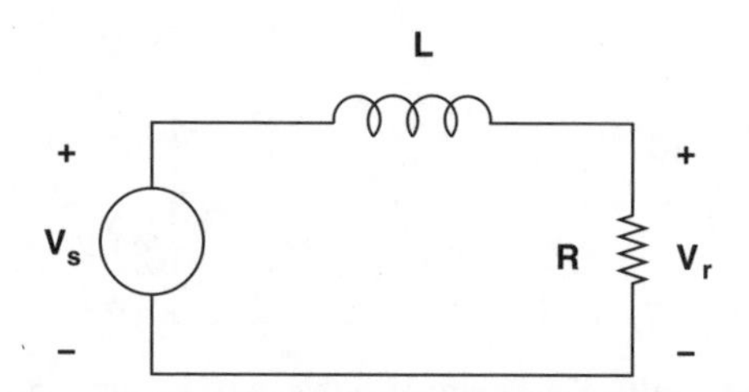

Figure 2.8 Power circuit

$$V_r = V_s \frac{R}{R + j\omega L}$$

This is shown in abstract form in Figure 2.9. Quite clearly, the 'receiving end' voltage is less than the load voltage, and the inequality will increase as the load increases (meaning current through the inductor increases). Consider what happens, however, when a capacitance is put in parallel with the resistor, as shown in Figure 2.10.

The output voltage is:

$$V_r = V_s \frac{\frac{\frac{R}{j\omega C}}{R + \frac{1}{j\omega C}}}{\frac{\frac{R}{j\omega C}}{R + \frac{1}{j\omega C}} + j\omega L} = V_s \frac{1}{\left(1 - \omega^2 LC\right) + \frac{j\omega L}{R}}$$

2.2.1.1 Example

Suppose the voltage source in the circuit of Figure 2.10 provides a sine wave with an RMS magnitude of $V_s = 10\,\text{kV}$, the load resistor is $R = 10\,\Omega$, the inductance is $L = 10\,\text{mH}$ and the system frequency is $\omega = 2\pi \times 60\,\text{Hz} = 377$ radians/second. The relative magnitude of the output voltage is readily calculated as a function of the capacitor value and is shown in Figure 2.11.

2.2.2 *Reactive Power Voltage Support*

Noting that the capacitor provides reactive power suggests that reactive power injection in can provide some amount of voltage control. Indeed this is the case, and in distribution systems

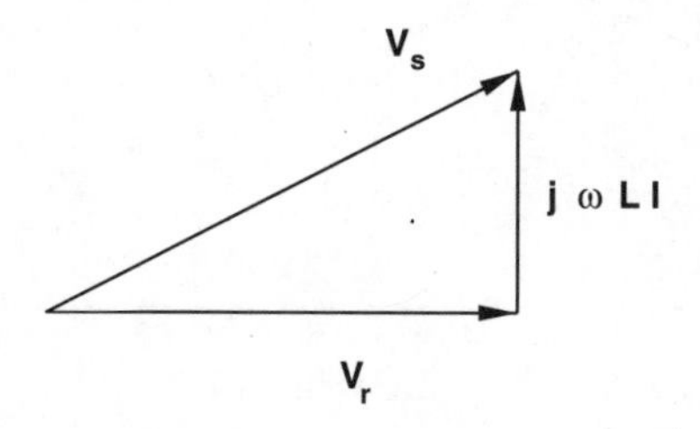

Figure 2.9 Voltage vectors

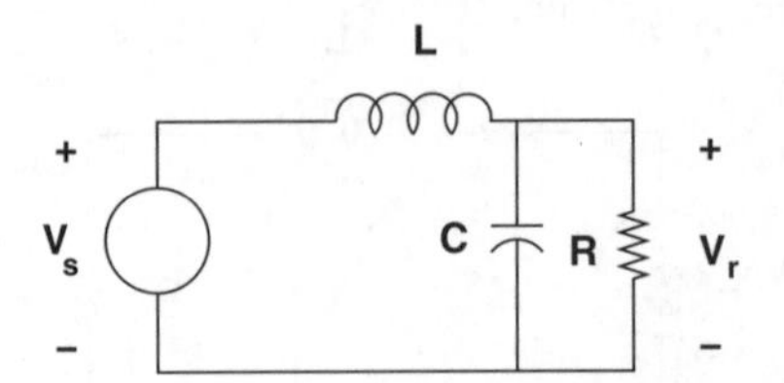

Figure 2.10 Power circuit with compensating capacitor

electric power utilities often used capacitors, usually switched in increments, to help control voltage profiles. To approach this problem, consider the arrangement shown in Figure 2.12. Here, the capacitor is replaced by a more general reactive admittance, defined by:

$$\mathrm{I} = jB\mathrm{V}$$

The voltage across the load resistor is found using the same calculation as for the capacitor case and is:

$$\frac{\mathrm{V_r}}{\mathrm{V_s}} = \frac{R}{(R - X_s B) + jX_s}$$

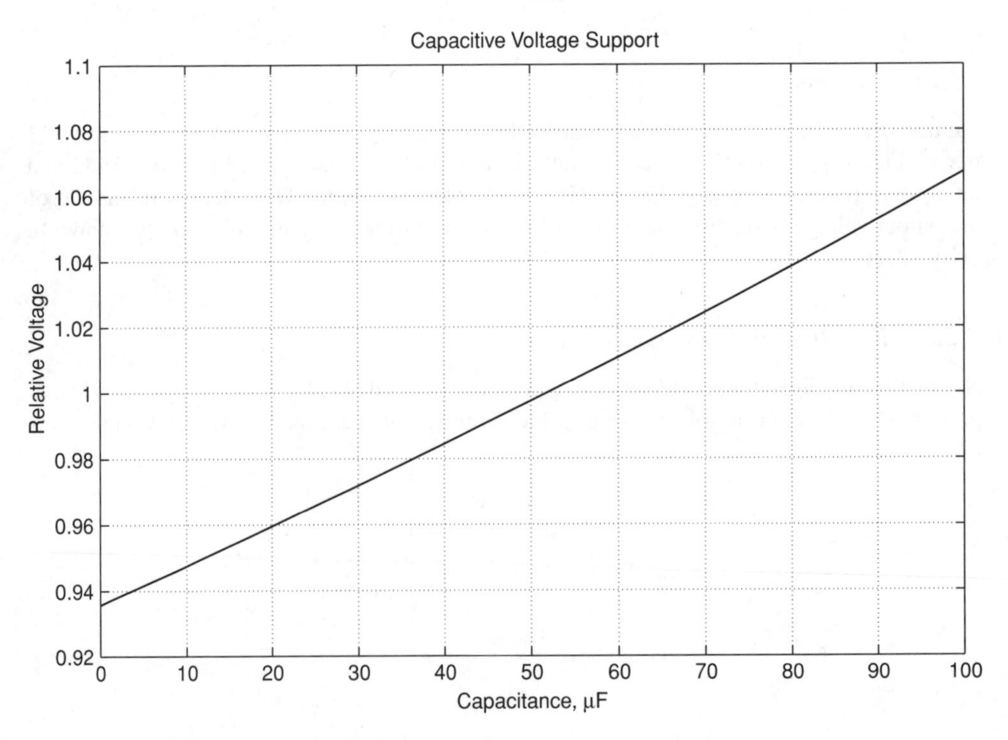

Figure 2.11 Voltage support with a capacitor

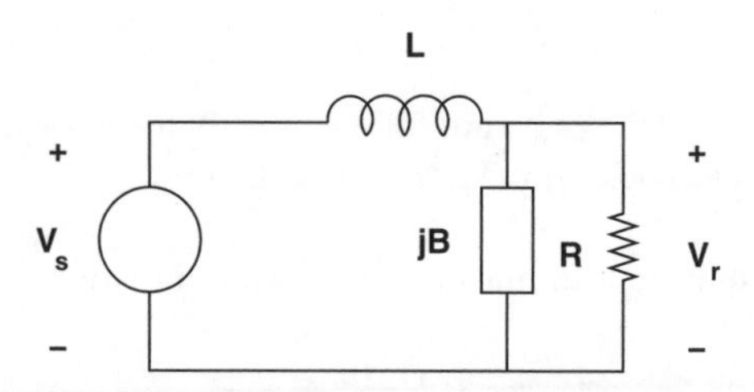

Figure 2.12 Power circuit with generalized reactive element

where the notation $X_s = \omega L$ has been used. Reactive power *produced* by the reactive element is:

$$Q = |\mathbf{V}_r|^2 B$$

These calculations may be carried out over a range of values of the reactive element B and then voltage $|\mathbf{V}_r|$ cross plotted against reactive power Q to get the effect of reactive power on voltage. This is done in Figure 2.13 to show how injection of reactive power affects voltage.

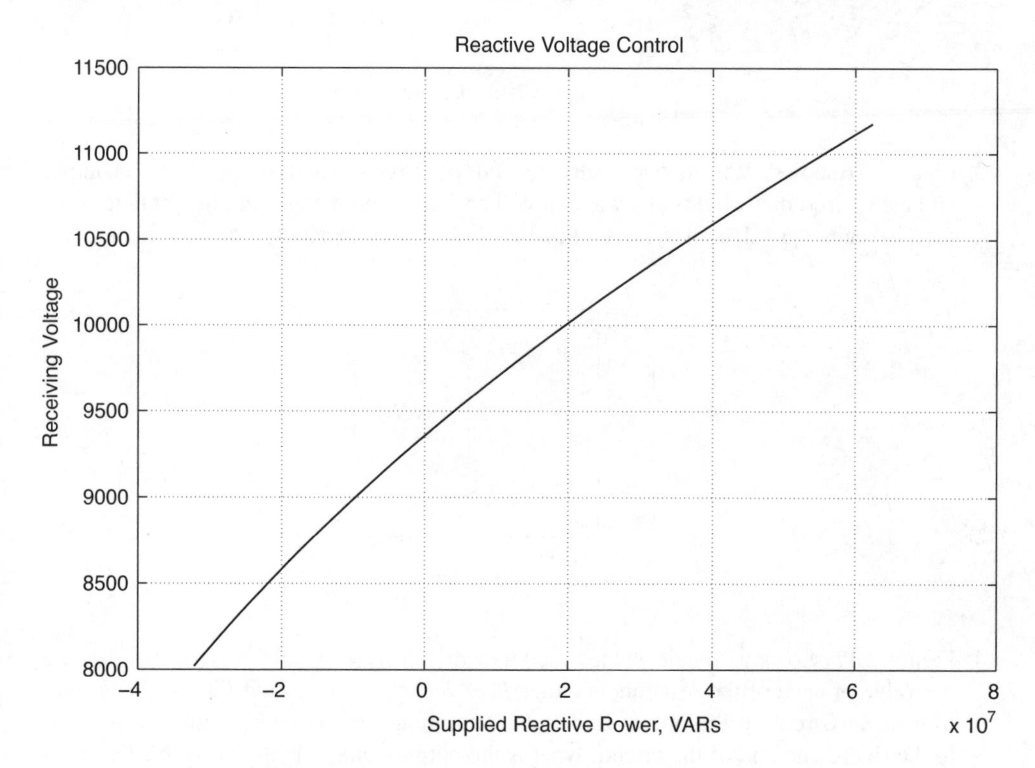

Figure 2.13 Voltage affected by reactive power injection

2.3 Problems

Note: the first few of these exercises do not follow directly the material of this chapter, but are here to help you verify your background in network theory.

1. Find the Thevenin equivalent for the circuit shown in Figure 2.14.

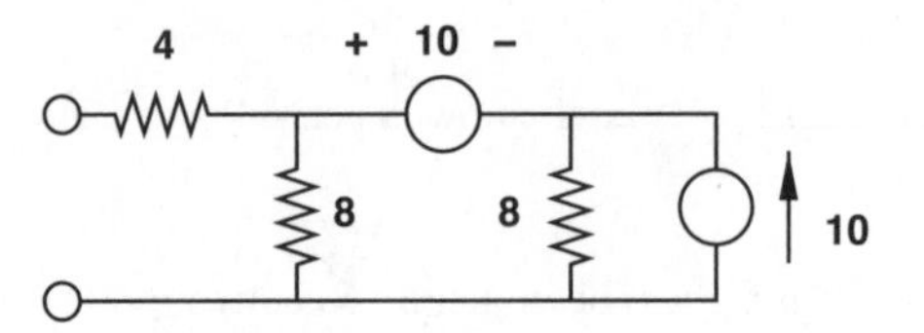

Figure 2.14 Circuit

2. Figure 2.15 shows two circuits, one with resistor values, the other with symbols. Show that these two circuits are equivalent if the values represented by the symbols are chosen correctly. Find the value of the symbols.

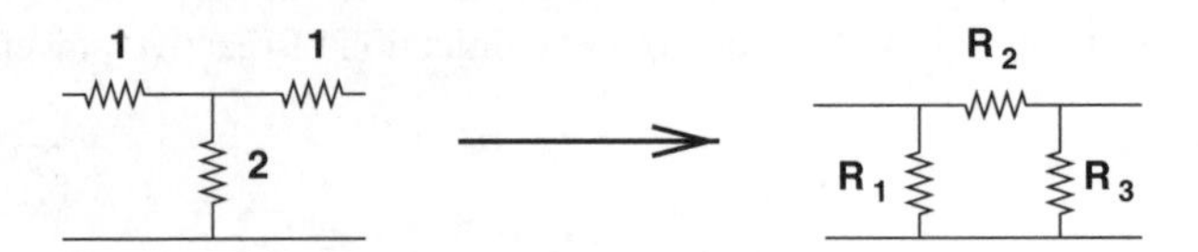

Figure 2.15 Circuit

3. Figure 2.16 shows a Wheatstone Bridge, loaded with a resistor at its output. Find the output voltage v_o. You may do this any way but will probably find it expedient to first determine the Thevenin equivalent of the circuit that excludes the horizontally oriented resistor.

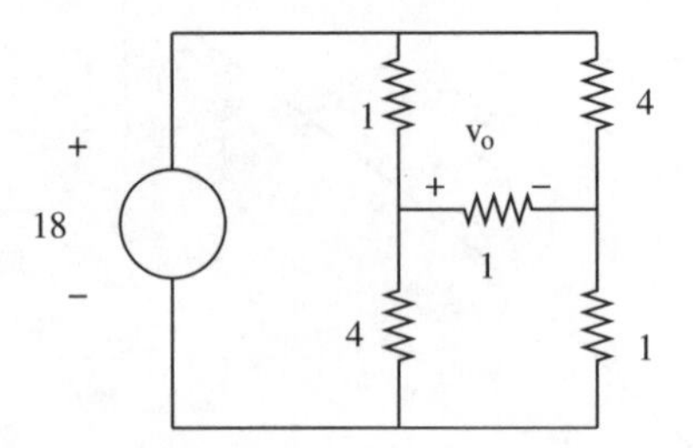

Figure 2.16 Loaded bridge

4. Figure 2.17 shows a 'magic ladder' network driven by two voltage sources. Assume the value of each of the resistors is either R or $2R$ where $R = 1\,\mathrm{k}\Omega$ Find the Thevenin Equivalent Circuit at the output terminals. Assuming there is nothing more connected to the right-hand end of the circuit, what is the output voltage V? What is the Thevenin equivalent resistance?

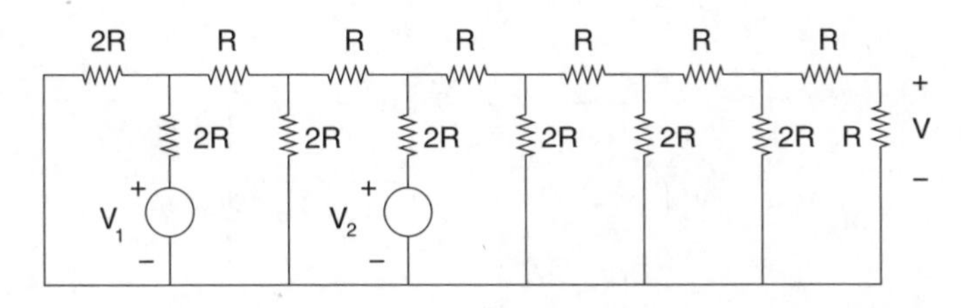

Figure 2.17 Magic ladder circuit

Hint: There is an easy way of doing this problem and there is a very hard way of working it. This circuit has some very nice properties, which is why it is called a 'magic ladder'. If you peer at it for a moment and consider the 'driving point' impedance at each of its nodes you can probably figure the easy way of working this problem.

5. Figure 2.18 shows a current source connected directly to a voltage source. Assume that the voltage and current are:

$$v = \mathrm{Re}\left\{V e^{j\omega t}\right\}$$
$$i = \mathrm{Re}\left\{I e^{j(\omega t-\psi)}\right\}$$

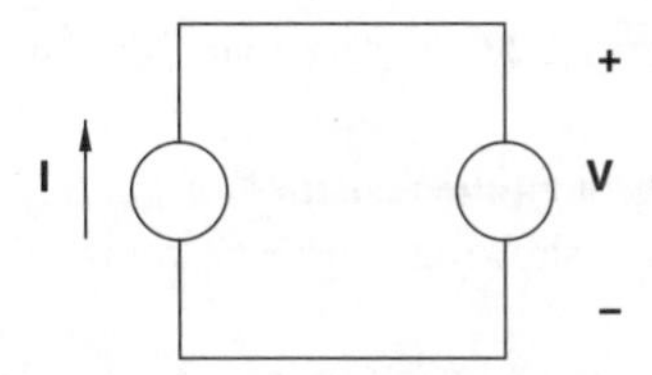

Figure 2.18 Current and voltage sources connected together

Assume that the values of voltage and current are $V_s = 120$ V, RMS and $I = 10$ A, also RMS.

(a) Calculate and sketch real and reactive power P and Q as a function of the angle ψ, which can range all around the unit circle (that is, from zero to 2π).
(b) Calculate and sketch instantaneous power for $\psi = 0$.
(c) Calculate and sketch instantaneous power for $\psi = \frac{\pi}{2}$.
(d) Sketch phasor diagrams for the two cases $\psi = 0$ and $\psi = \frac{\pi}{2}$.

6. Figure 2.19 shows a voltage source driving an inductor and a resistor in series. Assume the reactance of the inductor is $X = 20\,\Omega$ and the resistor is $R = 10\,\Omega$. The voltage source is 120 V, RMS.

(a) What is the voltage across the resistor?
(b) Draw a phasor diagram showing source voltage and voltages across the resistor and inductor.
(c) What are real and reactive power drawn from the source?

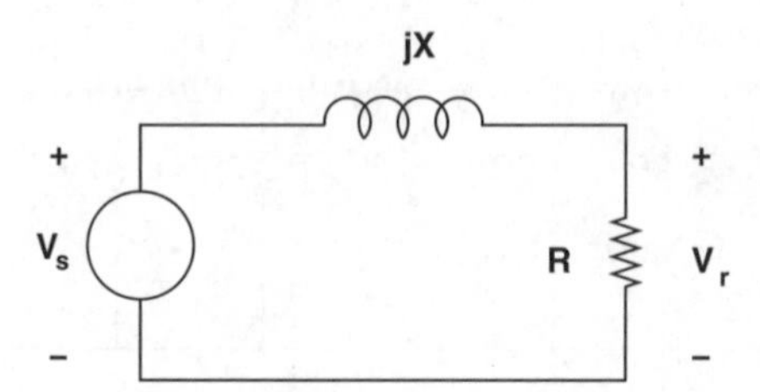

Figure 2.19 Voltage source driving L-R

7. Figure 2.20 shows a voltage source driving an inductor and a resistor in parallel. Assume the reactance of the inductor is $X = 20\,\Omega$ and the resistor is $R = 10\,\Omega$. The voltage source is 120 V, RMS.

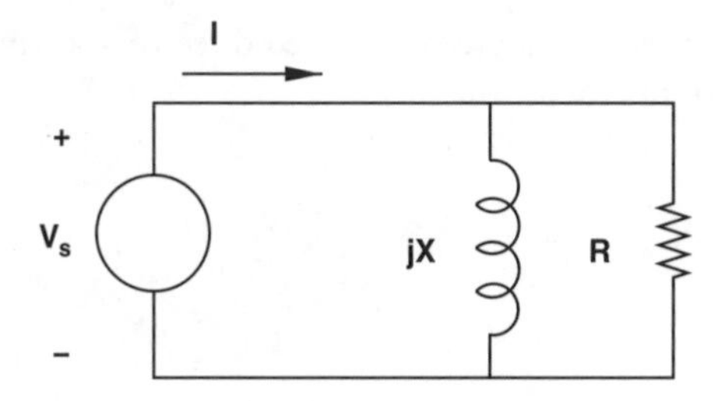

Figure 2.20 Voltage source driving parallel L-R

(a) What is the current through the resistor?
(b) Draw a phasor diagram showing source current and currents through the resistor and inductor.
(c) What are real and reactive power drawn from the source?

8. Figure 2.21 shows a current source driving a capacitor, an inductor and a resistor in parallel. Assume the reactance of the inductor is $X = 10\,\Omega$ and the resistor is $R = 10\,\Omega$. The current source is 10 A, RMS, and its frequency is 60 Hz.

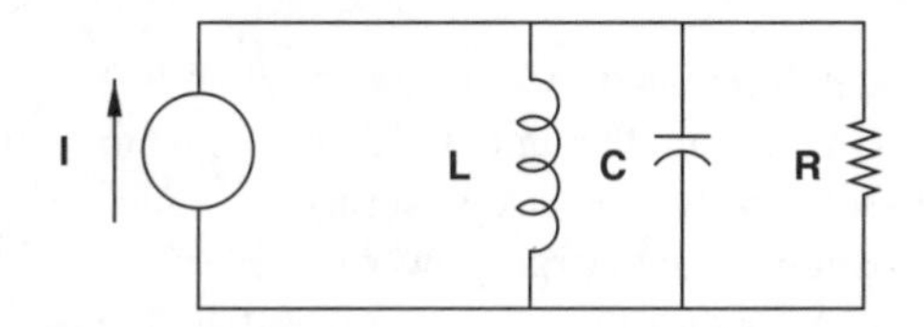

Figure 2.21 Current source driving parallel L-R-C

(a) What value should the capacitance have to maximize the voltage?
(b) Draw a phasor diagram showing source current and currents through the resistor, capacitor and inductor for the value of capacitance that maximizes voltage.
(c) What are real and reactive power drawn from the source for that value of capacitance?
(d) Calculate and sketch the magnitude of voltage for a range of capacitance from zero to twice the value you calculated to maximize voltage.

9. Figure 2.22 shows a voltage source driving a resistor through a capacitor and inductor connected in series. Assume the reactance of the inductor is $X = 10\,\Omega$ and the resistor is $R = 10\,\Omega$. The voltage source is 120 V, RMS, and its frequency is 60 Hz.

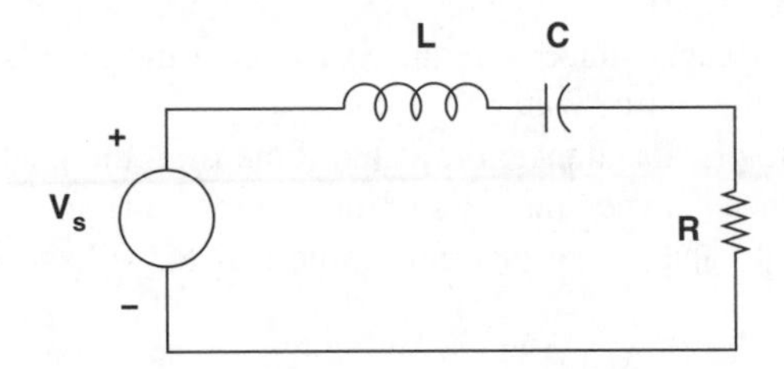

Figure 2.22 Voltage source driving series L-R-C

(a) What value should the capacitance have to maximize the resistor voltage?
(b) Draw a phasor diagram showing source voltage and voltages across the resistor, capacitor and inductor for the value of capacitance that maximizes voltage.
(c) What are real and reactive power drawn from the source for that value of capacitance?
(d) Calculate and sketch the magnitude of voltage for a range of capacitance from zero to twice the value you calculated to maximize voltage.

10. In the circuit of Figure 2.23, the voltage source is:

$$v_s = \text{Re}\left\{V_0 e^{j\omega t}\right\}$$

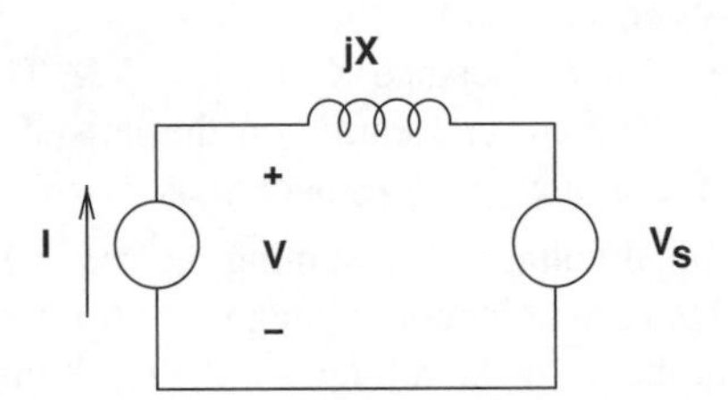

Figure 2.23 Inverter connected to power system

and the current source is:

$$i_s = \text{Re}\left\{I_0 e^{j(\omega t - \psi)}\right\}$$

Assume $V_0 = 100\,\text{V}$ and $I_0 = 1\text{A}$ and that the reactance $X = 10\,\Omega$.

(a) Draw a phasor diagram that shows the voltages V_s, V, and the voltage across the reactance for two cases: for $\psi = 0$ and $\psi = \frac{\pi}{2}$.
(b) Assuming arbitrary ψ, what is the locus of V?
(c) Assuming ψ can have any value, what are the maximum and minimum values of V? At what values of ψ do the maximum and minimum values of V occur?

11. Referring to the circuit of Figure 2.24, the voltage V_s is 120 V, RMS at 60 Hz. If the inductance L is 20 mH and the resistance R is 10 Ω.

(a) Assume the capacitance is omitted, draw the phasor diagram for voltages. What is the resistor voltage?
(b) What value of capacitance is required to make the resistor voltage have the same *magnitude* as the source voltage?
(c) For that value of capacitance, draw and dimension the phasor diagram. What is the phase difference between the resistor and source voltages?
(d) What value of capacitance maximizes the resistor voltage? What is that maximum voltage?

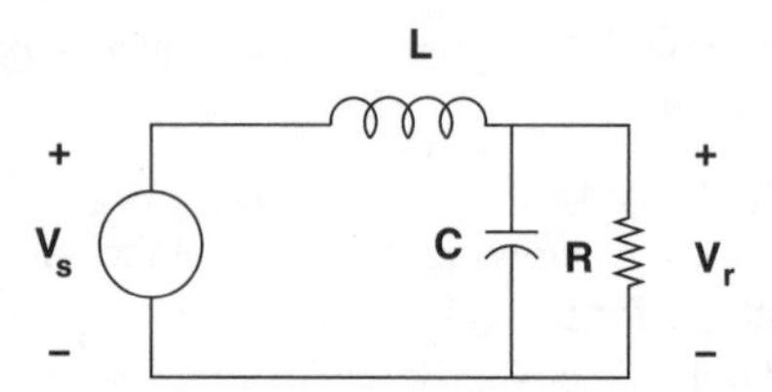

Figure 2.24 Resistive load with series inductance and parallel capacitance

12. Figure 2.23 shows, in highly simplified form, a situation similar to the inverters that are use for some renewable generation sources (wind, solar, ...). The current source on the left represents an inverter that is injecting current into the power system. The system itself is represented by the reactive impedance (representing a transmission line) and the voltage source on the right. For the moment we will assume a single phase system, although that is not the way most systems are built.
Assume the transmission line reactance $X = \omega L = 5\ \Omega$. The voltage source on the right is $V_s = 10$ kV (RMS). The current source on the left injects 1,000 A (RMS) into the system with a power factor angle *with respect to its measured terminal voltage*.

(a) Calculate the terminal voltage V, assuming the power factor of the current injection is: (a) unity, (b) 0.9, current lagging voltage, and (c) 0.9, current leading voltage.
(b) Calculate and plot the terminal voltage for the same three cases for current between $0 < I < 1000$A.
(c) Do the same calculation, but this time assume the *real* power varies between zero and 10 MW.

3

Transmission Lines

Transmission lines are open-air wires or insulated cables that carry electric power from one place to another. In the context of this book, 'transmission lines' are taken to mean not only the very high voltage lines that are referred to as 'transmission' by utility engineers, but also lower (but still high) voltage lines referred to as 'subtransmission' and 'distribution'. They all carry electric power and they all behave in roughly the same way. Transmission lines operate at high voltage to reduce losses, because loss is proportional to the square of current, and for a given amount of power carried, current is inversely proportional to voltage.

Transmission lines, being current carrying wires, produce magnetic fields and therefore exhibit inductance that can be described as being in series with the terminals of the line. Since they carry high voltage, they also produce electric fields that are terminated on the conductors, so that transmission lines also exhibit capacitance that is generally described as being in shunt with the terminals of the line. So a model of a transmission line that does a decent job of describing the behavior of the line at 'low' frequencies is as shown in Figure 3.1. This is often referred to as a 'pi' model for the line because of its form. Note that the series resistance of the line is neglected in this model. This is a limitation on the accuracy of line representation, but is usually justified by the notion (if not fact) that series resistance of a good transmission line is quite small compared with the reactance. For now, this assumption will be followed, but it is important to understand that it is only approximately true. A description of how to estimate transmission line inductance parameters is in the Appendix of Chapter 7.

It is necessary, in considering the model of Figure 3.1, to consider its range of applicability. That is, what does 'low frequency' mean in this context? In fact, the common sense approach would suggest that this model is probably reasonably good as long as the line is short (or at least not long) compared with a wavelength of the voltages and currents carried by the line. As it turns out, the critical length is one quarter wavelength, as will be shown shortly.

Electric Power Principles: Sources, Conversion, Distribution and Use James L. Kirtley

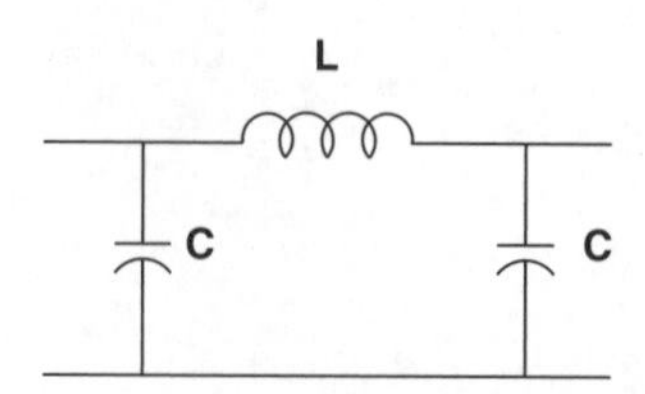

Figure 3.1 Standard 'pi' model for a transmission line

3.1 Modeling: Telegrapher's Equations

A transmission line that exhibits series inductance and parallel capacitance could be represented as a series of 'pi' models, each representing a suitably short part of the line. In fact, such representation has been used in analog computer models of transmission lines, with physical inductors and capacitors making up short line segments. Of course the frequency response of such line models is limited by the length represented by each section. This is shown, conceptually, in Figure 3.2.

If one takes the limit of short length for each section represented, it is possible to see that the series inductance produces a voltage drop proportional to the rate of change of current. Similarly, the capacitance produces a variation of line current proportional to the rate of change of voltage. The two sets of expressions that result are called the *Telegrapher's Equations*:

$$\frac{\partial v}{\partial x} = -L\frac{\partial i}{\partial t} \tag{3.1}$$

$$\frac{\partial i}{\partial x} = -C\frac{\partial v}{\partial t} \tag{3.2}$$

Here, the variable x is length along the line. Note that Equation 3.1 is the limit of KVL taken around a loop in Figure 3.2, while Equation 3.2 is the limit of KCL taken at a node in the same figure. The two parameters L and C are the inductance of the line per unit length (H/m) and the capacitance of the line per unit length (F/m), respectively. The two expressions are called the 'Telegrapher's Equations' because they were first derived to understand how pulses travel on telegraph lines. However, they are useful to power systems engineers understanding how not only pulses but also power frequency sine waves travel on power lines.

First we deal with a time domain description. Differentiating Equation 3.1 by $\frac{\partial}{\partial x}$, and Equation 3.2 by $\frac{\partial}{\partial t}$, the result is:

$$\frac{\partial^2 v}{\partial x^2} = -L\frac{\partial^2 i}{\partial x \partial t} = LC\frac{\partial^2 v}{\partial t^2} \tag{3.3}$$

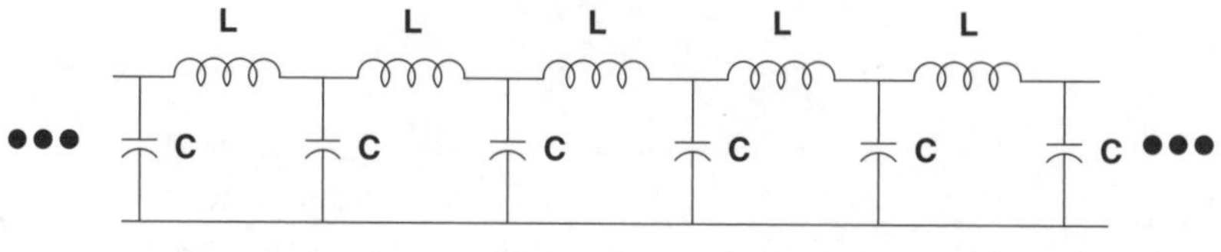

Figure 3.2 Transmission line model with multiple 'pi' sections

A similar procedure does the cross-differentiation in the opposite order and arrives at:

$$\frac{\partial^2 i}{\partial x^2} = -C\frac{\partial^2 v}{\partial x \partial t} = LC\frac{\partial^2 i}{\partial t^2} \tag{3.4}$$

Note that these two equations have general solutions. (3.3) is consistent with *any* solution of the form:

$$v(x, t) = v(x \pm ut)$$

if

$$u^2 = \frac{1}{LC}$$

Similarly, (3.4) is consistent with any solution of the form:

$$i(x, t) = i(x \pm ut)$$

3.1.1 Traveling Waves

A general solution for excitations on a transmission line is then:

$$v(x, t) = v_+(x - ut) + v_-(x + ut) \tag{3.5}$$

$$i(x, t) = i_+(x - ut) + i_-(x + ut) \tag{3.6}$$

The two expressions Equations 3.5 and 3.6 describe positive-going (v_+ and i_+) and negative-going (v_- and i_-) waves that travel together (that is, the pair of v and i form a *mode* that carries energy in either the forward (positive) or reverse (negative) direction. These modes can have arbitrary shape and, for a lossless line of the type postulated here, that shape does not change with time or distance along the line. Figure 3.3 shows pulses traveling along a transmission line.

3.1.2 Characteristic Impedance

Differentiating Equation 3.6 yields:

$$\frac{\partial i}{\partial t} = \mp u\frac{\partial i}{\partial x}$$

and then using Equation 3.1, one obtains:

$$\frac{\partial v}{\partial x} = \pm uL\frac{\partial i}{\partial x}$$

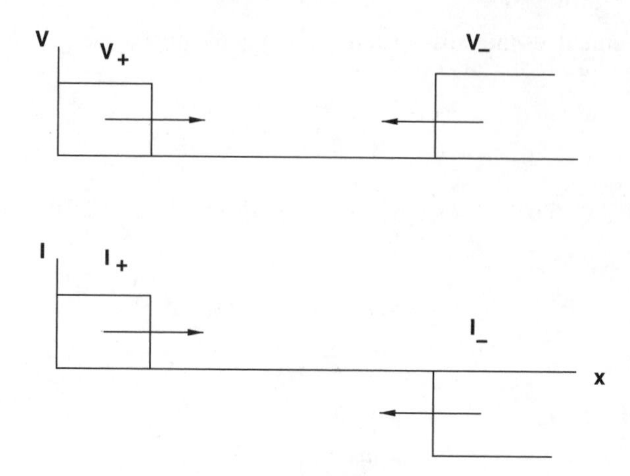

Figure 3.3 Traveling pulses on a transmission line

Integrating once with respect to x,

$$v_{\pm} = \pm u L i_{\pm}$$

and

$$uL = \frac{L}{\sqrt{LC}} = \sqrt{\frac{L}{C}} = Z_0$$

This defines what is called the *characteristic impedance* of the transmission line.

Note that the two transmission line parameters L and C combine in characteristic ways. $1/LC$ has units of m^2/HF. Since H is a V s/A and F is an A s/V a Henry times a Farad is a s^2, so the units of $1/LC$ are m/s. Similarly, the ratio of L/C is a $(\mathrm{v/A})^2$, so that $\sqrt{L/C}$ has units of Ω.

With the foregoing, a general solution for excitation on a transmission line becomes:

$$v(x,t) = v_+(x - ut) + v_-(x + ut)$$
$$i(x,t) = \frac{1}{Z_0} v_+(x - ut) - \frac{1}{Z_0} v_-(x + ut)$$

The excitations described here are what are called transverse electric and magnetic (TEM) modes. They are characterized by the fact that both electric and magnetic fields lie in a plane transverse to the direction of propagation of the mode (in this case, forward propagating in the +x direction or reverse, propagating in the −x direction. The magnetic and electric field lines lie in the $y - z$ plane. For modes such as this, the velocity of propagation is the speed of light in the medium of propagation. For open air (overhead) transmission lines, this is the speed of light ($u = c = 3 \times 10^8\mathrm{m/s}$). For insulated cables the speed of propagation will be modified by the dielectric constant of the insulation ($u = 1/\sqrt{\mu\epsilon}$).

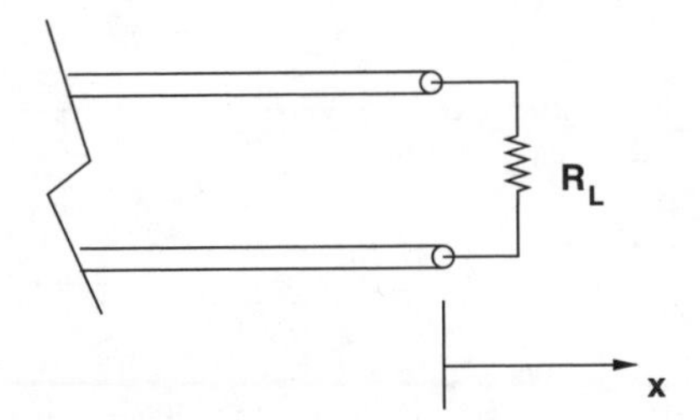

Figure 3.4 Resistive termination

3.1.3 Power

Traveling waves on transmission lines carry power: at a given spot on the transmission line, power flow along the line is:

$$P = P_+ - P_- = v_+ i_+ - v_- i_- = \frac{1}{Z_0}\left(v_+^2 - v_-^2\right) = Z_0\left(i_+^2 - i_-^2\right)$$

Traveling pulses on transmission lines are important for more than one reason. What are called 'surges' can be caused by switching operations, faults (short circuits) and by lightning, and so understanding how these excitations travel and interact with line terminations and other discontinuities is important. Consider the boundary condition that would be produced by a resistive termination at the end of a line, as shown in Figure 3.4. The line has a characteristic impedance Z_0 and the termination is assumed to have a resistance R_L.

3.1.4 Line Terminations and Reflections

Figure 3.4 shows a transmission line terminated by a resistor. At the termination, the ratio of voltage to current must, of course, be the resistance:

$$\frac{v_+ + v_-}{i_+ + i_-} = R$$

Since the voltage and current are related by the characteristic impedance,

$$\frac{v_+ + v_-}{v_+ - v_-} = \frac{R}{Z_0}$$

or, carrying out a little algebra, the reverse going voltage pulse (the reflection from the termination) is:

$$v_- = v_+ \frac{\frac{R}{Z_0} - 1}{\frac{R}{Z_0} + 1}$$

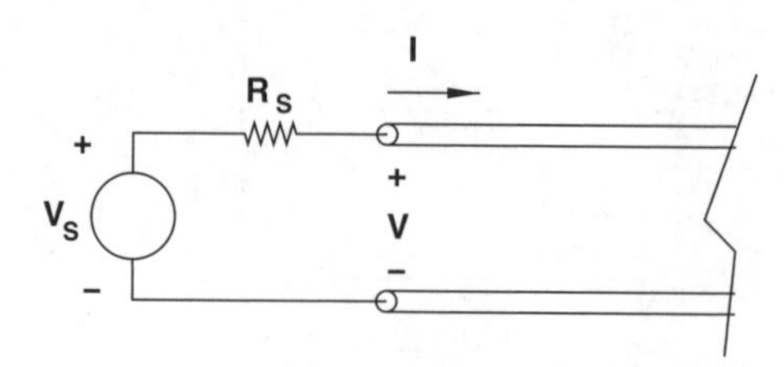

Figure 3.5 Termination with voltage source

Note that

1. If the termination resistance is equal to the characteristic impedance ($R = Z_0$), the termination is *matched* to the line and there is no reflection.
2. If the line is *open* ($R \rightarrow \infty$), the reflection is equal to and of the same sign as the incoming pulse and the voltage at the termination, being the sum of incoming and reflected pulses, is double the incoming pulse.
3. If the termination is *shorted* ($R = 0$), the reflected pulse is just the negative of the incoming pulse.

Figure 3.5 shows, schematically, a termination end of a line with a voltage source and resistance. If this termination is at the 'sending' end of the line, it is the positive going pulse that must be derived from the source voltage and the negative going pulse. It is straightforward to show that this is:

$$v_+ = \frac{v_s}{\frac{R_s}{Z_0} + 1} + v_- \frac{\frac{R_s}{Z_0} - 1}{\frac{R_s}{Z_0} + 1}$$

3.1.4.1 Examples

Consider the line situation shown in Figure 3.6. Further, assume that the line is terminated by its characteristic impedance ($R_L = Z_0$) and has an excitation source with a source impedance also equal to its characteristic impedance ($R_s = Z_0$). Suppose the source voltage is an ideal step: $v_s = V_s u(t)$.

The step of source voltage will launch a positive going wave of amplitude $v_+ = V_s/2$, and that pulse will propagate to the right. Since the line is 'matched', the reflected pulse will have

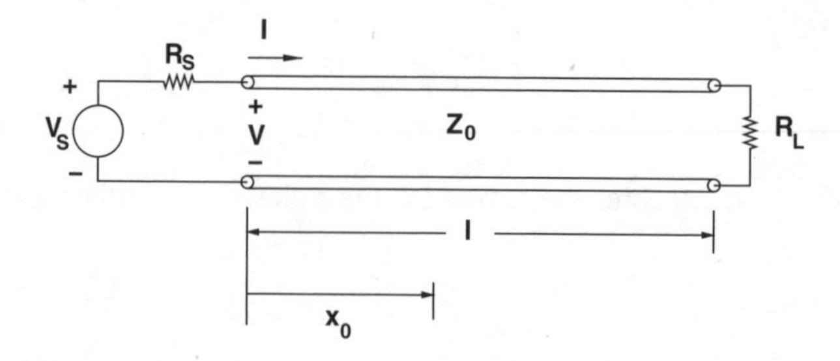

Figure 3.6 Example transmission line

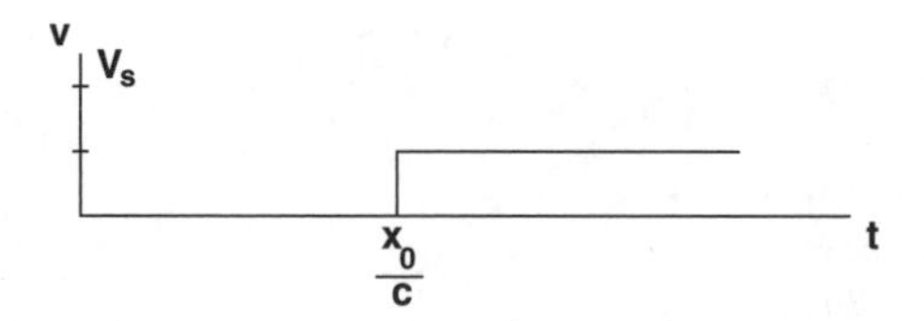

Figure 3.7 Voltage at point x_0, matched termination

an amplitude of zero, and so the voltage seen at a point $x = x_0$ along the line will be as shown in Figure 3.7.

Alternatively, consider the source impedance is still 'matched' but the termination end of the line is open ($R_L \to \infty$). The same positive going wave is launched by the step of voltage, but then there will be a positive reflection from the open termination end, and the voltage at point x_0 will be as shown in Figure 3.8.

3.1.4.2 Lightning

Transmission lines occasionally have to contend with lightning, which is often considered to be a current source, perhaps as shown in Figure 3.9. This is a pulse that rises with a time constant of roughly $1/2\mu s$ and then falls with a time constant of perhaps $8\mu s$. The crest amplitude varies but in this example is taken to be about 50 kA.

If a current source of the form shown in Figure 3.9 were to be imposed in the middle of a line as shown in Figure 3.10, two pulse waves would be launched: one going forward and one going back. Because of symmetry, the two pulses will be of equal amplitude:

$$v_+ = v_-$$

and the currents must add up to the current of the source:

$$i_+ - i_- = i_s$$

In the initial period, before there are any reflections to contend with, it is straightforward to match the voltage and current along the line to the source current, as shown in Figure 3.11.

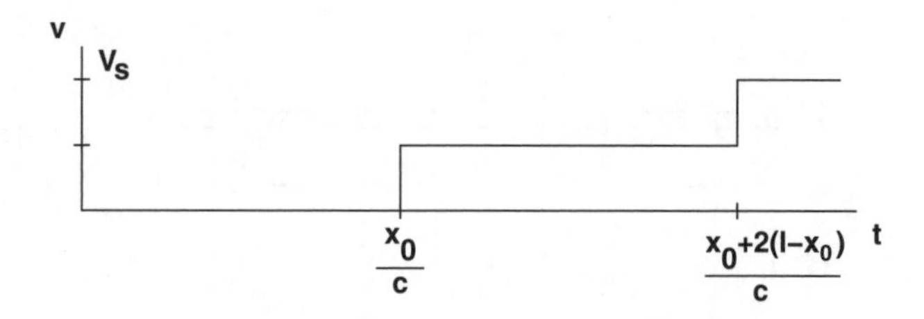

Figure 3.8 Voltage at point x_0, open termination

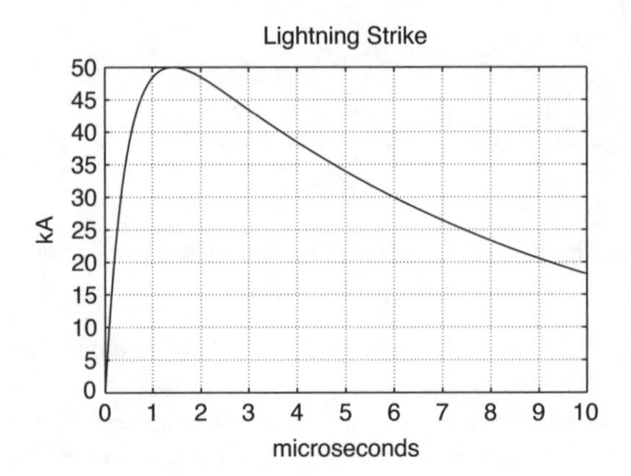

Figure 3.9 Idealized current from lightning

3.1.4.3 Inductive Termination

Consider a situation as shown in Figure 3.12. Here a line is terminated by an inductance.

In this circumstance, it is reasonable to think that the inductance initially 'looks' like an open circuit, because it is not possible to impose a step in current in an inductor. Over time, however, the inductor becomes a short circuit because it cannot support a DC voltage. A good way of examining this circuit is to realize that the line connected to the inductor can be represented as a Thevenin Equivalent circuit as shown in Figure 3.13, with open circuit voltage equal to twice the incoming (positive going) pulse and an equivalent impedance equal to the characteristic impedance of the line. This could be verified by examining the 'open circuit' and 'short circuit' behavior of the line.

The voltage at the line termination will then be, simply:

$$v(t) = 2v_+ e^{-t'/\tau}$$

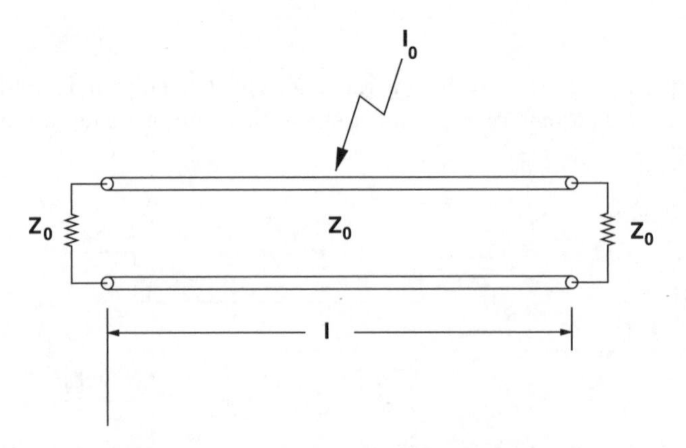

Figure 3.10 Lightning hits line in the middle

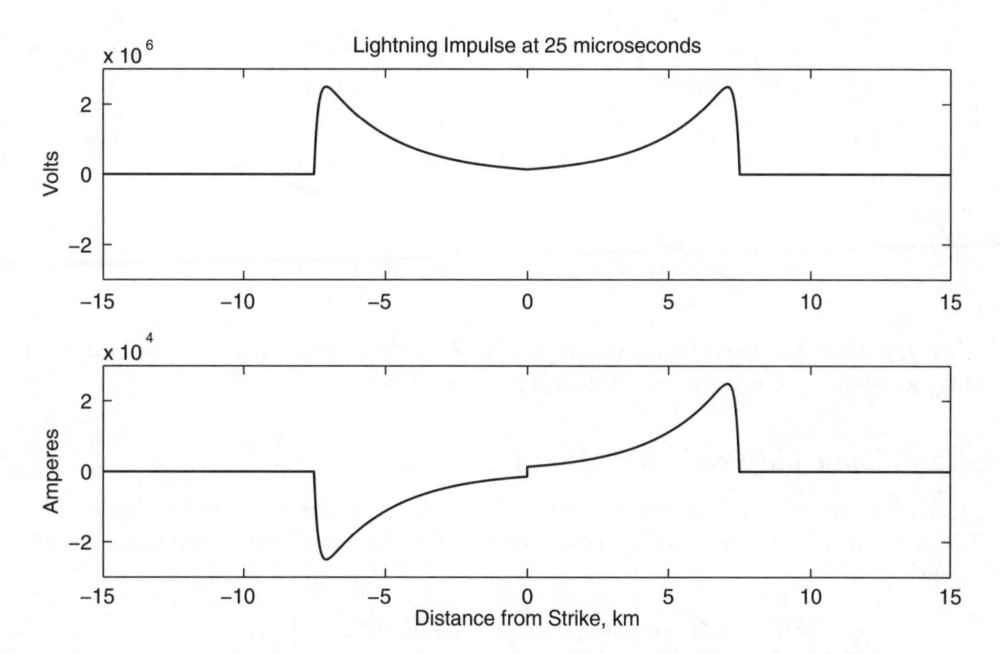

Figure 3.11 Voltage and current pulses from lightning strike

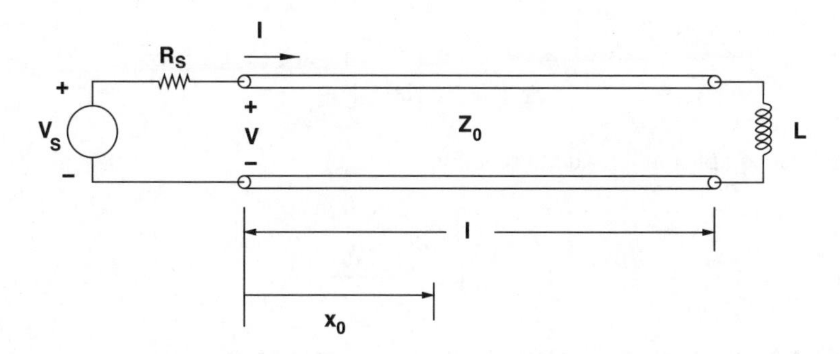

Figure 3.12 Line terminated by an inductance

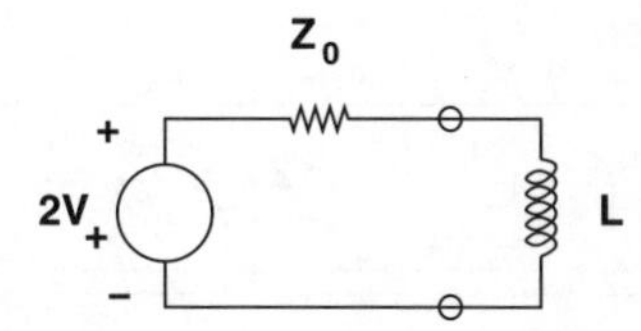

Figure 3.13 Thevenin equivalent circuit

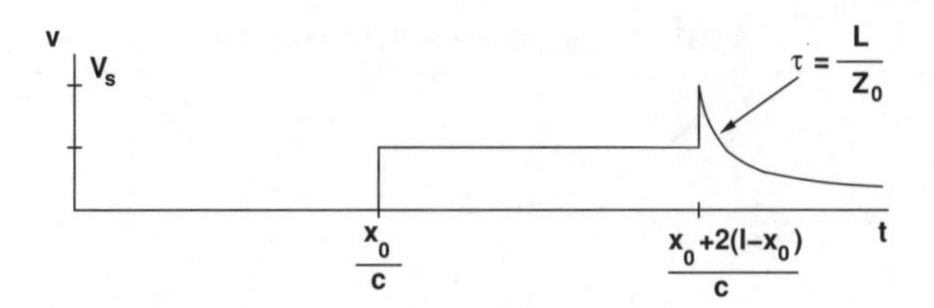

Figure 3.14 Voltage at x_0 with inductive termination

where t' is zero when the pulse arrives and $\tau = L/Z_0$ is the time constant. The voltage as seen at the location x_0 would then be as sketched in Figure 3.14.

3.1.5 Sinusoidal Steady State

Most operation of a power system involves voltages and currents varying sinusoidally. The forward- and reverse- going voltage and current functions can be expressed in the following fashion:

$$v(x,t) = \text{Re}\left\{\text{V}_+\text{e}^{j(\omega t - kx)} + \text{V}_-\text{e}^{j(\omega t + kx)}\right\}$$
$$i(x,t) = \text{Re}\left\{\text{I}_+\text{e}^{j(\omega t - kx)} + \text{I}_-\text{e}^{j(\omega t + kx)}\right\}$$

In the sinusoidal steady state it is possible to extract the time variation ($\text{e}^{j\omega t}$) so that

$$v = \text{Re}\left\{\text{V}\text{e}^{j\omega t}\right\}$$
$$i = \text{Re}\left\{\text{I}\text{e}^{j\omega t}\right\}$$

and then the complex amplitude (a function of space) is:

$$\mathbf{V} = \mathbf{V}_+\text{e}^{-jkx} + \mathbf{V}_-\text{e}^{jkx}$$
$$\mathbf{I} = \frac{\mathbf{V}_+}{Z_0}\text{e}^{-jkx} - \frac{\mathbf{V}_-}{Z_0}\text{e}^{jkx}$$

Consider the case shown in Figure 3.15. This is a general steady state case, with a (potentially) complex load and a (potentially) complex source impedance. Take the position of $x = 0$

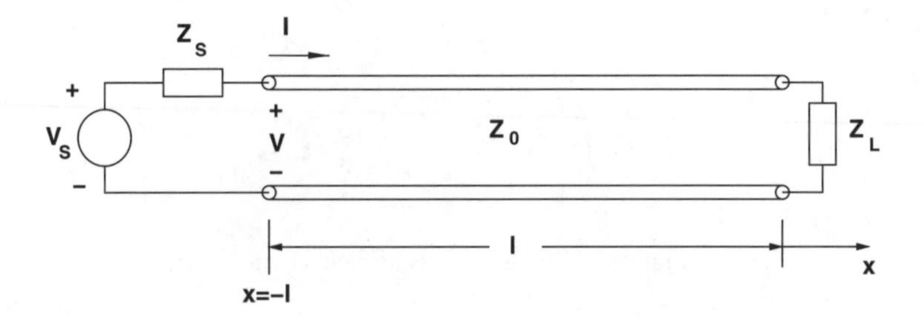

Figure 3.15 General steady state transmission line case

to be at the load end, so the source is at $x = -\ell$, The ratio of voltages is then:

$$\frac{\mathbf{V}_+ + \mathbf{V}_-}{\mathbf{V}_+ - \mathbf{V}_-} = \frac{Z_L}{Z_0}$$

Some algebra is required to find the two components of voltage along the line:

$$\mathbf{V}_+ = \frac{V_s\left(\frac{Z_L}{Z_0}+1\right)}{\left(\frac{Z_s}{Z_0}+1\right)\left(\frac{Z_L}{Z_0}+1\right)e^{jk\ell} - \left(\frac{Z_s}{Z_0}-1\right)\left(\frac{Z_L}{Z_0}-1\right)e^{-jk\ell}}$$

$$\mathbf{V}_- = \frac{V_s\left(\frac{Z_L}{Z_0}-1\right)}{\left(\frac{Z_s}{Z_0}+1\right)\left(\frac{Z_L}{Z_0}+1\right)e^{jk\ell} - \left(\frac{Z_s}{Z_0}-1\right)\left(\frac{Z_L}{Z_0}-1\right)e^{-jk\ell}}$$

Combined, these yield for voltage along the line:

$$\mathbf{V}(x) = V_s \frac{\frac{Z_L}{Z_0}\cos kx - j\sin kx}{\left(\frac{Z_s Z_L}{Z_0^2}+1\right) j\sin k\ell + \left(\frac{Z_s}{Z_0}+\frac{Z_L}{Z_0}\right)\cos k\ell}$$

This last expression is useful for examining the distribution of voltage along transmission lines. Generally, voltage tends to 'sag' for long lines that are heavily loaded and to rise for lines that are lightly loaded. Figure 3.16 shows this effect for a relatively long line (250 km)

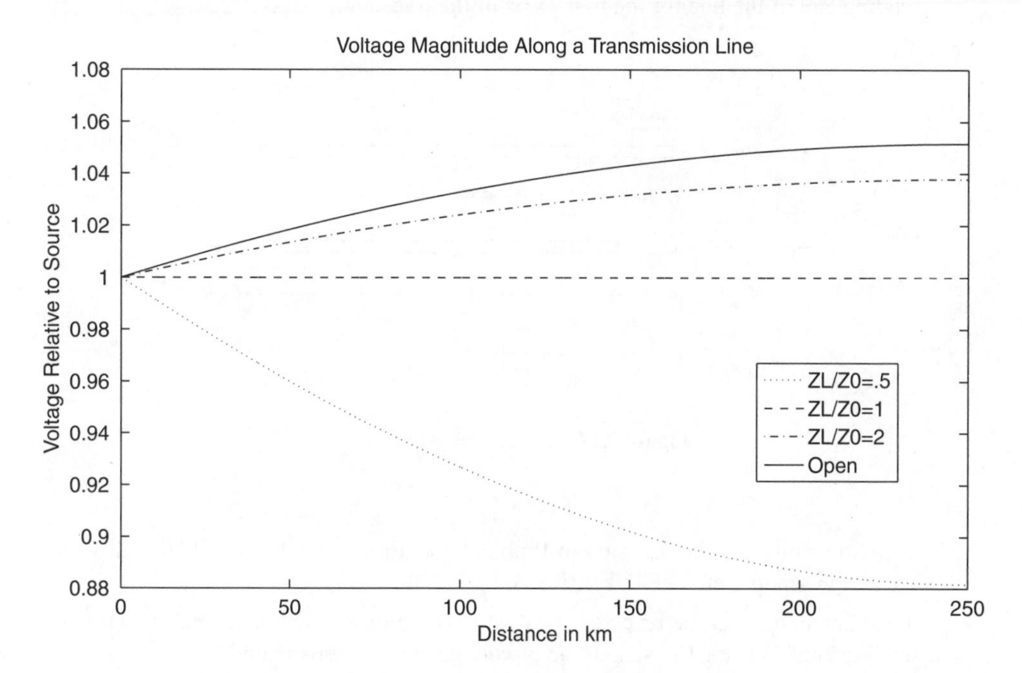

Figure 3.16 Voltage distribution along a long line

operating at 60 Hz. Note that for the case in which load impedance is a resistance with value equal to the characteristic impedance of the line, voltage magnitude is constant along the line. This is referred to as 'surge impedance loading' for reasons that should be clear from the discussion of the propagation of surges along the transmission line.

3.2 Problems

1. Shown in Figure 3.17 is a length of transmission line that is 50 km long. Actually, this is a coaxial cable with the following properties:

Rated current	325 A
Rated voltage	45 kV, RMS, line–neutral
Characteristic impedance	$Z_s = 30.3\ \Omega$
Inductance	$C = 0.18\ \mu\text{F/km}$

(a) What is the *inductance* per unit length of this line?

(b) What is the speed of propagation of signals in the line?

(c) This part is concerned only with one phase of this cable transmission line. It is carrying its rated current at rated voltage into a unity power factor load when, at time $t = 0$, which just happens to correspond with the peak of voltage (and current, since it is a unity power factor load) the current is interrupted at the load end. Everything happens very fast with respect to the line frequency, so assume the source voltage is constant at the peak value corresponding to 45 kV, RMS.

(d) Sketch, as a function of time, voltage at the receiving end of the line and current at the sending end of the line for the first 75 μs of the transient.

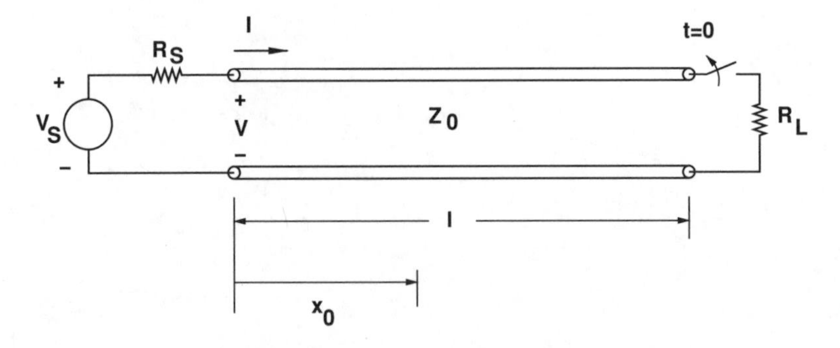

Figure 3.17 Transmission line example

2. The same transmission line as cited in Problem 1 is to be operated at 60 Hz and with a voltage, at the sending end, of 45 kV (RMS).

(a) If the line is *open* at the receiving end, what is the magnitude of current drawn at the sending end? What is the magnitude of voltage at the receiving end?

(b) Assuming the line is driving a load of unity power factor, how much current can it deliver to the load if the sending end current is limited to a magnitude of 325 A? Assume the sending end voltage is rated.

(c) Again assuming the conditions of the previous part, what is the *sending end* power factor?

3. Shown in Figure 3.18 is a length of transmission line that is 300 km long. Assume that the phase velocity for waves on the line is the speed of light: $C = 3 \times 10^8$ m/s. The line's characteristic impedance is 250 Ω.

(a) What are the inductance and capacitance per unit length of this line?

(b) The line is driven in the middle by a current source as shown: we assume this current source is a short pulse, 20,000 A in magnitude and of duration 20 μs. (This is a poor approximation to a lightning strike.) The line is shorted at one end and terminated by an impedance equivalent to it's characteristic, or *surge* impedance. Estimate and sketch the current in the short at the left-hand end of the line.

(c) For the same excitation, estimate and sketch the voltage across the terminating resistor at the right-hand end.

Your answers to the second and third parts should be labeled, dimensioned sketches. Assume the current surge starts at time $t = 0$.

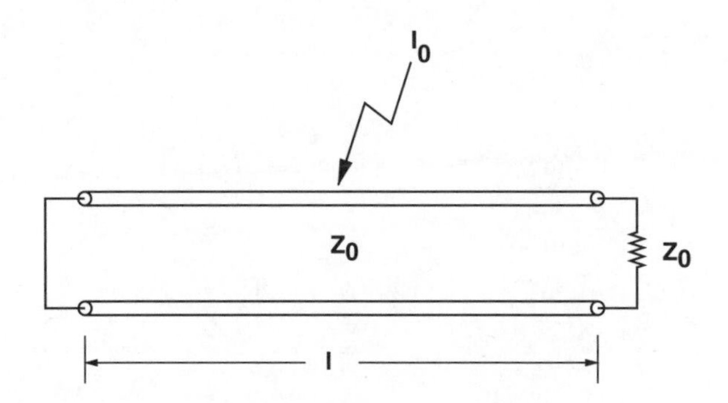

Figure 3.18 Transmission line example

4. The same transmission line as cited in Problem 3 is to be operated at 60 Hz and with a voltage, at the sending end, of 500 kV (RMS). (Note there are no real 500 kV, single phase lines, but just give us a little space here...)

(a) If the line is *open* at the receiving end, what is the magnitude of current drawn at the sending end? What is the magnitude of voltage at the receiving end?

(b) If the line is terminated with a resistance equivalent to the surge impedance: $R_L = Z_0$, what are the receiving end voltage and sending end current? What are real and reactive power at the sending end?

(c) Make the same estimates for a resistive load $R_L = \frac{Z_0}{0.8}$.

(d) Make the same estimates for a resistive load $R_L = \frac{Z_0}{1.2}$.

(e) Calculate and plot receiving end voltage for real power loads of between zero and surge impedance loading, assuming unity power factor at the receiving end. You will probably want to use MATLAB to do the heavy lifting for this part and the next.
(f) Now, assume that we can provide reactive compensation at the receiving end. Calculate and plot receiving end voltage for a reasonable range of compensating reactive VARs for real power loads of 80%, 100% and 120% of surge impedance loading. (You may do these for equivalent real admittance as the answer is really only important when voltage is nearly nominal.)

4

Polyphase Systems

Most electric power applications employ three phases. That is, three-separate power-carrying circuits, with voltages and currents staggered symmetrically in time are used. Two major reasons for the use of three-phase power are economical use of conductors and nearly constant power flow.

Systems with more than one phase are generally termed *polyphase*. Three-phase systems are the most common, but there are situations in which a different number of phases may be used. Two-phase systems have a simplicity that makes them useful for teaching vehicles and for certain servomechanisms. This is why two-phase machines show up in laboratories and textbooks. Systems with a relatively large number of phases are used for certain specialized applications such as controlled rectifiers for aluminum smelters. Six-phase systems have been proposed for very high power transmission applications.

Polyphase systems are qualitatively different from single-phase systems. In some sense, polyphase systems are more complex, but often much easier to analyze. This little paradox will become obvious during the discussion of electric machines. It is interesting to note that physical conversion between polyphase systems of different phase number is always possible.

4.0.1 Two-Phase Systems

The two-phase system is the simplest of all polyphase systems to describe. Consider a pair of voltage sources sitting side by side with:

$$v_1 = V\cos\omega t = \mathrm{Re}\{V\mathrm{e}^{j\omega t}\} \tag{4.1}$$

$$v_2 = V\sin\omega t = \mathrm{Re}\{V\mathrm{e}^{j(\omega t-\frac{\pi}{2})}\} \tag{4.2}$$

Suppose this system of sources is connected to a 'balanced load', as shown in Figure 4.1. The phasor diagram for this two phase voltage source is shown in Figure 4.2. Here, 'balanced' means both load impedances have the same value $\mathbf{Z}$. To compute the power flows in the system, it is convenient to re-write the voltages in complex form:

If each source is connected to a load with impedance:

$$\mathbf{Z} = |\mathbf{Z}|\mathrm{e}^{j\psi}$$

Electric Power Principles: Sources, Conversion, Distribution and Use James L. Kirtley

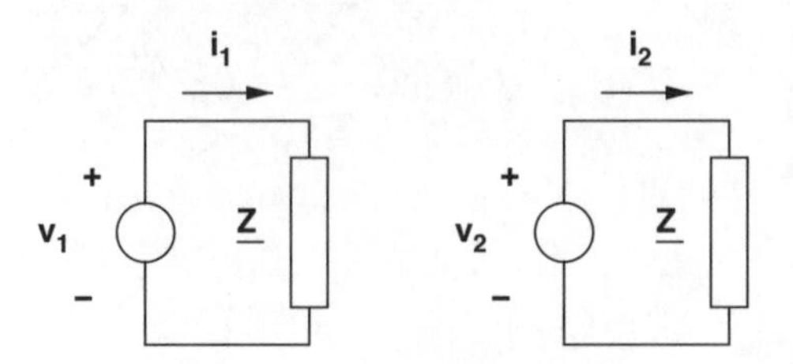

Figure 4.1 Two-phase system

then each of the two phase networks has the same value for real and reactive power:

$$P + jQ = \frac{V^2}{2|\mathbf{Z}|} e^{j\psi} \tag{4.3}$$

or:

$$P = \frac{V^2}{2|\mathbf{Z}|} \cos \psi \qquad\qquad Q = \frac{V^2}{2|\mathbf{Z}|} \sin \psi$$

It is straightforward to show that, for a system with voltage of the form described by Equations 4.1 and 4.2, instantaneous power is given by:

$$p_1 = \frac{V^2}{2|\mathbf{Z}|} \cos \psi \, [1 + \cos 2\omega t] + \frac{V^2}{2|\mathbf{Z}|} \sin \psi \sin 2\omega t$$

$$p_2 = \frac{V^2}{2|\mathbf{Z}|} \cos \psi \, [1 + \cos(2\omega t - \pi)] + \frac{V^2}{2|\mathbf{Z}|} \sin \psi \sin(2\omega t - \pi)$$

Note that the time-varying parts of these two expressions have opposite signs. Added together, they give instantaneous power:

$$p = p_1 + p_2 = \frac{|\mathbf{V}|^2}{|\mathbf{Z}|} \cos \psi$$

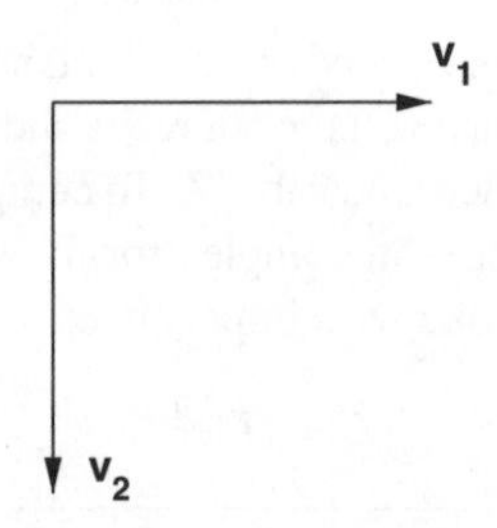

Figure 4.2 Phasor diagram for two-phase source

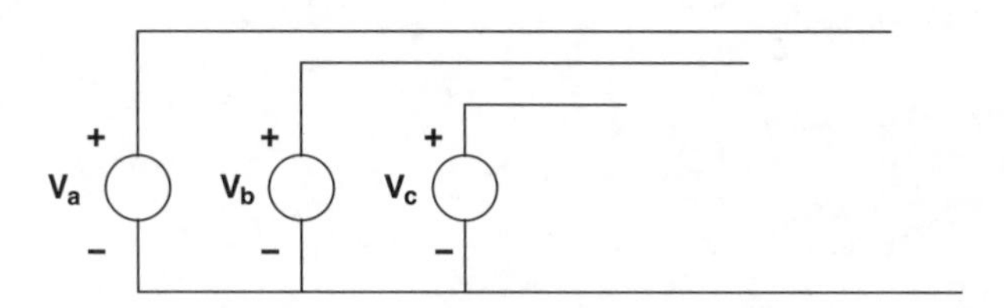

Figure 4.3 Three-phase voltage source

At least one of the advantages of polyphase power networks is now apparent. The use of a *balanced* polyphase system avoids the power flow pulsations due to ac voltage and current, and even the pulsations due to reactive energy flow. This has obvious benefits when dealing with motors and generators or, in fact, any type of source or load that would like to see constant power.

4.1 Three-Phase Systems

Now consider the arrangement of three voltage sources illustrated in Figure 4.3.

Assume that the three-phase voltages are:

$$v_a = V\cos\omega t = \mathrm{Re}\left[V e^{j\omega t}\right] \tag{4.4}$$

$$v_b = V\cos\left(\omega t - \frac{2\pi}{3}\right) = \mathrm{Re}\left[V e^{j\left(\omega t - \frac{2\pi}{3}\right)}\right] \tag{4.5}$$

$$v_c = V\cos\left(\omega t + \frac{2\pi}{3}\right) = \mathrm{Re}\left[V e^{j\left(\omega t + \frac{2\pi}{3}\right)}\right] \tag{4.6}$$

These three-phase voltages are illustrated in the time domain in Figure 4.4 and as complex phasors in Figure 4.5. Note the symmetrical spacing in time of the voltages. As in earlier

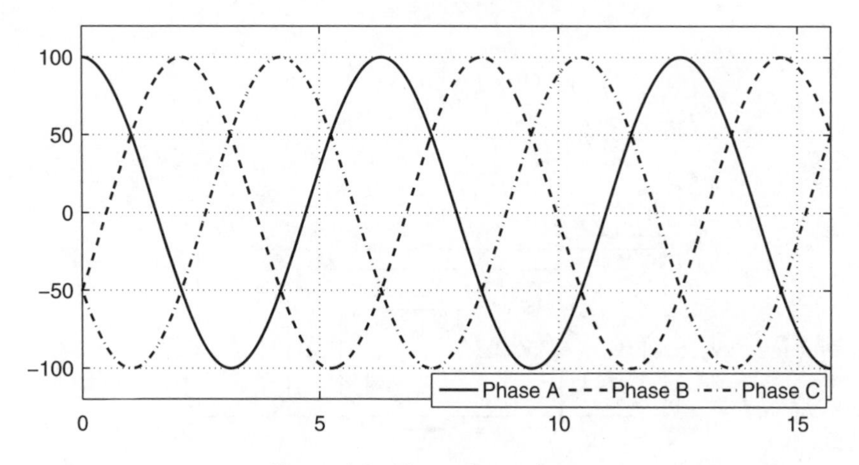

Figure 4.4 Three-phase voltages

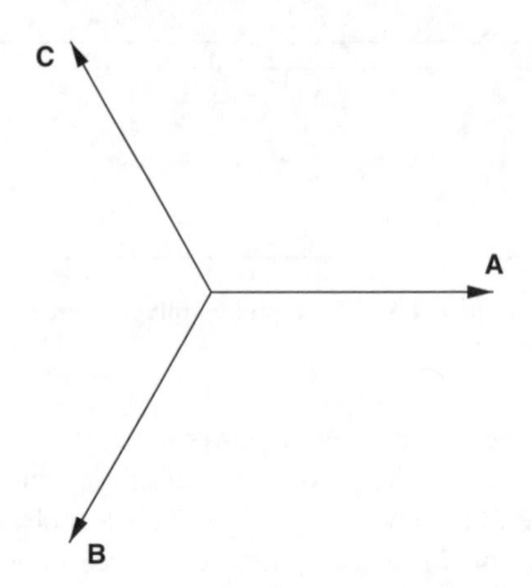

Figure 4.5 Phasor diagram: three-phase voltages

examples, the instantaneous voltages may be visualized by imagining Figure 4.5 spinning counterclockwise with angular velocity ω. The instantaneous voltages are just projections of the vectors of this 'pinwheel' onto the horizontal axis.

Consider connecting these three voltage sources to three identical loads, each with complex impedance $\mathbf{Z}$, as shown in Figure 4.6.

If voltages are as given by (4.4–4.6), then currents in the three phases are:

$$i_a = \mathrm{Re}\left\{\frac{\mathbf{V}}{\mathbf{Z}}e^{j\omega t}\right\} \qquad i_b = \mathrm{Re}\left\{\frac{\mathbf{V}}{\mathbf{Z}}e^{j(\omega t-\frac{2\pi}{3})}\right\} \qquad i_c = \mathrm{Re}\left\{\frac{\mathbf{V}}{\mathbf{Z}}e^{j(\omega t+\frac{2\pi}{3})}\right\}$$

Complex power in each of the three phases is:

$$P + jQ = \frac{V^2}{2|\mathbf{Z}|}(\cos\psi + j\sin\psi)$$

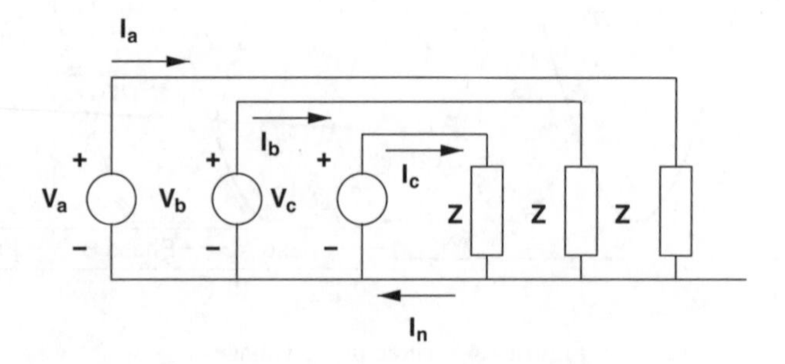

Figure 4.6 Three-phase source connected to balanced load

Then, remembering the time phase of the three sources, it is possible to write the values of instantaneous power in the three phases:

$$p_a = \frac{V^2}{2|\mathbf{Z}|}\{\cos\psi\,[1+\cos 2\omega t] + \sin\psi \sin 2\omega t\}$$
$$p_b = \frac{V^2}{2|\mathbf{Z}|}\left\{\cos\psi\left[1+\cos 2\left(\omega t - \frac{2\pi}{3}\right)\right] + \sin\psi \sin 2\left(\omega t - \frac{2\pi}{3}\right)\right\}$$
$$p_c = \frac{V^2}{2|\mathbf{Z}|}\left\{\cos\psi\left[1+\cos 2\left(\omega t + \frac{2\pi}{3}\right)\right] + \sin\psi \sin 2\left(\omega t + \frac{2\pi}{3}\right)\right\}$$

The sum of these three expressions is total instantaneous power, which is constant:

$$p = p_a + p_b + p_c = \frac{3}{2}\frac{V^2}{|\mathbf{Z}|}\cos\psi \tag{4.7}$$

It is useful, in dealing with three-phase systems, to remember that

$$\cos x + \cos\left(x - \frac{2\pi}{3}\right) + \cos\left(x + \frac{2\pi}{3}\right) = 0$$

regardless of the value of x.

Consider the current in the neutral wire, i_n in Figure 4.6. This current is given by:

$$i_n = i_a + i_b + i_c = \mathrm{Re}\left[\frac{\mathbf{V}}{\mathbf{Z}}\left(e^{j\omega t} + e^{j(\omega t - \frac{2\pi}{3})} + e^{j(\omega t + \frac{2\pi}{3})}\right)\right] = 0 \tag{4.8}$$

This shows the most important advantage of three-phase systems over two-phase systems: a wire with no current in it does not have to be very large. In fact, the neutral connection may be eliminated completely in many cases.

There is a fundamental difference between grounded and ungrounded systems if perfectly balanced conditions are not maintained. In effect, the ground wire provides isolation between the phases by fixing the neutral voltage at the star point to be zero. If the load impedances are not equal the load is said to be *unbalanced*. If an unbalanced system is grounded there may be current in the neutral. If an unbalanced load is not grounded, the star point voltage may not be zero, and the voltages will be different in the three phases at the load, even if the voltage sources all have the same magnitude.

4.2 Line–Line Voltages

A balanced three-phase set of voltages has a well-defined set of line–line voltages. If the line–neutral voltages are given by Equations 4.4–4.6, then line–line voltages are:

$$v_{ab} = v_a - v_b = \mathrm{Re}\left\{\mathbf{V}\left(1 - e^{-j\frac{2\pi}{3}}\right)e^{j\omega t}\right\}$$
$$v_{bc} = v_b - v_c = \mathrm{Re}\left\{\mathbf{V}\left(e^{-j\frac{2\pi}{3}} - e^{j\frac{2\pi}{3}}\right)e^{j\omega t}\right\}$$
$$v_{ca} = v_c - v_a = \mathrm{Re}\left\{\mathbf{V}\left(e^{j\frac{2\pi}{3}} - 1\right)e^{j\omega t}\right\}$$

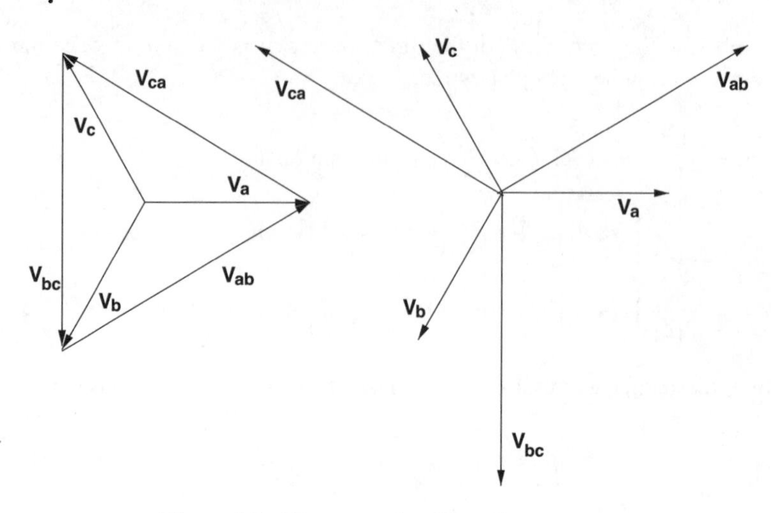

Figure 4.7 Line–neutral and line–line voltages

and these reduce to:

$$v_{ab} = \text{Re}\left\{\sqrt{3}\mathbf{V}e^{j\frac{\pi}{6}}e^{j\omega t}\right\}$$
$$v_{bc} = \text{Re}\left\{\sqrt{3}\mathbf{V}e^{-j\frac{\pi}{2}}e^{j\omega t}\right\}$$
$$v_{ca} = \text{Re}\left\{\sqrt{3}\mathbf{V}e^{j\frac{5\pi}{6}}e^{j\omega t}\right\}$$

The phasor relationship of line–neutral and line–line voltages is shown in Figure 4.7. Two things should be noted about this relationship:

- The line–line voltage set has a magnitude that is larger than the line–neutral voltage by a factor of $\sqrt{3}$.
- Line–line voltages are phase shifted by 30° with respect to the line–neutral voltages.

Clearly, line–line voltages themselves form a three-phase set just as do line–neutral voltages. This is shown conceptually in Figure 4.8. Power system components (sources, transformer windings, loads, etc.) may be connected either line–neutral or line–line. The former connection is often called *wye*, the latter is called *delta*, for obvious reasons. It should be noted that the *wye* connection is at least potentially a four-terminal connection, and could be grounded or ungrounded. The *delta* connection is inherently three-terminal and necessarily ungrounded.

4.2.1 Example: Wye and Delta Connected Loads

Loads may be connected in either line–neutral or line–line configuration. An example of the use of this flexibility is in a fairly commonly used distribution system with a line–neutral voltage of 120 V, RMS (about 170 V, peak). In this system the line–line voltage is 208 V, RMS (about 294 V, peak). Single–phase loads may be connected either line–line or line–neutral.

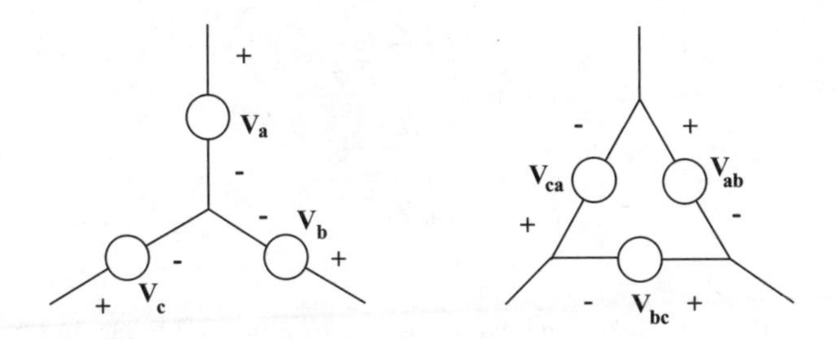

Figure 4.8 *Wye* and *delta* connected voltage sources

Suppose it is necessary to build a resistive heater to deliver 6 kW, to be made of three elements which may be connected in either *wye* or *delta*. Each of the three elements must dissipate 2000 W. Thus, since $P = \frac{V^2}{R}$, the *wye*-connected resistors would be:

$$R_y = \frac{120^2}{2000} = 7.2\,\Omega$$

while the *delta*-connected resistors would be:

$$R_\Delta = \frac{208^2}{2000} = 21.6\,\Omega$$

As is suggested by this example, *wye*- and *delta*-connected impedances are often directly equivalent. In fact, ungrounded connections are three-terminal networks which may be represented in two ways. The two networks shown in Figure 4.9, combinations of three passive impedances, are directly equivalent and identical in their terminal behavior if the relationships between elements are as given in Equations 4.9–4.14.

$$\mathbf{Z}_{ab} = \frac{\mathbf{Z}_a\mathbf{Z}_b + \mathbf{Z}_b\mathbf{Z}_c + \mathbf{Z}_c\mathbf{Z}_a}{\mathbf{Z}_c} \tag{4.9}$$

$$\mathbf{Z}_{bc} = \frac{\mathbf{Z}_a\mathbf{Z}_b + \mathbf{Z}_b\mathbf{Z}_c + \mathbf{Z}_c\mathbf{Z}_a}{\mathbf{Z}_a} \tag{4.10}$$

$$\mathbf{Z}_{ca} = \frac{\mathbf{Z}_a\mathbf{Z}_b + \mathbf{Z}_b\mathbf{Z}_c + \mathbf{Z}_c\mathbf{Z}_a}{\mathbf{Z}_b} \tag{4.11}$$

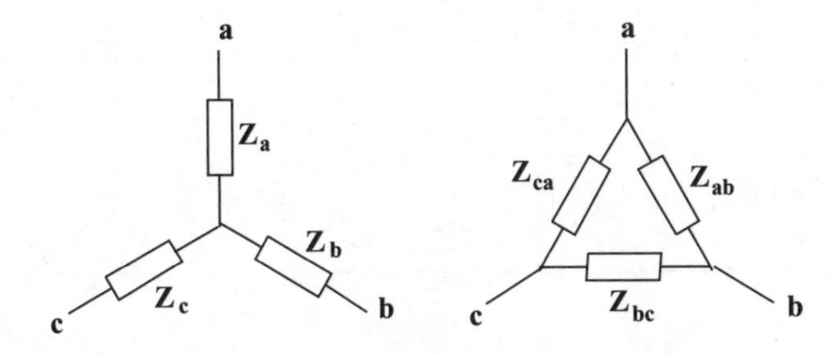

Figure 4.9 Wye and delta connected impedances

$$\mathbf{Z}_{\mathrm{a}} = \frac{\mathbf{Z}_{\mathrm{ab}}\mathbf{Z}_{\mathrm{ca}}}{\mathbf{Z}_{\mathrm{ab}} + \mathbf{Z}_{\mathrm{bc}} + \mathbf{Z}_{\mathrm{ca}}} \tag{4.12}$$

$$\mathbf{Z}_{\mathrm{b}} = \frac{\mathbf{Z}_{\mathrm{ab}}\mathbf{Z}_{\mathrm{bc}}}{\mathbf{Z}_{\mathrm{ab}} + \mathbf{Z}_{\mathrm{bc}} + \mathbf{Z}_{\mathrm{ca}}} \tag{4.13}$$

$$\mathbf{Z}_{\mathrm{c}} = \frac{\mathbf{Z}_{\mathrm{bc}}\mathbf{Z}_{\mathrm{ca}}}{\mathbf{Z}_{\mathrm{ab}} + \mathbf{Z}_{\mathrm{bc}} + \mathbf{Z}_{\mathrm{ca}}} \tag{4.14}$$

A special case of the *wye–delta* equivalence is that of *balanced* loads, in which:

$$\mathbf{Z}_{\mathrm{a}} = \mathbf{Z}_{\mathrm{b}} = \mathbf{Z}_{\mathrm{c}} = \mathbf{Z}_y$$

and

$$\mathbf{Z}_{\mathrm{ab}} = \mathbf{Z}_{\mathrm{bc}} = \mathbf{Z}_{\mathrm{ca}} = \mathbf{Z}_\Delta$$

in which case:

$$\mathbf{Z}_\Delta = 3\mathbf{Z}_y$$

4.2.2 Example: Use of Wye–Delta for Unbalanced Loads

The unbalanced load shown in Figure 4.10 is connected to a balanced voltage source. The problem is to determine the line currents. Note that his load is ungrounded (if it *were* grounded, this would be a trivial problem). The voltages are given by:

$$v_{\mathrm{a}} = V \cos \omega t$$

$$v_{\mathrm{b}} = V \cos \left(\omega t - \frac{2\pi}{3}\right)$$

$$v_{\mathrm{c}} = V \cos \left(\omega t + \frac{2\pi}{3}\right)$$

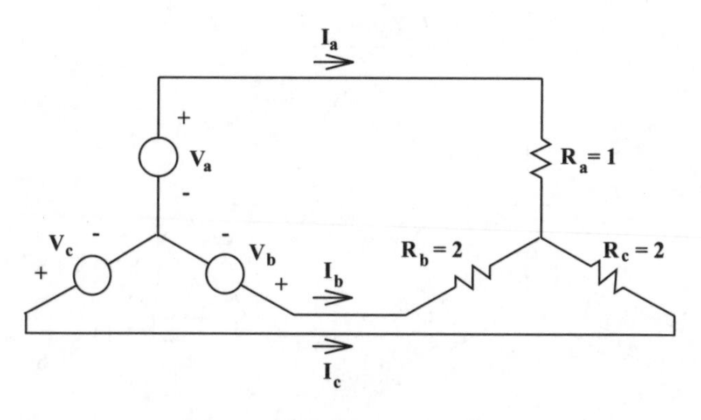

Figure 4.10 Unbalanced load

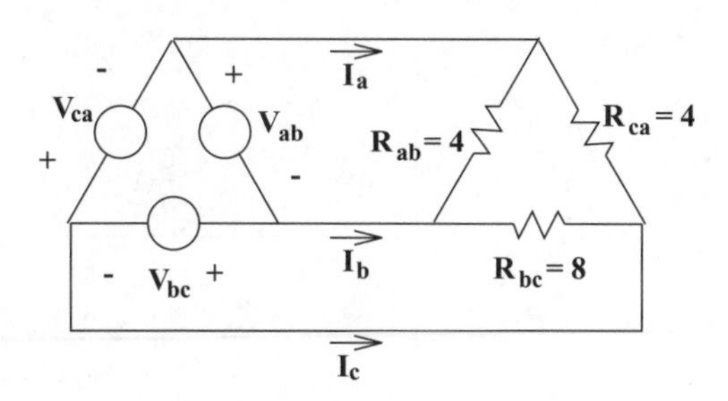

Figure 4.11 Delta equivalent

To solve this problem, convert both the source and load to delta equivalent connections, as shown in Figure 4.11. The values of the three resistors are:

$$r_{ab} = r_{ca} = \frac{2+4+2}{2} = 4$$

$$r_{bc} = \frac{2+4+2}{1} = 8$$

The complex amplitudes of the equivalent voltage sources are:

$$\begin{aligned}
\mathbf{V}_{ab} &= \mathbf{V}_a - \mathbf{V}_b = \mathbf{V}\left(1 - e^{-j\frac{2\pi}{3}}\right) &&= \mathbf{V}\sqrt{3}e^{j\frac{\pi}{6}} \\
\mathbf{V}_{bc} &= \mathbf{V}_b - \mathbf{V}_c = \mathbf{V}\left(e^{-j\frac{2\pi}{3}} - e^{j\frac{2\pi}{3}}\right) &&= \mathbf{V}\sqrt{3}e^{-j\frac{\pi}{2}} \\
\mathbf{V}_{ca} &= \mathbf{V}_c - \mathbf{V}_a = \mathbf{V}\left(e^{j\frac{2\pi}{3}} - 1\right) &&= \mathbf{V}\sqrt{3}e^{j\frac{5\pi}{6}}
\end{aligned}$$

Currents in each of the equivalent resistors are:

$$\mathbf{I}_{ab} = \frac{\mathbf{V}_{ab}}{r_{ab}} \quad \mathbf{I}_{bc} = \frac{\mathbf{V}_{bc}}{r_{bc}} \quad \mathbf{I}_{ca} = \frac{\mathbf{V}_{ca}}{r_{ca}}$$

The *line* currents are then just the difference between current in the legs of the *delta*:

$$\begin{aligned}
I_a &= I_{ab} - I_{ca} = \sqrt{3}V\left(\frac{e^{j\frac{\pi}{6}}}{4} - \frac{e^{j\frac{5\pi}{6}}}{4}\right) &&= \tfrac{3}{4}V \\
I_b &= I_{bc} - I_{ab} = \sqrt{3}V\left(\frac{e^{-j\frac{\pi}{2}}}{8} - \frac{e^{j\frac{\pi}{6}}}{4}\right) &&= -\left(\tfrac{3}{8} + j\tfrac{1}{4}\right)V \\
I_c &= I_{ca} - I_{bc} = \sqrt{3}V\left(\frac{e^{j\frac{5\pi}{6}}}{4} - \frac{e^{-j\frac{\pi}{2}}}{8}\right) &&= -\left(\tfrac{3}{8} - j\tfrac{1}{4}\right)V
\end{aligned}$$

These are shown in Figure 4.12.

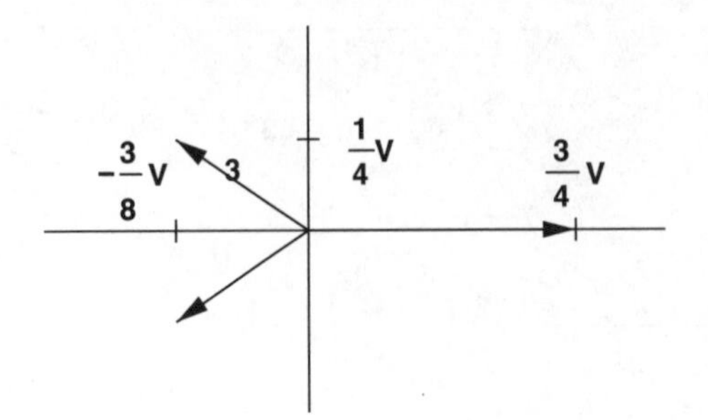

Figure 4.12 Line currents

4.3 Problems

1. Shown in Figure 4.13 is a three-phase voltage source. The three-phase voltages are:

$$v_a = \sqrt{2} \cdot 120 \cos(\omega t)$$
$$v_b = \sqrt{2} \cdot 120 \cos\left(\omega t - \frac{2\pi}{3}\right)$$
$$v_c = \sqrt{2} \cdot 120 \cos\left(\omega t + \frac{2\pi}{3}\right)$$

and note that the center point of this source is grounded.

For each of the six loads shown in Figure 4.14, find currents drawn from the three sources.

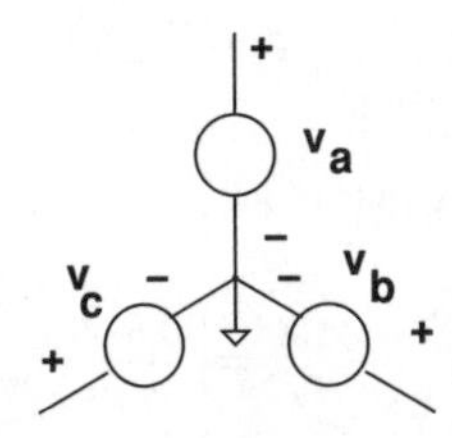

Figure 4.13 Three-phase voltage source

2. The situation is as shown in Figure 4.15. A three phase *current* source is feeding a three-phase resistive load. The currents are actually square waves, as shown in Figure 4.16. Assume the amplitude of the currents is 100 A and that each of the resistances in Figure 4.16 is 5 Ω. Estimate and draw a dimensioned sketch of each of the four voltages: v_a, v_b, v_c, v_g.

3. A three-phase *ungrounded* voltage source is shown in Figure 4.17 It is connected to an unbalanced *wye*-connected load as shown. Assume the source is a balanced, 120 V, line–neutral (208 V, line–line) sine wave voltage source.
Find the voltage between the neutral of the voltage source and the center point of the unbalanced load.

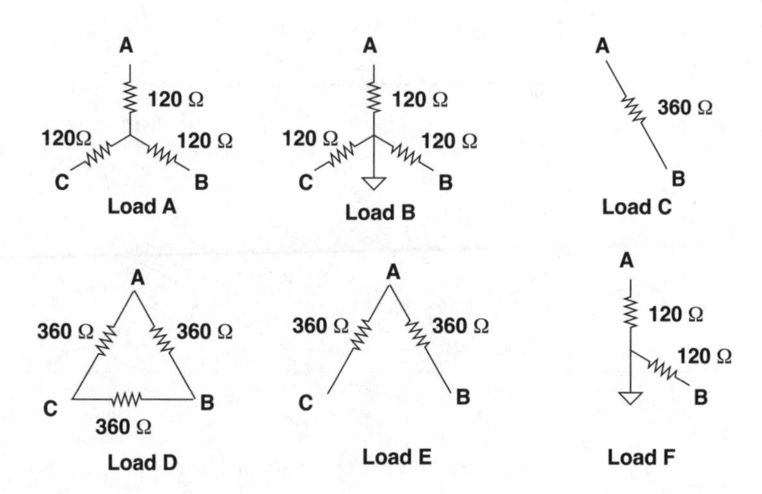

Figure 4.14 Resistive loads

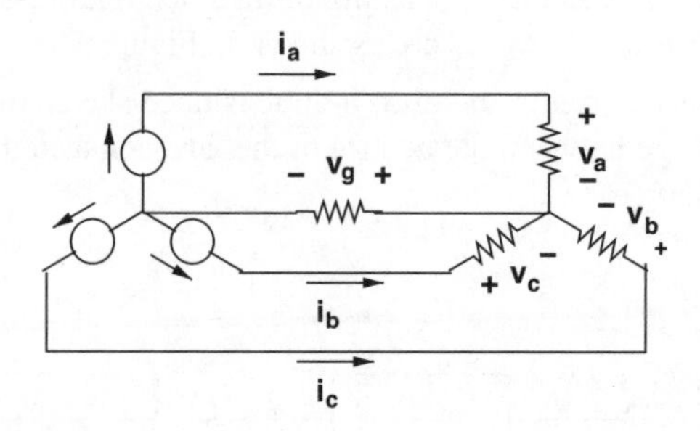

Figure 4.15 Current source feeding resistive load

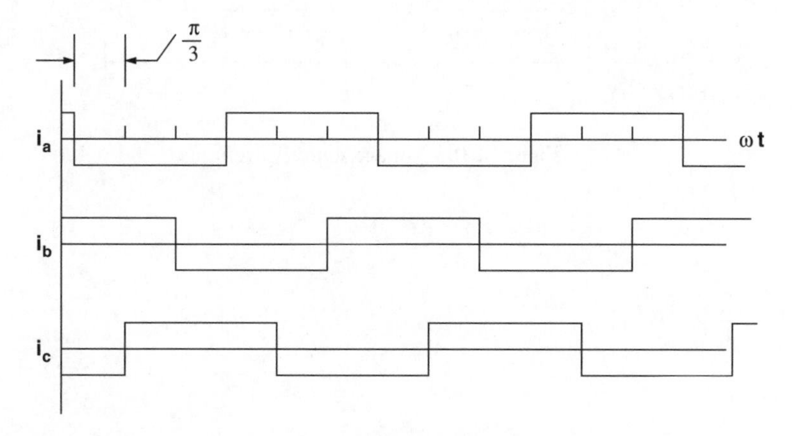

Figure 4.16 Currents

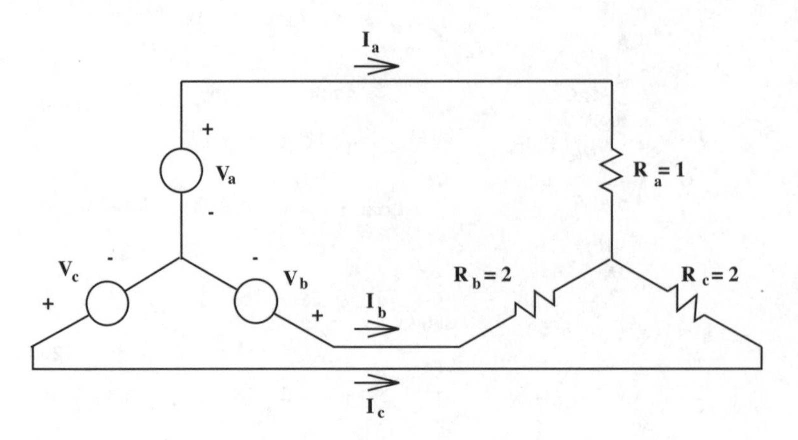

Figure 4.17 Voltage source and load

4. A voltage source consisting of three square-wave sources, connected in wye, is connected to a wye connected resistive load consisting of three ten ohm resistors. Neither source nor load is grounded. The square waves are as shown in Figure 4.18, with amplitude of 100 V.

(a) Find the three lead currents and draw a dimensioned sketch of their time behavior.

(b) What is the voltage between the neutral of the *wye*-connected resistors and the neutral of the voltage source?

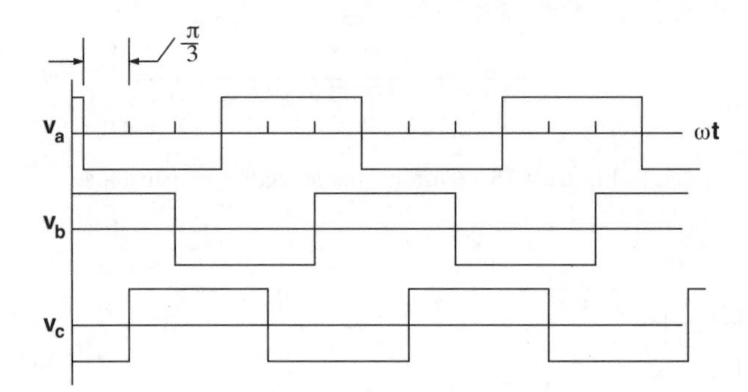

Figure 4.18 Voltage source waveforms

5

Electrical and Magnetic Circuits

Magnetic circuits offer, as do electric circuits, a way of simplifying the analysis of magnetic field systems which can be represented as having a collection of discrete elements. In electric circuits the elements are sources, resistors and so forth which are represented as having discrete currents and voltages. These elements are connected together with 'wires' and their behavior is described by network constraints (Kirchhoff's Voltage and Current Laws) and by constitutive relationships such as Ohm's Law. In magnetic circuits the lumped parameters are called 'reluctances' (the inverse of 'reluctance' is called 'permeance'). The analog to a 'wire' is referred to as a high permeance magnetic circuit element. Of course high permeability is the analog of high conductivity.

By organizing magnetic field systems into lumped parameter elements and using network constraints and constitutive relationships we can simplify the analysis of such systems.

5.1 Electric Circuits

Start with two conservation laws: conservation of charge and Faraday's Law. From these one can, with appropriate simplifying assumptions, derive the two fundamental circuit constraints embodied in Kirchhoff's laws.

5.1.1 Kirchoff's Current Law (KCL)

Conservation of charge could be written in integral form as:

$$\oiint \vec{J} \cdot \vec{n}\, \mathrm{d}a + \int_{\text{volume}} \frac{\mathrm{d}\rho_f}{\mathrm{d}t}\, \mathrm{d}v = 0 \tag{5.1}$$

This simply states that the sum of current *out* of some volume of space and rate of change of free charge in that space must be zero.

Now, if a discrete current is defined to be the integral of current density crossing through a part of the surface:

$$i_k = -\iint_{\text{surface}_k} \vec{J} \cdot \vec{n}\, \mathrm{d}a \tag{5.2}$$

and if it can be assumed that there is no accumulation of charge within the volume (in ordinary circuit theory the nodes are small and do not accumulate charge):

$$\oiint \vec{J} \cdot \vec{n}\, \mathrm{d}a = -\sum_k i_k = 0 \tag{5.3}$$

which holds if the sum over the index k includes all current paths into the node. This is Kirchhoff's Current Law (KCL).

5.1.2 *Kirchoff's Voltage Law (KVL)*

Faraday's Law is, in integral form:

$$\oint \vec{E} \cdot \mathrm{d}\vec{\ell} = -\frac{\mathrm{d}}{\mathrm{d}t} \iint \vec{B} \cdot \vec{n}\, \mathrm{d}a \tag{5.4}$$

where the closed loop in the left hand side of the equation is the edge of the surface of the integral on the right hand side.

If *voltage* is defined in the usual way, between points a and b for element k:

$$v_k = \int_{a_k}^{b_k} \vec{E} \cdot \mathrm{d}\vec{\ell} \tag{5.5}$$

Then, if the right-hand side of Faraday's Law (that is, magnetic induction) is negligible, the loop equation becomes:

$$\sum_k v_k = 0 \tag{5.6}$$

This works for circuit analysis because most circuits do not involve magnetic induction in the loops. However, it does form the basis for much head-scratching over voltages encountered by 'ground loops'.

5.1.3 *Constitutive Relationship: Ohm's Law*

Many of the materials used in electric circuits carry current through a linear conduction mechanism. That is, the relationship between electric field and electric current density is

$$\vec{J} = \sigma \vec{E} \tag{5.7}$$

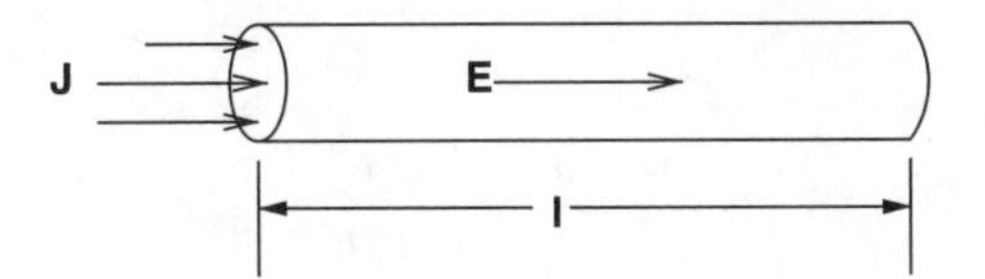

Figure 5.1 Simple rod-shaped resistor

Suppose, to start, we can identify a piece of material that has constant area and which is carrying current over some finite length, as shown in Figure 5.1. Assume this rod is carrying current density $\vec{J}$ (We won't say anything about how this current density managed to get into the rod, but assume that it is connected to something that can carry current – perhaps a wire). Total current carried by the rod is simply

$$I = |J|A$$

and then voltage across the element is:

$$v = \int \vec{E} \cdot d\ell = \frac{\ell}{\sigma A} I$$

from which we conclude the *resistance* is

$$R = \frac{V}{I} = \frac{\ell}{\sigma A}$$

Of course the lumped parameter picture can be used even with elements that are more complex. Consider the annular resistor shown in Figure 5.2. This is an end-on view of something that is uniform in cross-section and has depth D in the direction perpendicular to the page. Assume that the inner and outer elements are very good conductors, relative to the annular element in between. Assume further that this element has conductivity σ and inner and outer radii R_i and R_o, respectively.

Now, if this structure is carrying current from the inner to the outer electrode, current density at radius r would be:

$$\vec{J} = \vec{i}_r J_r(r) = \frac{I}{2\pi Dr}$$

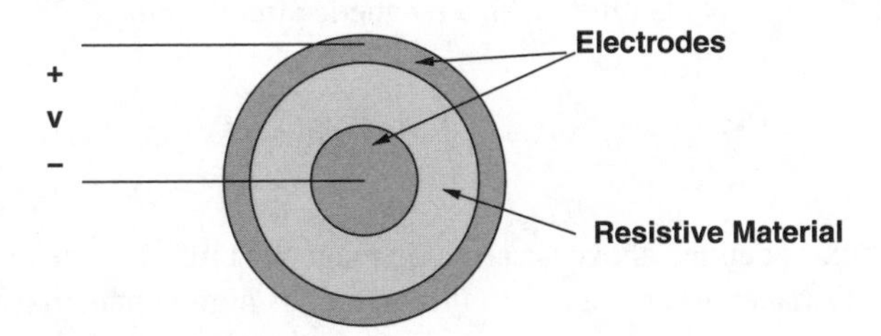

Figure 5.2 Annular resistor

Electric field is

$$E_r = \frac{J_r}{\sigma} = \frac{I}{2\pi D r \sigma}$$

Then voltage is

$$v = \int_{R_i}^{R_o} E_r(r) = \frac{I}{2\pi\sigma D}\log\frac{R_o}{R_i}$$

so that we conclude the resistance of this element is

$$R = \frac{\log\frac{R_o}{R_i}}{2\pi\sigma D}$$

5.2 Magnetic Circuit Analogies

In the electric circuit, elements for which voltage and current are defined are connected together by elements thought of as 'wires', or elements with zero or negligible voltage drop. The interconnection points are 'nodes'. In magnetic circuits the analogous thing occurs: elements for which magnetomotive force and flux can be defined are connected together by high permeability magnetic circuit elements (usually iron) which are the analog of wires in electric circuits.

5.2.1 Analogy to KCL

Gauss' Law is:

$$\oint\!\!\oint \vec{B} \cdot \vec{n}\, \mathrm{d}a = 0 \tag{5.8}$$

which means that the total amount of flux coming out of a region of space is *always* zero.

One can define a quantity which is sometimes called simply 'flux' or a 'flux tube'. This might be thought to be a collection of flux lines that can somehow be bundled together. Generally it is the flux that is identified with a magnetic circuit element. Mathematically it is:

$$\Phi_k = \iint \vec{B} \cdot \vec{n}\, \mathrm{d}a \tag{5.9}$$

In most cases, flux as defined above is carried in magnetic circuit elements which are made of high permeability material, analogous to the 'wires' of high conductivity material which carry current in electric circuits. It is possible to show that flux is largely contained in such high permeability materials.

If all of the flux tubes out of some region of space ('node') are considered in the sum, they must, according to Gauss' Law, add to zero:

$$\sum_k \Phi_k = 0 \tag{5.10}$$

5.2.2 *Analogy to KVL: Magnetomotive Force*

Ampere's Law is

$$\oint \vec{H} \cdot d\vec{\ell} = \iint \vec{J} \cdot \vec{n}\, da \tag{5.11}$$

Where, as for Faraday's Law, the closed contour on the left is the periphery of the (open) surface on the right. Now we define what we call *magnetomotive force* (MMF), in direct analogy to 'electromotive force', (voltage).

$$F_k = \int_{a_k}^{b_k} \vec{H} \cdot d\vec{\ell} \tag{5.12}$$

Further, define the current enclosed by a loop to be:

$$F_0 = \iint \vec{J} \cdot \vec{n}\, da \tag{5.13}$$

Then the analogy to KVL is:

$$\sum_k F_k = F_0$$

Note that the analogy is not exact as there is a source term on the right-hand side whereas KVL has no source term. Note also that *sign* counts here. The closed integral is taken in such direction so that the positive sense of the surface enclosed is positive (upwards) when the surface is to the left of the contour. (This is another way of stating the celebrated 'right-hand rule': if you wrap your right hand around the contour with your fingers pointing in the direction of the closed contour integration, your thumb is pointing in the positive direction for the surface.)

5.2.3 *Analogy to Ohm's Law: Reluctance*

Consider a 'gap' between two high permeability pieces as shown in Figure 5.3. If their permeability is high enough, there is negligible magnetic field H in them and so the MMF or 'magnetic potential' is essentially constant, just like in a wire. For the moment, assume that the gap dimension g is 'small' and uniform over the gap area A. Now, assume that some flux Φ is flowing from one of these to the other. That flux is

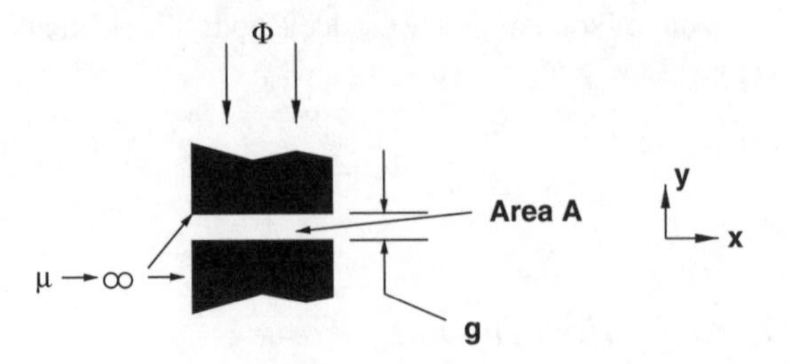

Figure 5.3 Air gap

$$\Phi = BA$$

where B is the flux density crossing the gap and A is the gap area. Note that this ignores 'fringing' fields in this simplified analysis. This neglect often requires correction in practice. Since the permeability of free space is μ_0, (assuming the gap is indeed filled with 'free space'), magnetic field intensity is

$$H = \frac{B}{\mu_0}$$

and gap MMF is just magnetic field intensity multiplied by gap dimension (g). This, of course, assumes that the gap is uniform and that so is the magnetic field intensity:

$$F = \frac{B}{\mu_0} g$$

Which means that the *reluctance* of the gap is the ratio of MMF to flux:

$$\mathcal{R} = \frac{F}{\Phi} = \frac{g}{\mu_0 A}$$

5.2.4 *Simple Case*

Consider the magnetic circuit situation shown in Figure 5.4. Here there is a piece of highly permeable material shaped to carry flux across a single air-gap. A coil is wound through the window in the magnetic material (this shape is usually referred to as a 'core'). The equivalent circuit is shown in Figure 5.5.

Note that in Figure 5.4, if the positive sense of the closed loop in a direction that goes vertically upwards through the leg of the core through the coil and then downwards through the gap, the current crosses the surface surrounded by the contour in the positive sense direction.

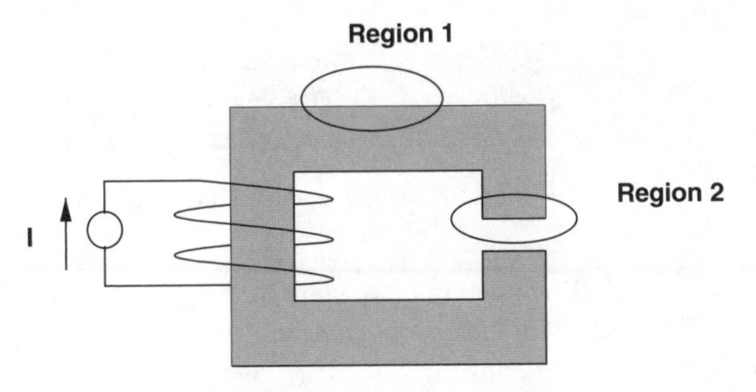

Figure 5.4 Single air-capped core

5.2.5 Flux Confinement

The gap in this case has the same reluctance as computed earlier, so that the flux in the gap is simply $\Phi = \frac{NI}{\mathcal{R}}$. Now, by focusing on the two regions indicated one might make a few observations about magnetic circuits. First, consider 'region 1' as shown in Figure 5.6.

In this picture, note that magnetic field $\vec{H}$ parallel to the surface must be the same inside the material as it is outside. Consider Ampere's Law carried out about a very thin loop consisting of the two arrows drawn at the top boundary of the material in Figure 5.6 with very short vertical paths joining them. If there is no current singularity inside that loop, the integral around it must be zero which means the magnetic field just inside must be the same as the magnetic field outside. Since the material is very highly permeable and $\vec{B} = \mu\vec{H}$, and 'highly permeable' means μ is very large, unless B is *really* large, $\vec{H}$ must be quite small. Thus the magnetic circuit has small magnetic field H and therefore flux densities parallel to and just outside its boundaries are also small.

At the surface of the magnetic material, since the magnetic field parallel to the surface must be very small, any flux lines that emerge from the core element must be perpendicular to the surface as shown for the gap region in Figure 5.7. This is true for region 1 as well as for region 2, but note that the total MMF available to drive fields across the gap is the same as would produce field lines from the area of region 1. Since any lines emerging from the magnetic material in region 1 would have very long magnetic paths, they must be very weak. Thus the magnetic circuit material largely confines flux, with only the relatively high permeance (low reluctance) gaps carrying any substantive amount of flux.

5.2.6 Example: C-Core

Consider a 'gapped' C-core as shown in Figure 5.8. This is two pieces of highly permeable material shaped generally like 'C's. They have uniform depth D in the direction perpendicular

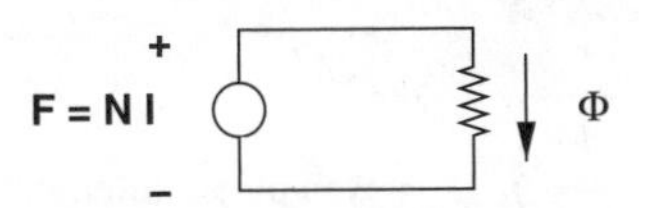

Figure 5.5 Equivalent circuit

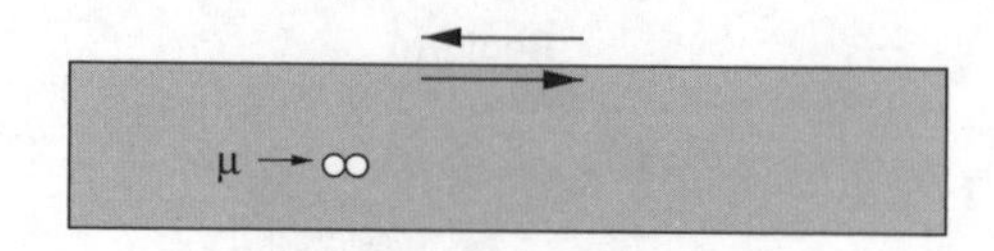

Figure 5.6 Flux confinement boundary: This is 'region 1'

to the page. The area $A = wD$, where w is the width at the gap. Assume the two gaps have the same area. Each of the gaps will have a reluctance:

$$\mathcal{R} = \frac{g}{\mu_0 A}$$

Suppose a coil with N turns is wound on this core as shown in Figure 5.9. Then a current I is made to flow in that coil. The magnetic circuit equivalent is shown in Figure 5.10. The two gaps are in series and, of course, in series with the MMF source. Since the two fluxes are the same and the MMFs add:

$$F_0 = NI = F_1 + F_2 = 2\mathcal{R}\Phi$$

and then

$$\Phi = \frac{NI}{2\mathcal{R}} = \frac{\mu_0 ANI}{2g}$$

and corresponding flux density in the gaps would be:

$$B_y = \frac{\mu_0 NI}{2g}$$

5.2.7 *Example: Core with Different Gaps*

As a second example, consider the perhaps oddly shaped core shown in Figure 5.11. Suppose the gap on the right has twice the area as the gap on the left. There are two gap reluctances:

$$\mathcal{R}_1 = \frac{g}{\mu_0 A} \qquad \mathcal{R}_2 = \frac{g}{2\mu_0 A}$$

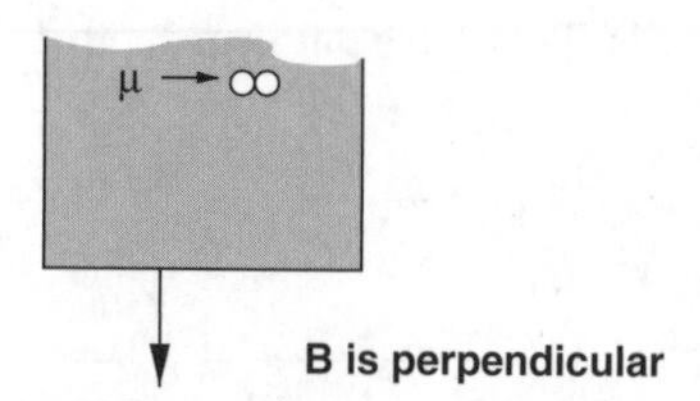

Figure 5.7 Gap boundary

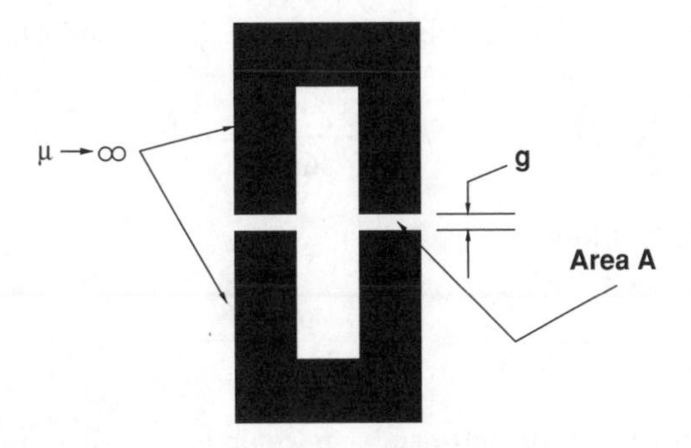

Figure 5.8 Gapped core

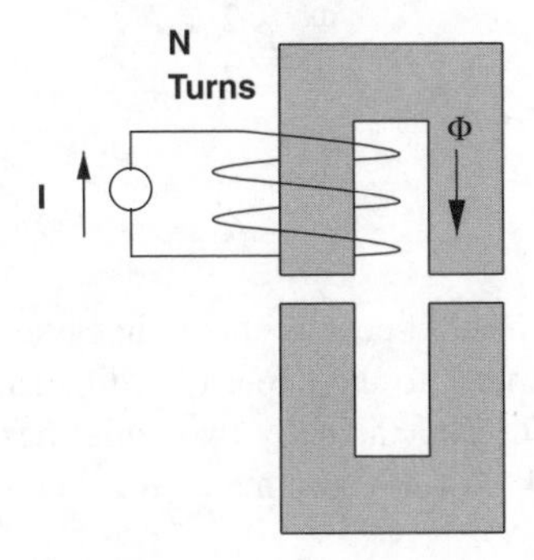

Figure 5.9 Wound, gapped core

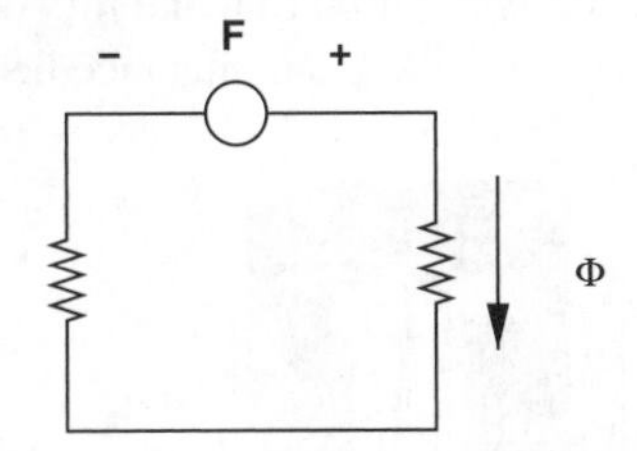

Figure 5.10 Equivalent magnetic circuit

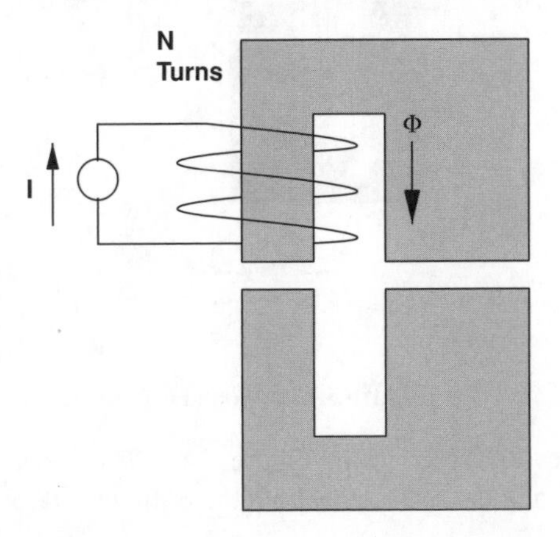

Figure 5.11 Wound, gapped core: different gaps

Since the two gaps are in series the flux is the same and the total reluctance is

$$\mathcal{R} = \frac{3}{2}\frac{g}{\mu_0 A}$$

Flux in the magnetic circuit loop is

$$\Phi = \frac{F}{\mathcal{R}} = \frac{2}{3}\frac{\mu_0 ANI}{g}$$

and the flux density across, say, the left-hand gap would be:

$$B_y = \frac{\Phi}{A} = \frac{2}{3}\frac{\mu_0 NI}{g}$$

5.3 Problems

1. An idealized picture of a gapped C-core is shown in Figure 5.12. In this picture the core leg width w is 2 cm, the depth into the paper (the dimension you cannot see) is 2.5 cm and the gap g is 0.5 mm. The coil (actually two coils) has a total of 100 turns (50 each side). Assume the permeability of the core material is very high so that its reluctance can be ignored.
   (a) What is the inductance of this device?
   (b) How much current in the coil is required to make flux density in the gap equal 1.8 T?
   (c) What gap would be required to make the inductance be 10 mH?

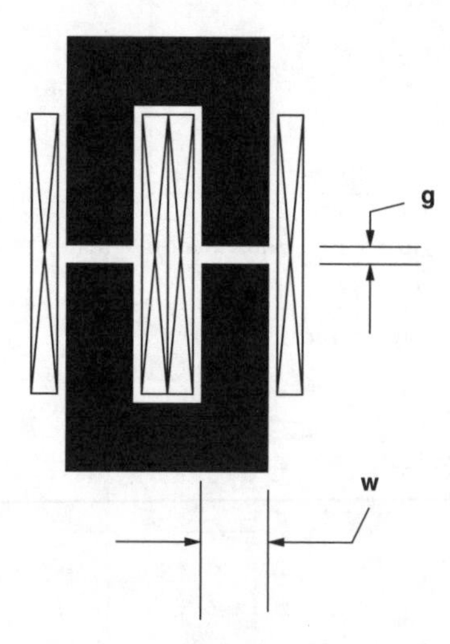

Figure 5.12 Wound C core

2. A rotary actuating device is shown in Figure 5.13. The inner part of this device can rotate about its center, and the surfaces of both the rotor and stator are cylindrical. The gap

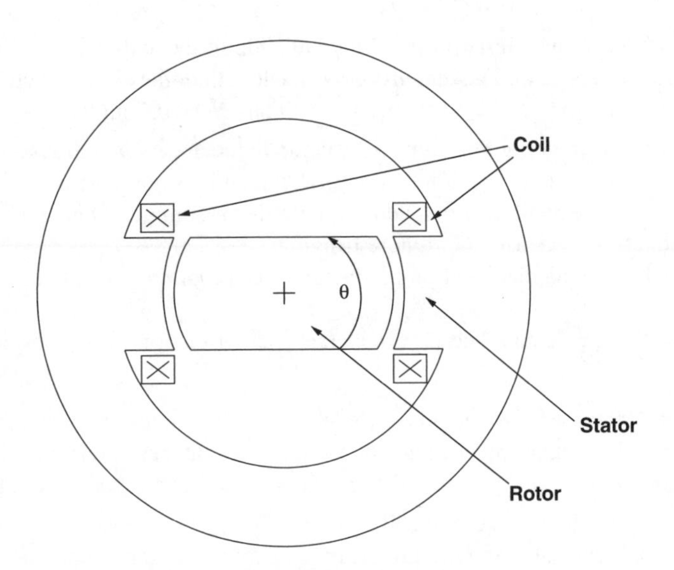

Figure 5.13 Rotary device

subtends an angle $\theta_g = 30°$. The gap radius is 5 cm and the gap dimension itself (clearance) is 0.1 mm. The coil wrapped around the stator poles has a total of 50 turns (in two coils, actually). The device is 10 cm long (in the dimension perpendicular to the paper).

(a) What is the maximum inductance?

(b) Ignoring fringing fields, calculate and sketch inductance of this device as a function of rotor position θ.

3. A solenoid actuator is shown in cartoon form in Figure 5.14. This is a cylindrical device, so if you were to look at it from either the right or left it would look like a number of circles.

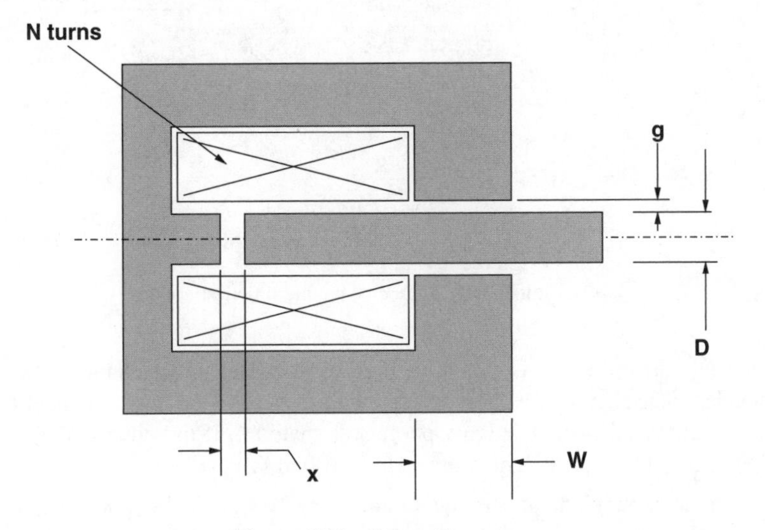

Figure 5.14 Solenoid actuator

The rod is fixed in its lateral position but can slide in the x direction. In this picture, the radius of that rod is 1 cm. The dimension W, the length of the air-gap section is 2 cm. (This picture is not drawn to accurate scale.) The coil has $N = 100$ turns.

(a) If the clearance gap $g = 1$ mm, what is the inductance when the rod is all the way in ($x = 0$). (Use the assumption that $g \ll R$, which is almost true here.)
(b) If the clearance gap $g = 1$ cm, you can't use that assumption, but you can still calculate the inductance when $x = 0$. Ignore fringing.
(c) If $g = 1$ mm, calculate and plot inductance as a function of rod position for $0 < x < 1$ cm.
(d) If $g = 1$ cm, calculate and plot inductance as a function of rod position for $0 < x < 1$ cm.

4. A coil with 100 turns is wound on a toroidal core as shown in Figure 5.15. The core has an inner radius of 2 cm, an outer radius of 5 cm and a thickness of 1 cm. The core material has a uniform permeability of $\mu = 200\mu_0$ and saturates at a flux density of 1.2 T.

(a) What is the inductance of the coil?
(b) How much current is required to bring the core to the start of saturation? (That is, so that the material at the inner diameter of the core reaches a flux density of 1.2 T.)
(c) How much current is required to fully saturate the core? (That is, to make all of the material in the core have a flux density of at least 1.2 T.)

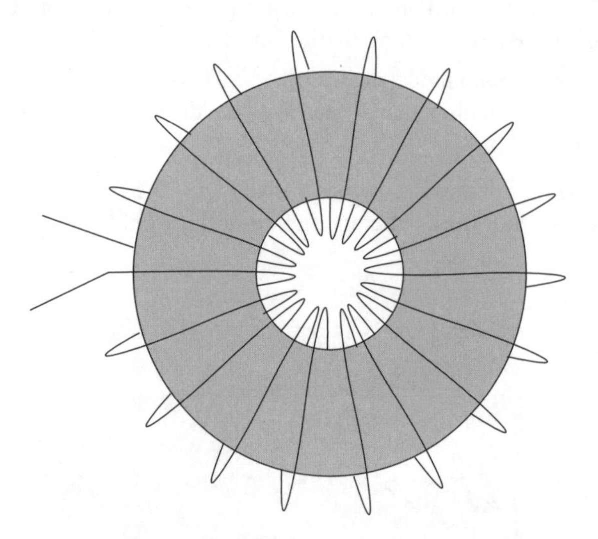

Figure 5.15 Inductor wound on circular core

5. An unusual sort of sensor is shown in Figure 5.16. The upper element, made of highly permeable material can move horizontally with respect to the lower element (a sort of 'E' core), which is also made of highly permeable material. If the outer coils are driven by a 2000 Hz current source with magnitude $I_0 = 100$ mA.

(a) When the upper element is in the middle ($x = 0$), what is the inductance of the driving coils? What voltage is required from the current source?

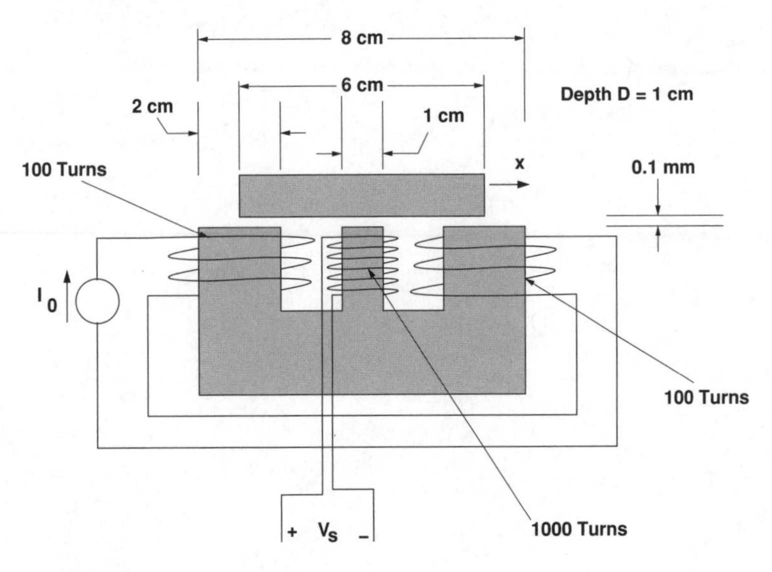

Figure 5.16 Sensor

(b) If the value of displacement x is small, what is the relationship between x and output voltage V_s?

(c) Ignoring fringing fields, what is the magnitude of output voltage V_s as a function of x for $0 < x < 1$ cm?

6

Transformers

Transformers are essential parts of most power systems. Their role is to convert electrical energy at one voltage to some other voltage. Transmission lines are typically operated at voltages that are substantially higher than either generation or utilization voltage, for at least two reasons.

- Because power is voltage multiplied by current, and because transmission line losses are proportional to the square of current, higher voltages generally produce lower transmission line losses.
- Transmission line capacity to carry power is roughly proportional to the square of voltage, so to make intensive use of transmission line right-of-way: that is to enable transmission lines to carry high power levels, high transmission voltages are required. This is true for underground cables as well as for above the ground transmission lines.

For these reasons, a transformer is usually situated right at the output of each generating unit to transform the power from generation voltage, which is in the range, usually, between 10 and 30 kV and transmission voltage which is in the high or extra high voltage range, typically between 138 and 765 kV. At substations that connect transmission lines to distribution circuits the power is stepped down in voltage. Distribution circuits generally operate in the range of 6 kV to 35 kV. (However, there are still lower voltage distribution primaries and some higher voltage circuits might be classified as 'distribution'.) Then before electric power is connected to customer loads there is yet another transformer to transform it from the distribution primary voltage to the customer at a voltage that is typically on the order of 240 V. In the United States that last transformer is center tapped to give the familiar 120 V, RMS, single phase voltage used for most appliances. Figure 6.1 is an annotated cutaway drawing of a large power transformer manufactured by ABB, illustrating some of the major parts of such a transformer unit.

6.1 Single-phase Transformers

The transformers used in electric power systems may be built as three-phase units and may have multiple and complicated winding patterns. But three-phase transformers are at the core

Electric Power Principles: Sources, Conversion, Distribution and Use James L. Kirtley

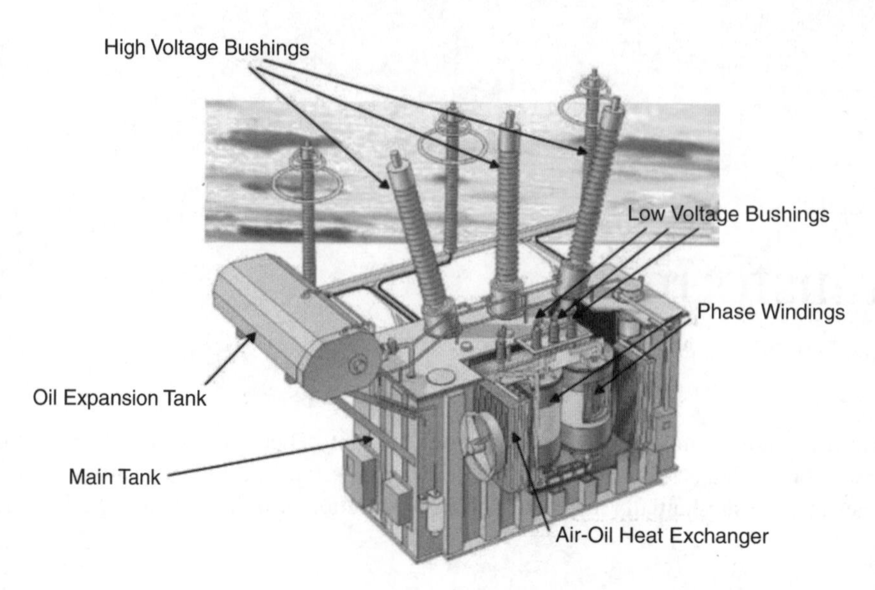

Figure 6.1 ABB Trafo-Star^{TM} transformer. Reproduced by permission of ABB Ltd. Annotation by Author

of things, understandable as three single-phase units, and there are many applications for transformers that work with only single-phase electric power, so this discussion starts with single-phase transformers.

6.1.1 Ideal Transformer

Transformers are actually fairly complex electromagnetic devices, subject to involved analysis, but for the purpose of learning about electric power systems, it will be sufficient to start with a simplified model for the transformer which is called the *ideal* transformer. This is a two-port circuit element, shown in Figure 6.2.

The *ideal transformer* as a network element constrains its terminal variables in the following way:

$$\frac{v_1}{N_1} = \frac{v_2}{N_2} \tag{6.1}$$

$$N_1 i_1 = -N_2 i_2 \tag{6.2}$$

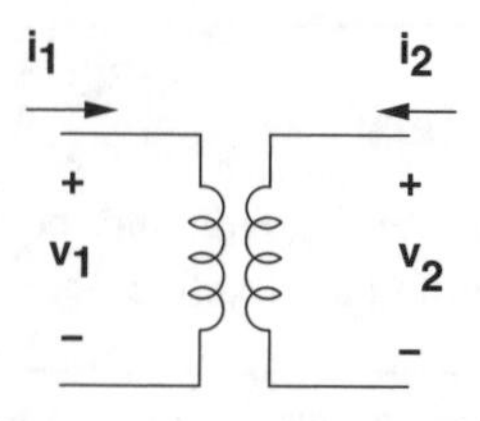

Figure 6.2 Ideal transformer

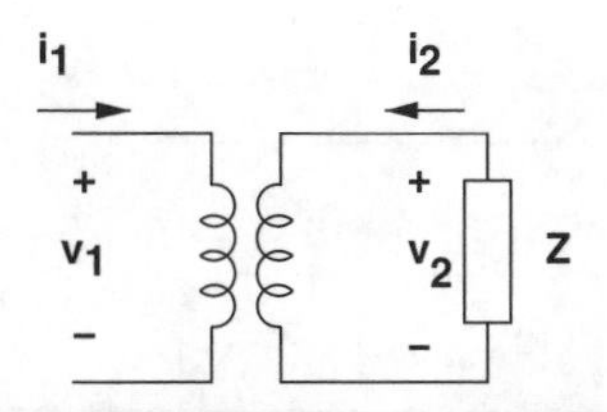

Figure 6.3 Impedance transformation

1. In normal operation, a transformer *turns ratio* $\frac{N_1}{N_2}$ is selected so that the desired voltages appear at the proper terminals. For example, to convert 13.8 kV distribution voltage to the 120/240 V level suitable for residential or commercial single phase service, one would use a transformer with turns ratio of $\frac{13800}{240} = 57.5$. To split the low voltage in half, a *center tap* on the low voltage winding would be used.
2. The transformer, at least in its *ideal* form, does not consume, produce nor store energy. Note that, according to Equations 6.1 and 6.2, the *sum* of power flows into an ideal transformer is identically zero:

$$p_1 + p_2 = v_1 i_1 + v_2 i_2 = 0 \tag{6.3}$$

3. The transformer also transforms impedances. To show how this is, look at Figure 6.3. Here, some impedance is connected to one side of an ideal transformer. See that it is possible to find an equivalent impedance viewed from the other side of the transformer.
   Noting that:

$$\mathbf{I}_2 = -\frac{N_1}{N_2}\mathbf{I}_1$$

   and that

$$\mathbf{V}_2 = -\mathbf{Z}\mathbf{I}_2$$

   Then the ratio between input voltage and current is:

$$\mathbf{V}_1 = \frac{N_1}{N_2}\mathbf{V}_2 = \left(\frac{N_1}{N_2}\right)^2 \mathbf{Z}\mathbf{I}_1 \tag{6.4}$$

6.1.2 Deviations from Ideal Transformer

As it turns out, the ideal transformer model is not too bad for showing the behavior of a real transformer under most circumstances. In transformers operating at power frequencies there are two meaningful deviations from ideal behavior of real transformers, and these are related to the transformer cores themselves, and because the transformer windings have resistance and leakage reactance.

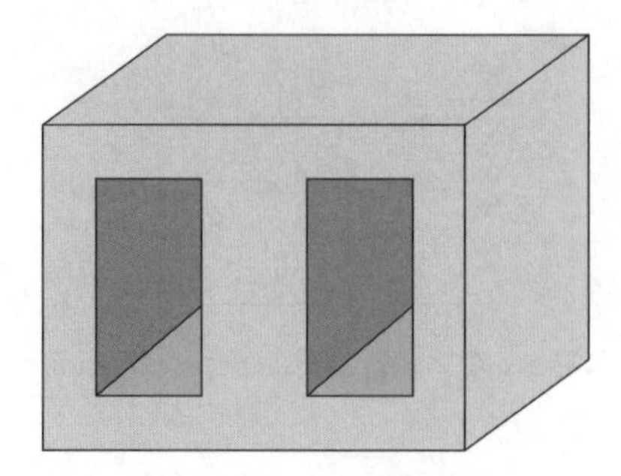

Figure 6.4 Transformer core

Transformers are wound on magnetic circuits (cores) made of thin sheets, called laminations, of steel. For power transformers, grain-oriented silicon steel is used so that the transformer has relatively low loss when energized. The transformer core itself might look like the cartoon shown in Figure 6.4. In this figure, the steel laminations would be oriented in the plane of the paper and the windings of the transformer would be located in the 'windows' in that core.

Assuming that the transformer has N_1 turns in one winding, connected to a voltage of V_1, and that the core itself has an area A_c, then the flux density in the core is

$$B_c \approx \frac{V_1}{\omega N_1}$$

The core material will draw some reactive power and will, of course, have losses from hysteresis and eddy currents. That reactive power and loss might be approximated by shunt reactive and resistive elements. If real and reactive power drawn by the core are P_c and Q_c, respectively, the resistance and reactance elements that describe transformer behavior are:

$$R_\phi \approx \frac{V_1^2}{P_c}$$

$$X_\phi \approx \frac{V_1^2}{Q_c}$$

Methods of estimating real power loss in magnetic circuits will be described in Chapter 8. Generally, it is possible to use tabulated data on core materials in conjunction with the flux density B_c and frequency ω to deduce a 'watts per kilogram' and a 'VARs per kilogram' number for a given core material. The values of P_c and Q_c are then calculated by using those two numbers and multiplying by the mass of the core. It is important to realize here that the behavior of magnetic iron is nonlinear with respect to flux density, so these resistive and reactive core elements are only approximations to actual transformer behavior, and their values will be functions of operating point. Core loss is a very important element of transformer behavior since power transformers in utility systems are virtually always energized, so that 'no-load' loss in power transformers is relatively expensive.

The windings in the power transformer, of course, have resistance. They also make magnetic flux, not all of which is mutually coupled. (The flux in the core is all mutually coupled, but there will be some 'leakage' flux in the air around the core, and this is what gives rise to the

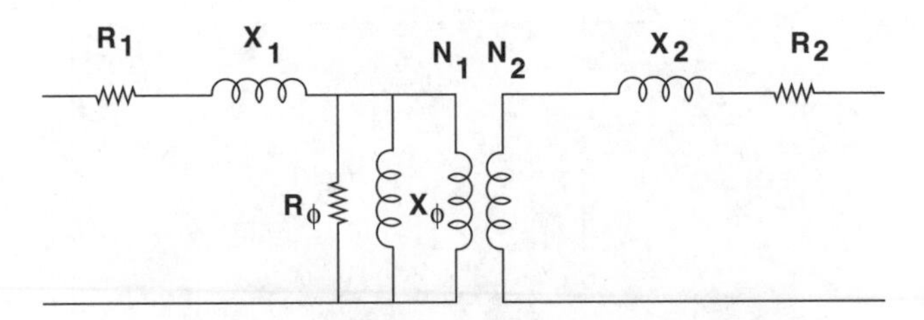

Figure 6.5 Equivalent circuit model of a power transformer

leakage flux elements.) An approximately complete model of a power transformer is shown in Figure 6.5. The leakage reactance elements and resistances are associated with each of the windings. Because impedances can be transformed across the ideal transformer, a somewhat more compact model can be derived, as shown in Figure 6.6. Here:

$$R = R_1 + \left(\frac{N_1}{N_2}\right)^2 R_2$$

$$X = X_1 + \left(\frac{N_1}{N_2}\right)^2 X_2$$

Typically, the core elements of the power transformer are fairly large, so they have little effect on the behavior of the power system, although the core loss element is economically important. Also, the series resistance tends to be quite small, so in system studies it is often found that only the series leakage reactance need be represented. As will be shown in Section 7.2, the ideal transformer part of the transformer equivalent circuit does not appear in most system studies.

6.2 Three-Phase Transformers

Polyphase transformers can be implemented as three single-phase transformers or as three sets of winding on a suitable core. Figure 6.7 shows a cartoon of a three-phase core. A set of windings, each constituting a single phase transformer, would be wound around each leg of this three-legged core.

A three-phase transformer operates as three single-phase transformers. The complication is that there are a number of ways of winding them, and a number of ways of interconnecting

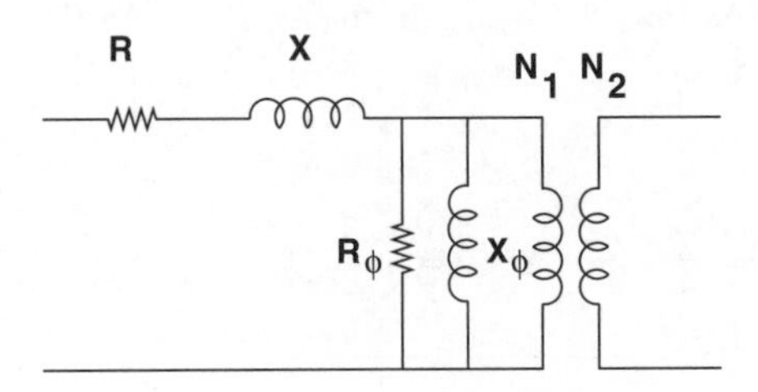

Figure 6.6 Equivalent circuit model of a power transformer with all leakage and winding resistance referred to one side

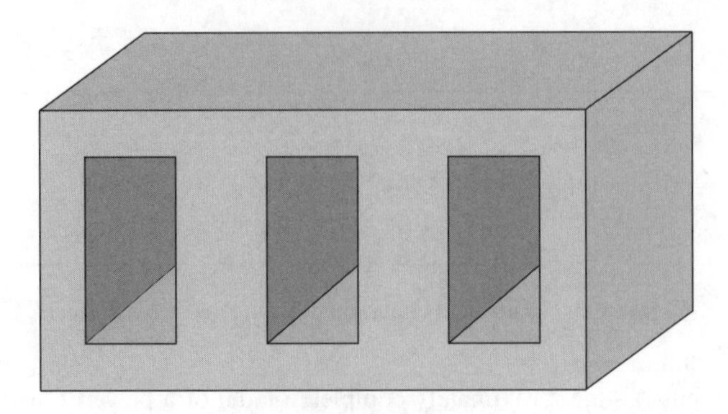

Figure 6.7 Three-phase, three-leg transformer core

them. On either 'side' of a transformer connection (i.e. the *high-voltage* and *low-voltage* sides), it is possible to connect transformer windings either line to neutral (*wye*), or line to line (*delta*). Thus we may speak of transformer connections being *wye–wye*, *delta–delta*, *wye–delta*, or *delta–wye*.

Connection of transformers in either *wye–wye* or *delta–delta* is reasonably easy to understand. Each of the line–neutral (in the case of *wye–wye*), or line–line (in the case of *delta–delta*) voltages is transformed by one of the three transformers. On the other hand, the interconnections of a *wye–delta* or *delta–wye* transformer are a little more complex. Figure 6.8 shows a *delta–wye* connection, in what might be called 'wiring diagram' form. A more schematic (and more common) form of the same picture is shown in Figure 6.9. In that picture, winding elements that *appear* parallel are wound on the same core segment, and so constitute a single phase transformer.

Assume that N_Δ and N_Y are numbers of turns. If the individual transformers are considered to be ideal, the following voltage and current constraints exist:

$$v_{aY} = \frac{N_Y}{N_\Delta}(v_{a\Delta} - v_{b\Delta}) \qquad i_{a\Delta} = \frac{N_Y}{N_\Delta}(i_{aY} - i_{cY})$$

$$v_{bY} = \frac{N_Y}{N_\Delta}(v_{b\Delta} - v_{c\Delta}) \qquad i_{b\Delta} = \frac{N_Y}{N_\Delta}(i_{bY} - i_{aY})$$

$$v_{cY} = \frac{N_Y}{N_\Delta}(v_{c\Delta} - v_{a\Delta}) \qquad i_{c\Delta} = \frac{N_Y}{N_\Delta}(i_{cY} - i_{bY})$$

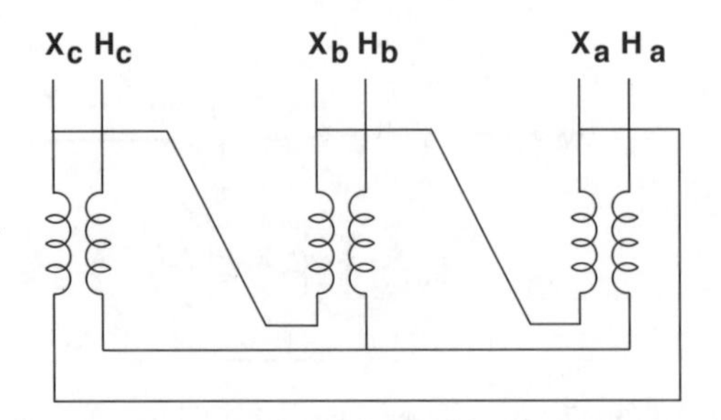

Figure 6.8 *Delta–wye* transformer connection

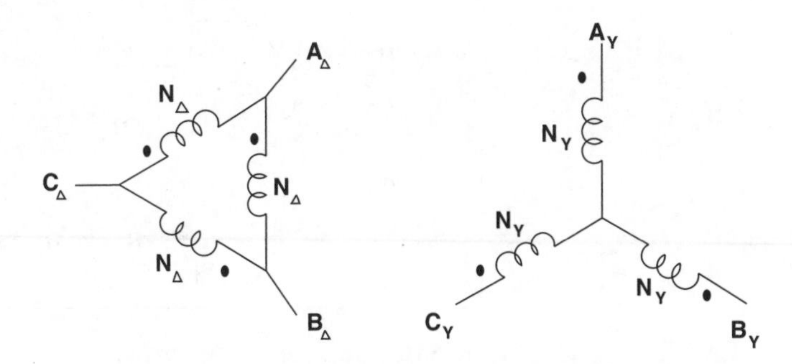

Figure 6.9 Schematic of *Delta–wye* transformer connection

where each of the *voltages* are line–neutral and the *currents* are in the lines at the transformer terminals.

Consider what happens if a *delta–wye* transformer is connected to a balanced three-phase voltage source, so that:

$$v_{a\Delta} = \mathrm{Re}\left(\mathbf{V}\,\mathrm{e}^{j\omega t}\right)$$
$$v_{b\Delta} = \mathrm{Re}\left(\mathbf{V}\,\mathrm{e}^{j(\omega t - \frac{2\pi}{3})}\right)$$
$$v_{c\Delta} = \mathrm{Re}\left(\mathbf{V}\,\mathrm{e}^{j(\omega t + \frac{2\pi}{3})}\right)$$

Then, complex amplitudes on the *wye* side are:

$$\mathbf{V}_{aY} = \frac{N_Y}{N_\Delta}\mathbf{V}\left(1 - \mathrm{e}^{-j\frac{2\pi}{3}}\right) = \sqrt{3}\frac{N_Y}{N_\Delta}\mathbf{V}\,\mathrm{e}^{j\frac{\pi}{6}}$$
$$\mathbf{V}_{bY} = \frac{N_Y}{N_\Delta}\mathbf{V}\left(\mathrm{e}^{-j\frac{2\pi}{3}} - \mathrm{e}^{j\frac{2\pi}{3}}\right) = \sqrt{3}\frac{N_Y}{N_\Delta}\mathbf{V}\,\mathrm{e}^{-j\frac{\pi}{2}}$$
$$\mathbf{V}_{cY} = \frac{N_Y}{N_\Delta}\mathbf{V}\left(\mathrm{e}^{j\frac{2\pi}{3}} - 1\right) = \sqrt{3}\frac{N_Y}{N_\Delta}\mathbf{V}\,\mathrm{e}^{j\frac{5\pi}{6}}$$

Two observations should be made here:

- The ratio of voltages (that is, the ratio of either *line-line* or *line-neutral*) is different from the *turns ratio* by a factor of $\sqrt{3}$.
- All *wye* side voltages are shifted in *phase* by 30° with respect to the *delta* side voltages.

6.2.1 *Example*

Suppose we have the following problem to solve:

> A balanced three-phase *wye*-connected resistor is connected to the *delta* side of a *wye–delta* transformer with a nominal *voltage* ratio of
>
> $$\frac{v_\Delta}{v_Y} = N$$

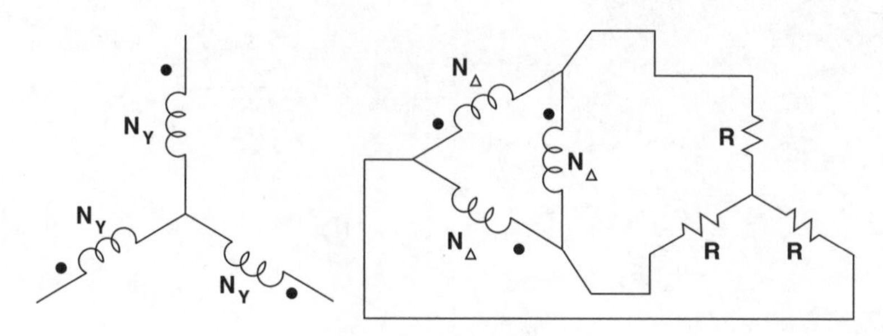

Figure 6.10 Example

What is the impedance looking into the *wye* side of the transformer, assuming drive with a balanced source?

The situation is shown in Figure 6.10.

It is important to remember the relationship between the *voltage* ratio and the *turns* ratio, which is:

$$\frac{v_\Delta}{v_Y} = N = \frac{N_\Delta}{\sqrt{3}N_Y}$$

so that:

$$\frac{N_\Delta}{N_Y} = \sqrt{3}N$$

Next, the *wye–delta* equivalent transform for the load makes the picture look like Figure 6.11

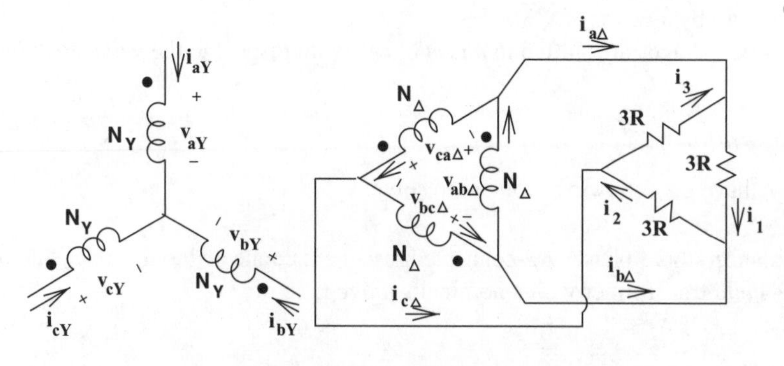

Figure 6.11 Equivalent situation

In this situation, each transformer secondary winding is connected directly across one of the three resistors. Currents in the resistors are given by:

$$i_1 = \frac{v_{ab\Delta}}{3R}$$
$$i_2 = \frac{v_{bc\Delta}}{3R}$$
$$i_3 = \frac{v_{ca\Delta}}{3R}$$

Line currents are:

$$i_{a\Delta} = i_1 - i_3 = \frac{v_{ab\Delta} - v_{ca\Delta}}{3R} = i_{1\Delta} - i_{3\Delta}$$
$$i_{b\Delta} = i_2 - i_1 = \frac{v_{bc\Delta} - v_{ab\Delta}}{3R} = i_{2\Delta} - i_{1\Delta}$$
$$i_{c\Delta} = i_3 - i_2 = \frac{v_{ca\Delta} - v_{bc\Delta}}{3R} = i_{3\Delta} - i_{2\Delta}$$

Solving for currents in the legs of the transformer Δ, subtract, for example, the second expression from the first:

$$2i_{1\Delta} - i_{2\Delta} - i_{3\Delta} = \frac{2v_{ab\Delta} - v_{bc\Delta} - v_{ca\Delta}}{3R}$$

Now, taking advantage of the fact that the system is balanced:

$$i_{1\Delta} + i_{2\Delta} + i_{3\Delta} = 0$$
$$v_{ab\Delta} + v_{bc\Delta} + v_{ca\Delta} = 0$$

to find:

$$i_{1\Delta} = \frac{v_{ab\Delta}}{3R}$$
$$i_{2\Delta} = \frac{v_{bc\Delta}}{3R}$$
$$i_{3\Delta} = \frac{v_{ca\Delta}}{3R}$$

Finally, the ideal transformer relations give:

$$v_{ab\Delta} = \frac{N_\Delta}{N_Y} v_{aY} \qquad i_{aY} = \frac{N_\Delta}{N_Y} i_{1\Delta}$$
$$v_{bc\Delta} = \frac{N_\Delta}{N_Y} v_{bY} \qquad i_{bY} = \frac{N_\Delta}{N_Y} i_{2\Delta}$$
$$v_{ca\Delta} = \frac{N_\Delta}{N_Y} v_{cY} \qquad i_{cY} = \frac{N_\Delta}{N_Y} i_{3\Delta}$$

so that:

$$i_{aY} = \left(\frac{N_\Delta}{N_Y}\right)^2 \frac{1}{3R} v_{aY}$$
$$i_{bY} = \left(\frac{N_\Delta}{N_Y}\right)^2 \frac{1}{3R} v_{bY}$$
$$i_{cY} = \left(\frac{N_\Delta}{N_Y}\right)^2 \frac{1}{3R} v_{cY}$$

The apparent resistance (that is, apparent were it to be connected in *wye*) at the *wye* terminals of the transformer is:

$$R_{eq} = 3R\left(\frac{N_Y}{N_\Delta}\right)^2$$

Expressed in terms of *voltage* ratio, this is:

$$R_{eq} = 3R\left(\frac{N}{\sqrt{3}}\right)^2 = R\left(\frac{v_Y}{v_\Delta}\right)^2$$

It is important to note that this solution took the long way around. Taken consistently (uniformly on a line–neutral or uniformly on a line–line basis), impedances transform across transformers by the square of the *voltage* ratio, no matter what connection is used.

6.3 Problems

1. An ideal transformer is shown in Figure 6.12. Use this model for a distribution transformer from $V_H = 24$ kV to $V_X = 240$ V (RMS). This is, of course, a single phase transformer and is of the type commonly used to provide the 240 V service commonly used in the United States. It is center-tapped to provide 120 V for branch circuits. Assume this transformer is rated at 24 kVA.

(a) What are the current ratings I_X and I_H?

(b) If the secondary ('X') side has 26 turns, how many turns does the primary ('H') side have?

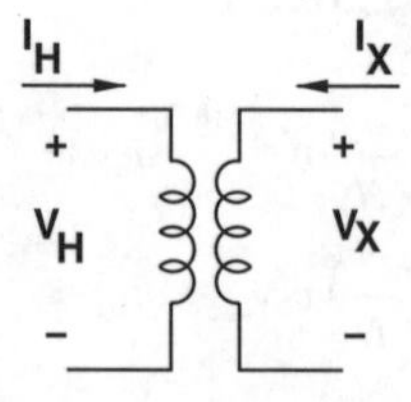

Figure 6.12 Ideal transformer

(c) If this transformer is loaded by a resistor of value R, what value of R results in the transformer being loaded to its rating?
(d) If the secondary is loaded by a resistor of value R, what is the impedance looking into the primary side?

2. Two alternative models for a single phase transformer are shown in Figure 6.13. 'Model A' shows the leakage and core elements referred to the 'H' side, while 'Model B' shows them on the 'X' side. This is a 24 kVA, 8 kV to 240 V, RMS transformer.

(a) If the core absorbs 100 W and 1000 VARs, what are the core elements R_c and X_c if they are located:
- On the 'H' side (Model A)?
- On the 'X' side (Model B)?

(b) If, when running at rated current, the copper loss is 1,200 W, what is the series resistance element:
- If it is on the 'H' side: R_H
- If it is on the 'X' side: R_X

(c) If $X_H = 133\Omega$ in Model A, what is X_X in Model B?

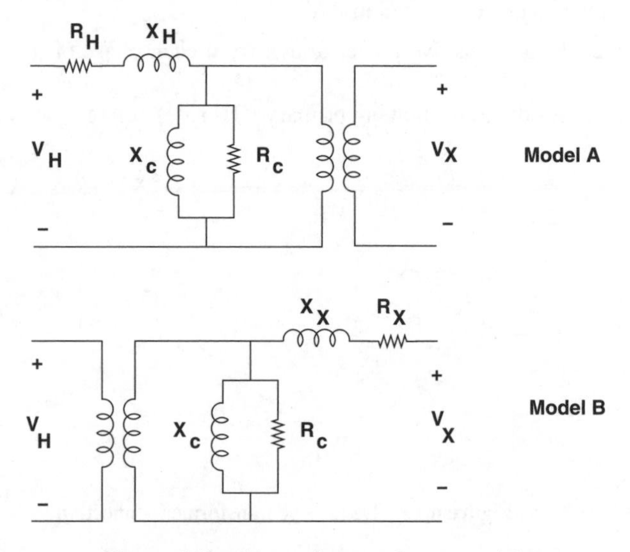

Figure 6.13 Alternative transformer models

3. A *delta–wye* connected transformer is shown in Figure 6.14. This is a 128 kV to 345 kV transformer, line–line, RMS. The 'X' side is the 128 kV side.

(a) What is the turns ratio $N_X : N_Y$?
(b) Draw phasor diagrams of the primary and secondary voltages. Be sure to get relative phase angles right.
(c) If a *balanced* load draws 100 MVA at 0.8 power factor from the 'H' side, what are the currents on both primary and secondary. Assume that Phase A voltage on the 'X' side has zero phase angle (that is, use V_{aX} as the phase reference).

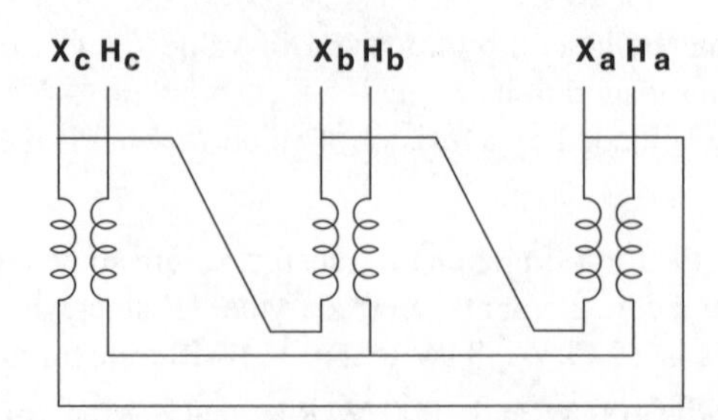

Figure 6.14 *Delta–wye* transformer connection

4. A *delta–wye* connected transformer is shown in schematic format in Figure 6.15. This transformer is from 600 V, RMS on the 'H' side to 208 V, RMS on the 'X' side.
   (a) What is the turns ratio between the 'H' side to the 'X' side?
   (b) If the transformer is open on the 'X' side but connected to a balanced 600 V, line–line source on the 'H' side, draw a phasor diagram showing voltages on the two sides.
   (c) Now the transformer is loaded by a single 10Ω resistor connected from terminal A_X to the neutral (star point) of the 'X' side. The 'H' side is connected to a 600 V (line–line, RMS) voltage source connected in *wye*.
   - What are real and reactive power drawn from each phase of the source on the 'H' side?
   - Draw a phasor diagram showing primary ('H' side) voltages and currents.

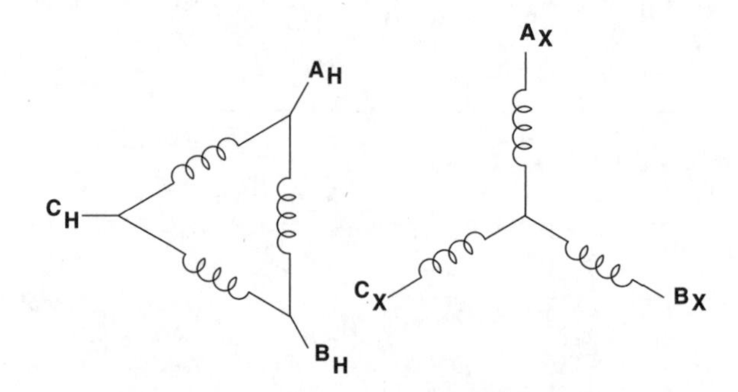

Figure 6.15 Delta–wye transformer connection

5. A *delta–wye* transformer connection is shown in schematic form in Figure 6.16, This transformer is a step-down arrangement from 13.8 kV, line-line, to 480 V, line–line, both RMS. See Figure 6.17. The resistors on the wye side are each drawing 100 A.
   (a) How much power is being drawn on the wye side from each of the three transformer secondaries?
   (b) What is the turn's ratio between the physical transformers that make up the three-phase transformer bank?
   (c) What are the three currents in the primary side of each of the three transformers? Show them in a reasonably proportioned vector diagram. Show the voltages across those transformer windings as well.

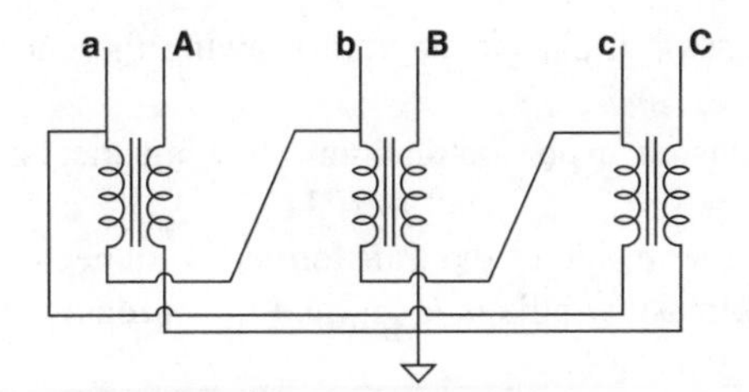

Figure 6.16 Transformer hookup

(d) Now, what are the three currents into the terminals of the *delta*-connected transformer primary? Show them in a well-proportioned vector diagram.

(e) Assuming the source on the 13.8 kV side is *wye*-connected, how much real and reactive power are being drawn from each phase? Does this correspond with power drawn by the load?

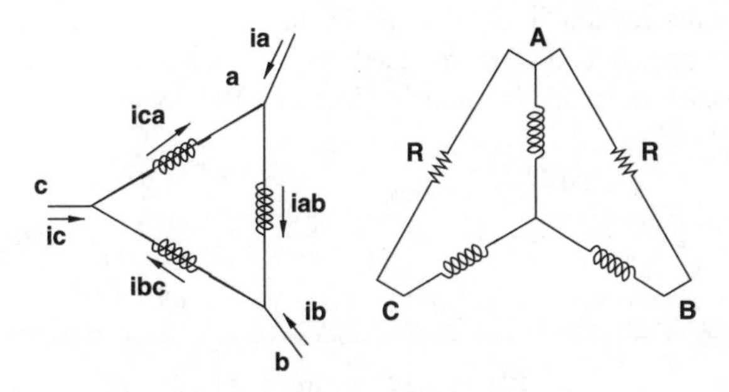

Figure 6.17 Unbalanced loaded transformer connection

6. Two transformers are connected as shown in Figure 6.18. This corresponds to two phases of a *wye*-connected primary (grounded, of course), with the secondaries in 'open *delta*'. The transformers are wound for a 13.8 kV/480 V ratio. The primaries are connected to a 13.8 kV (line–line) source and the open *delta* secondary is driving a *wye*-connected, balanced load that draws 100 A per phase.

(a) What is the physical turn's ratio of these transformers?

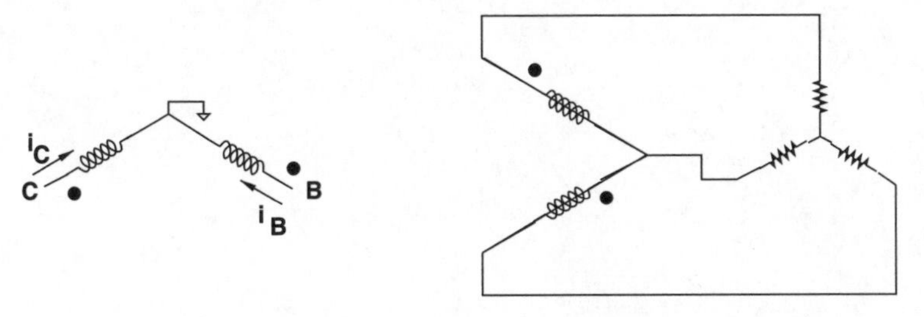

Figure 6.18 Two-transformer open delta

(b) Sketch and dimension a phasor diagram showing the currents in the two secondary windings.
(c) Sketch and dimension a phasor diagram showing the currents in the two primary windings.
(d) Assuming that the *wye* side of the transformer is connected to a voltage source, draw currents in the primary windings i_A, i_B and i_C, in relationship to the voltages on the *wye* side.
(e) Show that the sum of real and reactive powers on the *wye* side matches the same quantities on the *delta* side.
(f) What would happen if the ground on the *wye* side is removed?

7. Figure 6.19 shows a *wye*-connected load connected to a power source through a *delta-wye*-connected transformer. Assume that the load is two 30 Ω resistors and one 50 Ω resistor connected in *wye*. The source is balanced, 4160 V, RMS, line–line. The transformer is 4160/480, *delta* primary, solidly grounded *wye* secondary. Calculate the phase currents in the three phase load and the phase currents at the 4160 volt source if:

(a) The *wye*-connected load is solidly grounded,
(b) The *wye*-connected load is ungrounded, and
(c) The *wye*-connected load is grounded through a 10 Ω resistor.

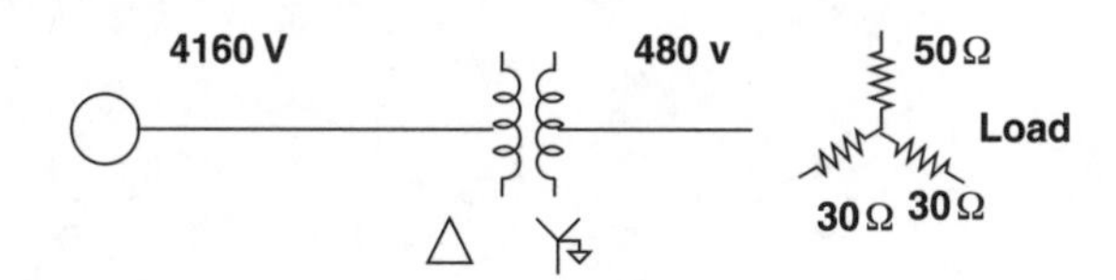

Figure 6.19 Source and load

7

Polyphase Lines and Single-Phase Equivalents

By now, one might suspect that a balanced polyphase system may be regarded simply as three single-phase systems, even though the three phases are physically interconnected. This feeling is reinforced by the equivalence between *wye*- and *delta*-connected sources and impedances. One more step is required to show that single phase equivalence is indeed useful, and this concerns situations in which the phases have mutual coupling.

7.1 Polyphase Transmission and Distribution Lines

Lines are such system elements as transmission or distribution lines: overhead wires, cables or even in-plant buswork. Such elements have impedance, so that there is some voltage drop between the *sending* and *receiving* ends of the line. This impedance is more than just conductor resistance: the conductors have both *self* and *mutual* inductance, because currents in the conductors make magnetic flux which, in turn, is linked by all conductors of the line. A cartoon of one format of overhead transmission line is shown in Figure 7.1. The phase conductors are arranged as parallel conductors in bundles (in this case, bundles of two conductors). These wires are supported by insulators. In many cases those insulators are actually strings of individual plates, often made of ceramic material, arranged to have a very long surface between the conductor and ground. Usually there are also accompanying ground wires that serve to carry the inevitable but usually small neutral currents that result from system unbalance, fault currents and that partially shield the active conductors from lightning.

A schematic view of a *line* is shown in Figure 7.2. Actually, only the inductance components of line impedance are shown, since they are the most interesting parts of line impedance.

Working in complex amplitudes, it is possible to write the voltage drops for the three phases by:

$$\begin{aligned}
\mathbf{V}_{a1} - \mathbf{V}_{a2} &= j\omega L\mathbf{I}_a + j\omega M(\mathbf{I}_b + \mathbf{I}_c) \\
\mathbf{V}_{b1} - \mathbf{V}_{b2} &= j\omega L\mathbf{I}_b + j\omega M(\mathbf{I}_a + \mathbf{I}_c) \\
\mathbf{V}_{c1} - \mathbf{V}_{c2} &= j\omega L\mathbf{I}_c + j\omega M(\mathbf{I}_a + \mathbf{I}_b)
\end{aligned}$$

Electric Power Principles: Sources, Conversion, Distribution and Use James L. Kirtley

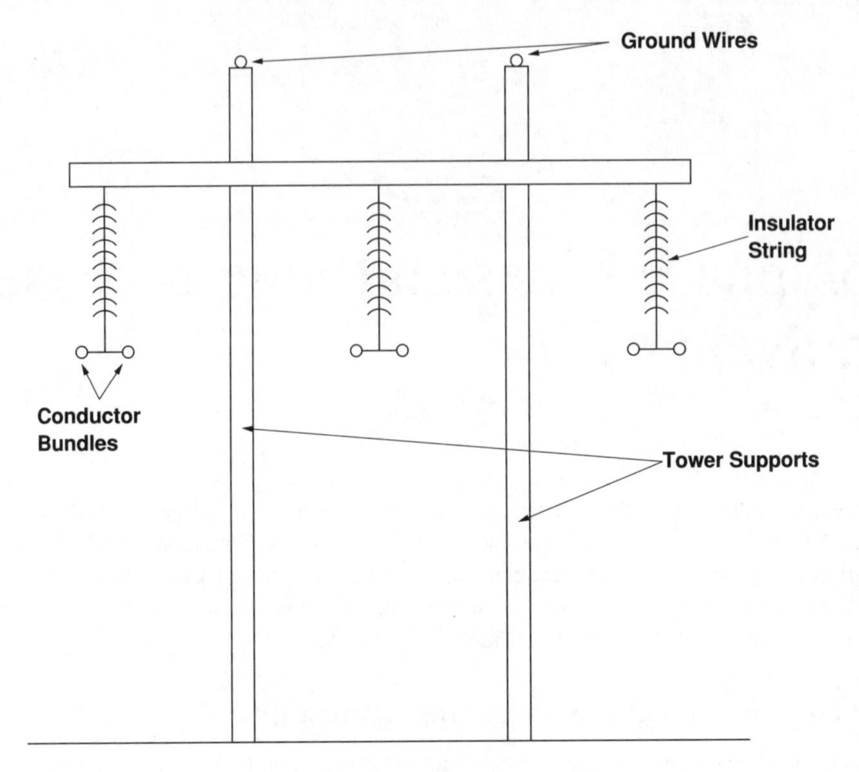

Figure 7.1 Three phase overhead line

If the currents form a balanced set:

$$\mathbf{I}_a + \mathbf{I}_b + \mathbf{I}_c = 0 \tag{7.1}$$

Then the voltage drops are simply:

$$\begin{aligned} \mathbf{V}_{a1} - \mathbf{V}_{a2} &= j\omega(L - M)\,\mathbf{I}_a \\ \mathbf{V}_{b1} - \mathbf{V}_{b2} &= j\omega(L - M)\,\mathbf{I}_b \\ \mathbf{V}_{c1} - \mathbf{V}_{c2} &= j\omega(L - M)\,\mathbf{I}_c \end{aligned}$$

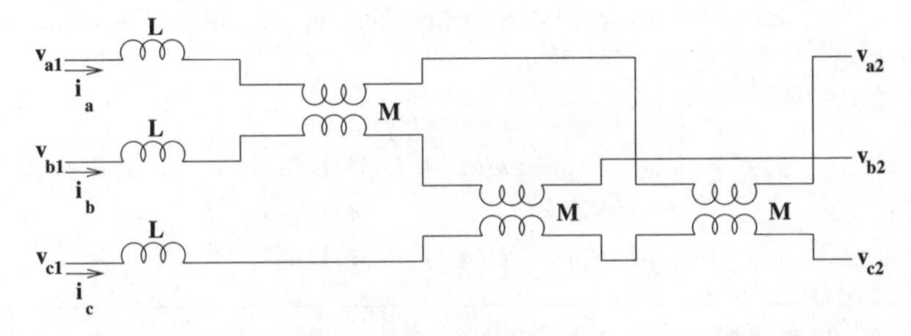

Figure 7.2 Schematic of a balanced three-phase line with mutual coupling

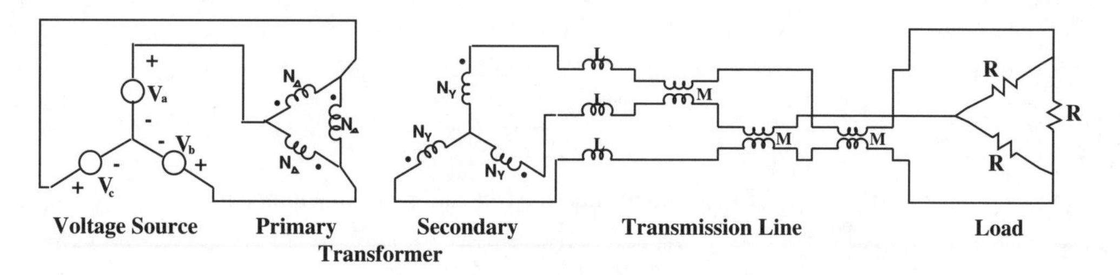

Figure 7.3 Example

In this case, an apparent inductance, suitable for the balanced case, has been defined:

$$L_1 = L - M \tag{7.2}$$

which describes the behavior of one phase in terms of its own current. It is most important to note that this inductance is a valid description of the line only if Equation 7.1 holds, which it does, of course, in the *balanced* case.

7.1.1 Example

To show how the analytical techniques which come from the network simplification resulting from single phase equivalents and *wye–delta* transformations, consider the following problem:

> A three-phase resistive load is connected to a balanced three-phase source through a transformer connected in *delta–wye* and a polyphase line, as shown in Figure 7.3. The problem is to calculate power dissipated in the load resistors.

The three-phase voltage source has:

$$v_a = \text{Re}\left[\sqrt{2}V_{RMS}e^{j\omega t}\right]$$
$$v_b = \text{Re}\left[\sqrt{2}V_{RMS}e^{j(\omega t - \frac{2\pi}{3})}\right]$$
$$v_c = \text{Re}\left[\sqrt{2}V_{RMS}e^{j(\omega t + \frac{2\pi}{3})}\right]$$

This problem is worked by a succession of simple transformations. First, the *delta*-connected resistive load is converted to its equivalent *wye* with $R_Y = \frac{R}{3}$.

Next, since the problem is balanced, the self- and mutual inductances of the line are directly equivalent to self inductances in each phase of $L_1 = L - M$.

Now, the transformer secondary is facing an impedance in each phase of:

$$\mathbf{Z}_{Ys} = j\omega L_1 + R_Y$$

The *delta–wye* transformer has a *voltage* ratio of:

$$\frac{v_p}{v_s} = \frac{N_\Delta}{\sqrt{3}N_Y}$$

so that, on the primary side of the transformer, the line and load impedance is:

$$\mathbf{Z}_p = j\omega L_{eq} + R_{eq}$$

where the equivalent elements are:

$$L_{eq} = \frac{1}{3}\left(\frac{N_\Delta}{N_Y}\right)^2 (L - M)$$

$$R_{eq} = \frac{1}{3}\left(\frac{N_\Delta}{N_Y}\right)^2 \frac{R}{3}$$

Magnitude of current flowing in each phase of the source is:

$$|\mathbf{I}| = \frac{\sqrt{2}V_{RMS}}{\sqrt{(\omega L_{eq})^2 + R_{eq}^2}}$$

Dissipation in one phase is:

$$P_1 = \frac{1}{2}|\mathbf{I}|^2 R_{eq} = \frac{V_{RMS}^2 R_{eq}}{(\omega L_{eq})^2 + R_{eq}^2}$$

And, of course, total power dissipated is just three times the single-phase dissipation.

7.2 Introduction To Per-Unit Systems

Strictly speaking, *per-unit* systems are nothing more than normalizations of voltage, current, impedance and power. These normalizations of system parameters because they provide simplifications in many network calculations. As we will discover, while certain *ordinary* parameters have very wide ranges of value, the equivalent *per-unit* parameters fall in a much narrower range. This helps in understanding how certain types of system behave.

7.2.1 Normalization Of Voltage And Current

The basis for the per-unit system of notation is the expression of voltage and current as fractions of *base* levels. Thus the first step in setting up a per-unit normalization is to pick *base* voltage and current.

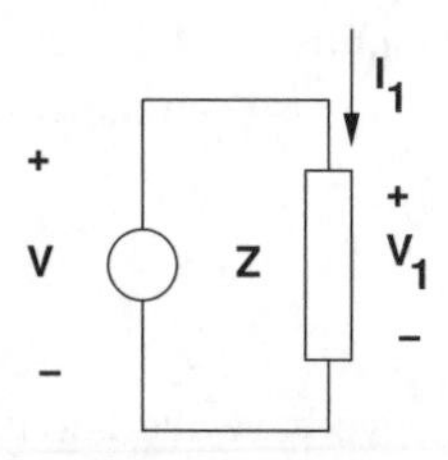

Figure 7.4 Example

Consider the simple situation shown in Figure 7.4. For this network, the complex amplitudes of voltage and current are:

$$\mathbf{V}_1 = \mathbf{I}_1 \mathbf{Z} \tag{7.3}$$

Start by defining two *base* quantities, V_B for voltage and I_B for current. In many cases, these will be chosen to be nominal or rated values. For generating plants, for example, it is common to use the rated voltage and rated current of the generator as *base* quantities. In other situations, such as system stability studies, it is common to use a standard, system wide base system.

The *per-unit* voltage and current are then simply:

$$\mathbf{v}_1 = \frac{\mathbf{V}_1}{\mathbf{V}_B} \tag{7.4}$$

$$\mathbf{i}_1 = \frac{\mathbf{I}_1}{I_B} \tag{7.5}$$

Applying equation 7.4 and 7.5 to Equation 7.3

$$\mathbf{v}_1 = \mathbf{i}_1 \mathbf{z}$$

where the *per-unit* impedance is:

$$\mathbf{z} = \mathbf{Z} \frac{I_B}{V_B}$$

This leads to a definition for a *base impedance* for the system:

$$Z_B = \frac{V_B}{I_B}$$

Of course there is also a *base power*, which for a single phase system is:

$$P_B = V_B I_B$$

as long as V_B and I_B are expressed in RMS. It is interesting to note that, as long as normalization is carried out in a consistent way, there is no ambiguity in per-unit notation. That is, *peak* quantities normalized to *peak* base quantities will be the same, in per-unit, as RMS quantities normalized to RMS bases. This advantage is even more striking in polyphase systems, as we are about to see.

7.2.2 Three-Phase Systems

When describing polyphase systems, we have the choice of using either line–line or line–neutral voltage and line current or current in delta equivalent loads. In order to keep straight analysis in *ordinary* variables, it is necessary to carry along information about which of these quantities is being used. There is no such problem with *per-unit* notation.

We may use as base quantities either line to neutral voltage V_{Bl-n} or line to line voltage V_{Bl-l}. Taking the base *current* to be line current I_{Bl}, we may express base *power* as:

$$P_B = 3V_{Bl-n}I_{Bl}$$

Because line–line voltage is, under normal operation, $\sqrt{3}$ times line–neutral voltage, an equivalent statement is:

$$P_B = \sqrt{3}V_{Bl-l}I_{Bl}$$

If base *impedance* is expressed by line–neutral voltage and line current (this is the common convention, but is not required),

$$Z_B = \frac{V_{Bl-n}}{I_{Bl}}$$

Then, base impedance is, written in terms of base power:

$$Z_B = \frac{P_B}{3I_B^2} = 3\frac{V_{Bl-n}^2}{P_B} = \frac{V_{Bl-l}^2}{P_B}$$

Note that a single per-unit voltage applies equally well to line–line, line–neutral, peak and RMS quantities. For a given situation, each of these quantities will have a different *ordinary* value, but there is only one *per-unit* value.

7.2.3 Networks With Transformers

One of the most important advantages of the use of per-unit systems arises in the analysis of networks with transformers. Properly applied, a per-unit normalization will cause nearly all ideal transformers to disappear from the per-unit network, thus greatly simplifying the analysis.

To show how this comes about, consider the ideal transformer as shown in Figure 7.5.

The ideal transformer imposes the constraints that:

$$\mathbf{V}_2 = N\mathbf{v}_1$$

$$\mathbf{I}_2 = \frac{1}{N}\mathbf{I}_1$$

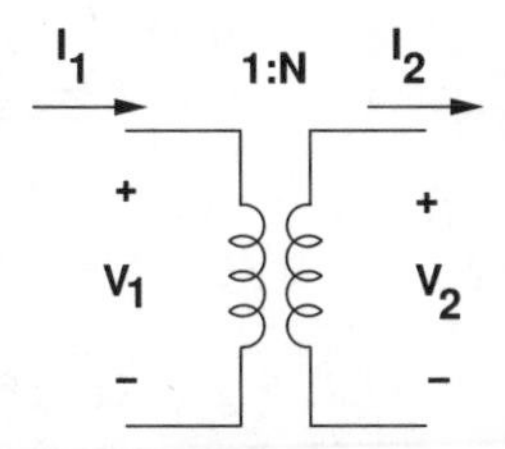

Figure 7.5 Ideal transformer with voltage and current conventions noted

Normalized to base quantities on the two sides of the transformer, the per-unit voltage and current are:

$$\mathbf{v}_1 = \frac{\mathbf{V}_1}{V_{B1}}$$
$$\mathbf{i}_1 = \frac{\mathbf{I}_1}{I_{B1}}$$
$$\mathbf{v}_2 = \frac{\mathbf{V}_2}{V_{B2}}$$
$$\mathbf{i}_2 = \frac{\mathbf{I}_2}{I_{B2}}$$

Now: note that if the *base* quantities are related to each other as if *they* had been processed by the transformer:

$$V_{B2} = N V_{B1} \tag{7.6}$$
$$I_{B2} = \frac{I_{B1}}{N} \tag{7.7}$$

then $\mathbf{v}_1 = \mathbf{v}_2$ and $\mathbf{i}_1 = \mathbf{i}_2$, as if the ideal transformer were not there (that is, consisted of an ideal wire).

Equation 7.6 and 7.7 reflect a general rule in setting up per-unit normalizations for systems with transformers. Each segment of the system should have the same base *power*. Base *voltages* transform according to transformer *voltage* ratios. For three-phase systems, of course, the *voltage* ratios differ from the physical turns ratios by a factor of $\sqrt{3}$ if *delta–wye* or *wye–delta* connections are used. It is, however, the *voltage* ratio that must be used in setting base voltages.

7.2.4 *Transforming From One Base To Another*

Very often data such as transformer leakage inductance is given in per-unit terms, on some base (perhaps the unit's rating), while in order to do a system study it is necessary to express the same data in per-unit in some other base (perhaps a unified system base). It is always possible to do this by the two step process of converting the per-unit data to its *ordinary* form, then re-normalizing it in the new base. However, it is easier to just convert it to the new base in the following way.

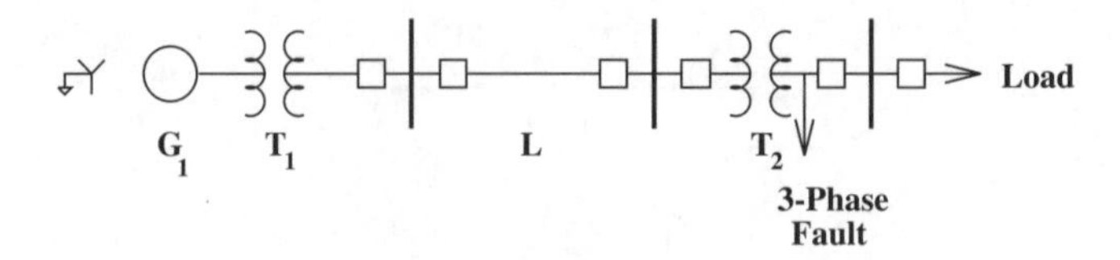

Figure 7.6 One-line diagram of faulted system

Note that impedance in Ohms (*ordinary* units) is given by:

$$\mathbf{Z} = \mathbf{z}_1 Z_{\mathrm{B}1} = \mathbf{z}_2 Z_{\mathrm{B}2} \tag{7.8}$$

Here, of course, $\mathbf{z}_1$ and $\mathbf{z}_2$ are the same *per-unit* impedance expressed in different *bases*. This could be written as:

$$\mathbf{z}_1 \frac{V_{\mathrm{B}1}^2}{P_{\mathrm{B}1}} = \mathbf{z}_2 \frac{V_{\mathrm{B}2}^2}{P_{\mathrm{B}2}} \tag{7.9}$$

This yields a convenient rule for converting from one base system to another:

$$\mathbf{z}_1 = \frac{P_{\mathrm{B}1}}{P_{\mathrm{B}2}} \left(\frac{V_{\mathrm{B}2}}{V_{\mathrm{B}1}} \right)^2 \mathbf{z}_2 \tag{7.10}$$

7.2.5 Example: Fault Study

To illustrate some of the concepts with which we have been dealing, we will do a short circuit analysis of a simple power system. This system is illustrated, in one-line diagram form, in Figure 7.6.

7.2.5.1 One-Line Diagram of the Situation

A one-line diagram is a way of conveying a lot of information about a power system without becoming cluttered with repetitive pieces of data. Drawing all three phases of a system would involve quite a lot of repetition that is not needed for most studies. Further, the three phases *can* be reconstructed from the one-line diagram if necessary. It is usual to use special symbols for different components of the network. For our network, we have the following pieces of data:

A three-phase fault is assumed to occur on the 34.5 kV side of the transformer T_2. This is a symmetrical situation, so that only one phase must be represented. The per-unit impedance diagram is shown in Figure 7.7. It is necessary to proceed now to determine the value of the components in this circuit.

First, it is necessary to establish a uniform base per-unit value for each of the system components. In this case, all of the parameters are put into a *base power* of 100 MVA and

Table 7.1 Per-unit data for fault example

Symbol	Component	Base P (MVA)	Base V (kV)	Impedance (per-unit)
G_1	Generator	200	13.8	$j0.18$
T_1	Transformer	200	13.8/138	$j0.12$
L_1	Transmission line	100	138	$0.02 + j0.05$
T_2	Transformer	50	138/34.5	$j0.08$

voltage bases of 138 kV on the line, 13.8 kV at the generator, and 34.5 kV at the fault. Using Equation 7.9:

$$
\begin{aligned}
x_g &= \tfrac{100}{200} \times 0.18 \ = 0.09 \text{ per-unit} \\
x_{T1} &= \tfrac{100}{200} \times 0.12 \ = 0.06 \text{ per-unit} \\
x_{T2} &= \tfrac{100}{50} \times 0.08 \ = 0.16 \text{ per-unit} \\
r_l &= 0.02 \text{ per-unit} \\
x_l &= 0.05 \text{ per-unit}
\end{aligned}
$$

Total impedance is:

$$
\begin{aligned}
\mathbf{z} &= j\left(x_g + x_{T1} + x_l + x_{T2}\right) + r_l \\
&= j0.36 + 0.02 \text{ per-unit} \\
|\mathbf{z}| &= 0.361 \text{ per-unit}
\end{aligned}
$$

Now, if e_g is equal to one per-unit (generator internal voltage equal to base voltage), then the per-unit *current* is:

$$
|\mathbf{i}| = \frac{1}{0.361} = 2.77 \text{ per-unit}
$$

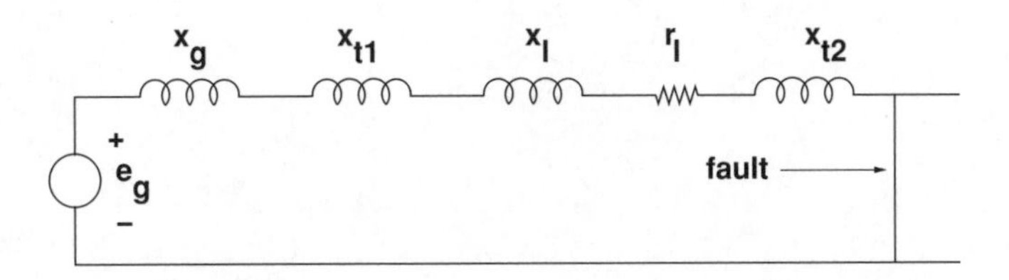

Figure 7.7 Impedance diagram for fault example

This may be translated back into ordinary units by getting base current levels. These are:

- On the base at the generator:

$$I_{\mathrm{B}} = \frac{100\mathrm{MVA}}{\sqrt{3} \times 13.8\mathrm{kV}} = 4.18\,\mathrm{kA}$$

- On the line base:

$$I_{\mathrm{B}} = \frac{100\mathrm{MVA}}{\sqrt{3} \times 138\mathrm{kV}} = 418\,\mathrm{A}$$

- On the base at the fault:

$$I_{\mathrm{B}} = \frac{100\mathrm{MVA}}{\sqrt{3} \times 34.5\mathrm{kV}} = 1.67\,\mathrm{kA}$$

Then the actual fault currents are:

- At the generator $|\mathbf{I}_{\mathrm{f}}| = 11,595\,\mathrm{A}$
- On the transmission line $|\mathbf{I}_{\mathrm{f}}| = 1159\,\mathrm{A}$
- At the fault $|\mathbf{I}_{\mathrm{f}}| = 4633\,\mathrm{A}$

7.3 Appendix: Inductances of Transmission Lines

Transmission lines are inductive, and the inductance of a line is the most important parameter in establishing how power flows through that line. This section is devoted to the derivation of inductance in transmission line-like structures.

7.3.1 Single Wire

To start, consider a wire and associated coordinate system as shown in Figure 7.8. Assume that this wire is carrying some current I in the axial (z) direction, which in this case would be *out* of the page. It is clear that the magnetic field produced by this current will be in the azimuthal (ϕ) direction. Using Ampere's Law, the magnetic field in the region *outside* the wire will be:

$$H_{\phi} = \frac{I}{2\pi r} \qquad r > R$$

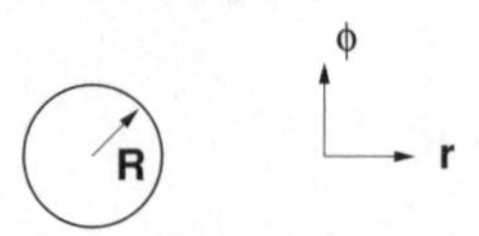

Figure 7.8 Wire and coordinate system for first inductance calculation

As it turns out, it is also necessary to consider the magnetic field inside the wire. If one assumes that current density is uniform in the wire, it is:

$$\vec{J} = \vec{i}_z \frac{I}{\pi R^2}$$

The magnetic field inside the wire is then:

$$H_\phi = J_z \frac{\pi r^2}{2\pi r} = \frac{r}{2\pi R^2} I$$

To compute inductance, it is appropriate to use energy stored in the magnetic fields:

$$w_m = \frac{1}{2} L I^2$$

In the case of the wire carrying current, it is possible to estimate the energy contained in a volume around the wire bounded by an outer radius R_o, using the expression for energy per unit volume, $W = \frac{\mu_0}{2}|H|^2$, energy stored in magnetic fields per unit length is:

$$\begin{aligned} w_m &= \int_0^R \frac{\mu_0}{2} \left(\frac{rI}{2\pi R^2} \right)^2 2\pi r \mathrm{d}r + \int_R^{R_o} \frac{\mu_0}{2} \left(\frac{I}{2\pi r} \right)^2 2\pi r \mathrm{d}r \\ &= \frac{\mu_0}{16\pi} I^2 + \frac{\mu_0}{4\pi} \log \frac{R_o}{R} I^2 \end{aligned}$$

From which one may conclude that the inductance per unit length associated within the region for $r < R_o$ is:

$$L = \frac{\mu_o}{8\pi} + \frac{\mu_0}{2\pi} \log \frac{R_o}{R}$$

The issue here, of course, is this expression is unbounded as $R_o \to \infty$. This difficulty will disappear once all of the currents in the system, which must add to zero, are considered.

The expression for inductance can be simplified a bit by noting that $\frac{1}{4} = \log \mathrm{e}^{\frac{1}{4}}$, so that:

$$L = \frac{\mu_0}{2\pi} \left(\log \mathrm{e}^{\frac{1}{4}} + \log \frac{R_o}{R} \right) = \frac{\mu_o}{2\pi} \log \left(\mathrm{e}^{\frac{1}{4}} \frac{R_o}{R} \right) = \frac{\mu_0}{2\pi} \log \frac{R_o}{R'}$$

where the effective radius of the wire is:

$$R' = R\mathrm{e}^{-\frac{1}{4}} \approx 0.78R$$

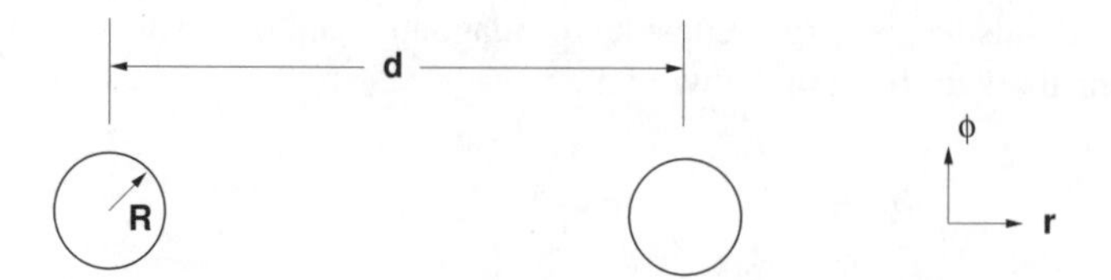

Figure 7.9 Two wires for mutual inductance calculation

7.3.2 Mutual Inductance

Consider the situation shown in Figure 7.9. A second wire runs parallel to the first and with spacing d. Flux linked by this second wire with respect to that outer radius is:

$$\lambda = \int_{\mathrm{d}}^{R_o} \mu_0 \frac{I}{2\pi r} \mathrm{d}r = I \frac{\mu_0}{2\pi} \log \frac{R_o}{d}$$

Mutual inductance is then:

$$M = \frac{\mu_0}{2\pi} \log \frac{R_o}{d}$$

Now: consider that the two conductors are carrying the same current, but in opposite directions, as one would expect for a circuit. That is, one wire is carrying current in the $+z$ direction, the other in the $-z$ direction. The flux linked would then be:

$$\lambda = LI - MI$$

and this becomes:

$$\lambda = \left(\frac{\mu_0}{2\pi} \log \frac{R_o}{R'} - \frac{\mu_0}{2\pi} \log R_o d \right) I = \frac{\mu_0}{2\pi} \log \frac{d}{R'}$$

Thus with closed circuits, the difficulty with unbounded radii goes away.

7.3.3 Bundles of Conductors

In some transmission line situations multiple wires are run in parallel. This serves (as will be shown here) to reduce the line inductance. It also reduces the electric field concentration around the wire, reducing corona discharges. Consider the situation shown in Figure 7.10. A number N of conductors runs in parallel with distances R_k from the first conductor. Here take $R_1 = R'$. Then, if currents are equal in the conductors of a bundle, flux linked by wire number one is:

$$\lambda = \frac{I}{N} \frac{\mu_0}{2\pi} \sum_{k=1}^{N} \log \frac{R_o}{R_k}$$

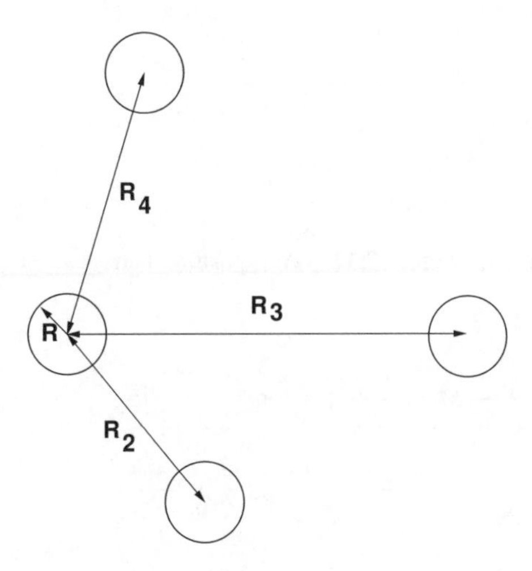

Figure 7.10 Multiple wires in a bundle

Note that

$$\frac{1}{N}\sum_{k=1}^{N}\log\frac{R_o}{R_k}=\log\frac{R_o}{\sqrt[N]{R'R_2R_3...R_N}}$$

This expression demonstrates that the bundle of conductors has the same inductance as a single conductor with an effective radius that is the *geometric mean distance* between the conductors:

$$\mathrm{GMD}=\sqrt[N]{\prod_{k=1}^{N}R_k}$$

From the foregoing, it is apparent that the effective radius, or self geometric mean distance of a round conductor is 0.78 times its actual radius, assuming of course that current flow is uniform within the conductor.

7.3.4 Transposed Lines

One might notice that the phase-to-phase spacing, and hence mutual inductances between different pairs of phases in situations such as the line illustrated in Figure 7.1 are not the same, leading to the possibility of unbalanced impedances. To bring lines into closer balance, transmission lines are often *transposed* as shown in Figure 7.11. At spots roughly one third of the way along the line the conductors are routed to occupy different lateral positions, so that each conductor occupies each lateral position for the same fraction as each of the other conductors. The line inductance then would be:

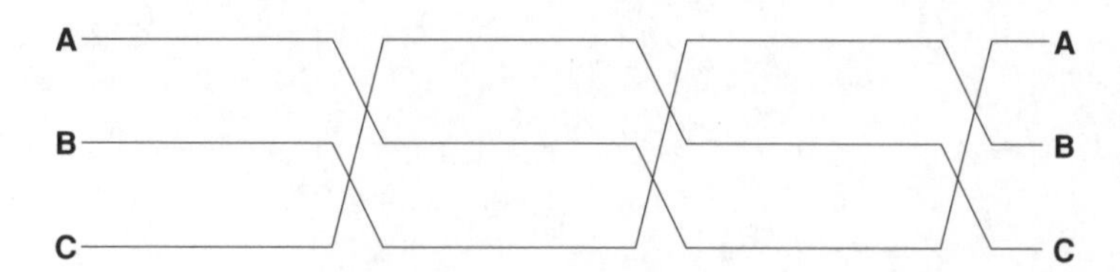

Figure 7.11 Transposition Pattern

$$
\begin{aligned}
L - M &= \frac{\mu_0 \ell}{2\pi}\left(\frac{2}{3}\log\frac{d}{R'} + \frac{1}{3}\log\frac{2d}{R'}\right) \\
&= \frac{\mu_0 \ell}{2\pi}\left(\frac{1}{3}\log(\frac{d}{R'})^2 + \frac{1}{3}\log\frac{2d}{R'}\right) \\
&= \frac{\mu_0 \ell}{2\pi}\frac{1}{3}\log\frac{2d^3}{R'^2} \\
&= \frac{\mu_0 \ell}{2\pi}\log\frac{\sqrt[3]{2}d}{R'}
\end{aligned}
$$

7.4 Problems

1. A 60 Hz transmission line has average parameters of:

Conductor self inductance	$L = 13.6 \times 10^{-7}$ H/m
Average mutual inductance	$M = 4.6 \times 10^{-7}$ H/m
Conductor resistance	$R = 1.2 \times 10^{-6}$ Ω/m

(a) What is the series reactance of a the line if it is 100 km long?
(b) What is the total series impedance of this line?

2. A system is to be normalized to a base condition of 138 kV (1-1) and 100 MVA.

(a) What is the base *current*?
(b) What is the base *impedance*?

3. What is the per-unit series impedance of the transmission line of Question 1 on the basis of Question 2 ?

4. The situation is shown in Figure 7.12. Consider the source to be a 138 kV, line-line, RMS three-phase voltage source. A fault occurs at the far end of a line that is 50 kM long with a series reactance of 0.4 Ω/km. Ignore line resistance. What is the fault current?

Figure 7.12 Fault situation for Problem 4

5. Using the same situation as in Figure 7.12, but with a series line impedance of $Z = j0.35 + 0.02\Omega/\text{km}$, what is the fault current?

6. A fault situation is shown in Figure 7.13. The generator can be considered to be a three-phase voltage source of 13.8 kV, line-line, RMS, in series with a reactance of 0.25 per-unit, on a base of 200 MVA, 13.8 kV. The transformer is 13.8 to 138 kV and has a series reactance of 0.05 per-unit on a base of 100 MVA and its rated voltages. The transmission line is 50 km long and has series impedance of $Z = j0.35 + 0.02\Omega/\text{km}$. A solid three-phase fault occurs at the end of the line. What are the currents:

(a) At the fault?
(b) In the generator leads?

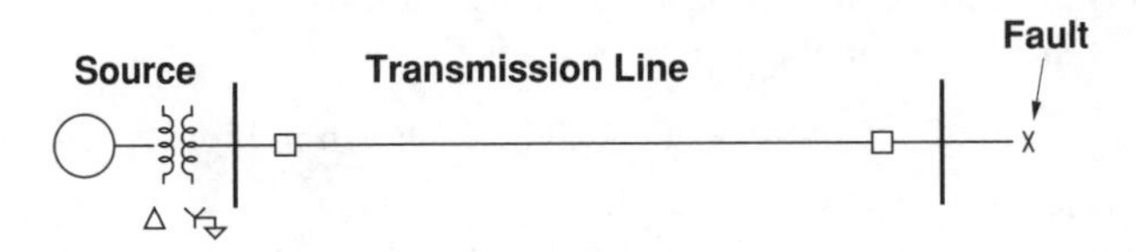

Figure 7.13 Fault situation for Problem 6

7. A fault situation is shown in Figure 7.14. The line is 50 km of overhead line with series impedance of $Z = j0.35 + 0.02\Omega/\text{km}$. The generator is a 13.8 kV, 200 MVA unit with a series reactance of 0.25 per-unit on its own base. Transformer T_1 is from 13.8 kV to 138 kV and has a series reactance of 0.05 per-unit on the system base of 100 MVA and consistent voltage units (13.8 kV to 138 kV). Transformer T_2 is 138 kV to 2.4 kV and is rated at 20 MVA, 138 kV to 2.4 kV and has a series reactance of 0.07 per-unit on a base consistent with its voltage and VA rating. The fault occurs at the 2.4 kV side of the transformer, and is a solid three-phase fault. The transmission line is 50 km long and has series impedance of $Z = j0.35 + 0.02\Omega/\text{km}$. What are:

(a) current in the fault?
(b) current in the line?
(c) current in the generator leads?

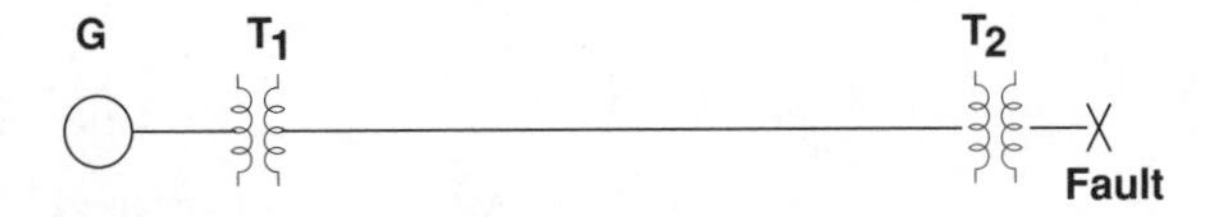

Figure 7.14 Fault situation for Problem 7

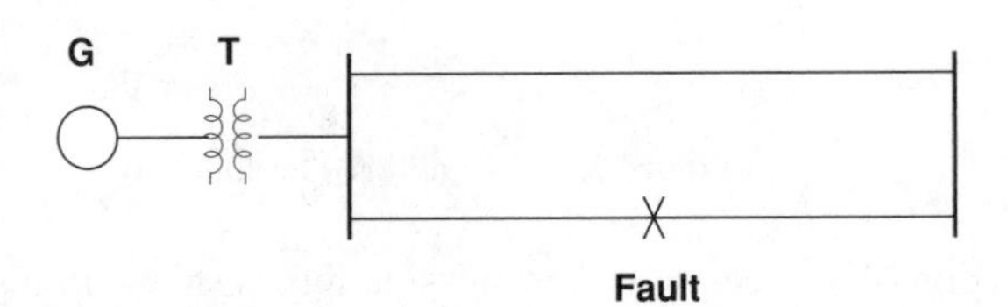

Figure 7.15 Fault situation for Problem 8

8. This fault situation is as shown in Figure 7.15. Both lines are 100 km long and have series impedance of $Z = j0.35 + 0.02\Omega/\text{km}$. The generator is rated at 500 MVA and has a terminal voltage of 24 kV, line–line, RMS. It has a series impedance of 0.25 per-unit on its rating as base. The transformer has a series reactance of 0.05 per-unit on the same base as the generator, with a secondary (high) voltage of 345 kV. The bus on the right is a load bus and can be considered to have no connections other than the two transmission lines. The fault occurs right in the middle of the lower line. The currents asked for are real currents, in amperes.

(a) What is the current in the fault?
(b) What is the current in the upper line?
(c) What is the current in the lower line, to the left of the fault?
(d) What is the current in the transformer, high side?
(e) What is the current in the generator leads?

9. The cross-section of a two-conductor bundle is shown in Figure 7.16.

(a) What is its GMD?
(b) Assuming a return radius $R_o = 100$ m, what is the inductance of this bundle in H/m?

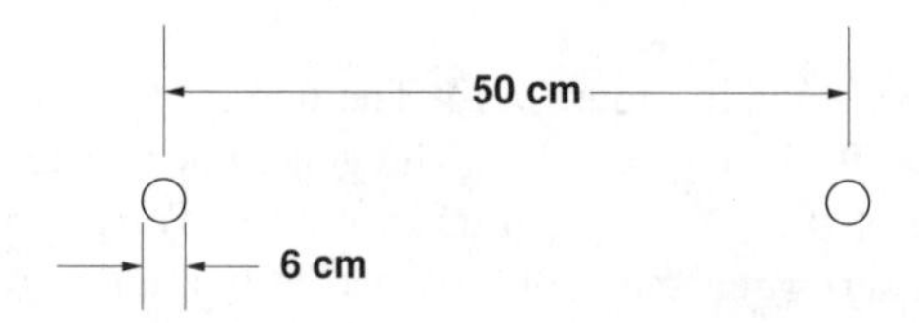

Figure 7.16 Bundle for Problem 9

10. The cross-section of a four-conductor bundle is shown in Figure 7.17. What is its GMD?

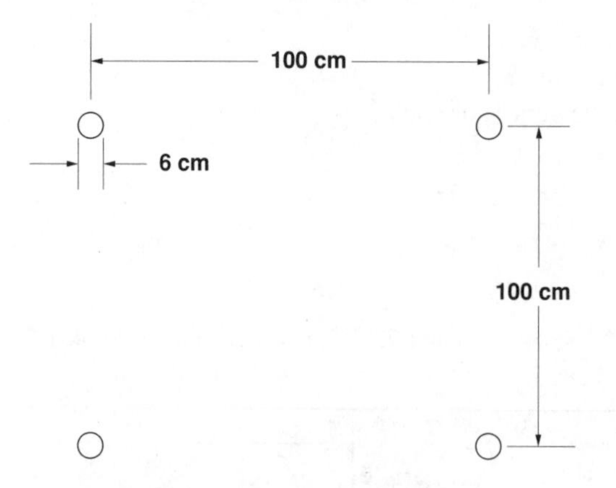

Figure 7.17 Bundle for Problem 10

11. The cross-section of a three-phase transmission line is shown in Figure 7.18. It consists of 6 cm diameter conductors in two conductor bundles with center-to-center spacing of 0.5 m. The line has bundle to bundle spacing of 10 m, center-to-center.

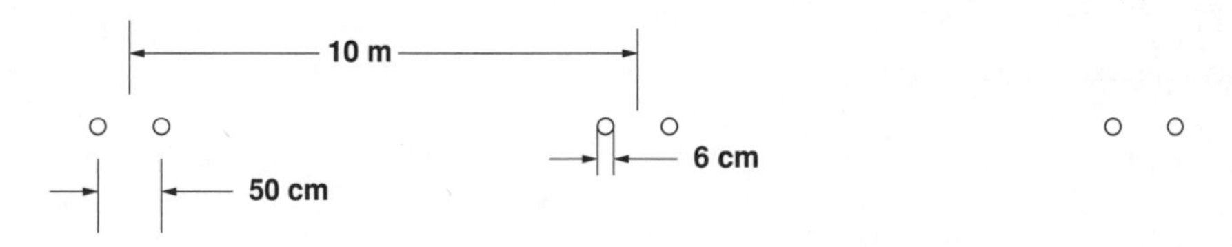

Figure 7.18 Three-phase transmission line for problem 11

(a) What is the series inductance $L - M$ in H/m for the two phases near each other? That is, center phase to outer phase.
(b) What is the series inductance $L - M$ for the two outside phases?
(c) If the line is transposed, what is the average $L - M$ in H/m?
(d) If the conductors are aluminum with conductivity of $\sigma = 3 \times 10^7$ S/m, and each bundle has two conductors with a diameter of 6 cm, what is the series resistance of the line?
(e) In a 60 Hz system, what would be the total line impedance of 10 km of this line?

8

Electromagnetic Forces and Loss Mechanisms

This chapter reviews some of the fundamental processes involved in electric machinery. Motors and generators convert electrical power to and from mechanical form, and to understand how they work one must understand the basics of electromechanical energy conversion. In the section on energy conversion processes we examine two fundamental ways of estimating electromagnetic forces: those involving thermodynamic arguments (conservation of energy) and field methods (Maxwell's Stress Tensor and Poynting's Theorem) are considered here.

This is a brief review of the fundamentals of electromechanical energy conversion. A more thorough explication of this area can be found in the three-volume set *Electromechanical Dynamics* (Woodson and Melcher, 1968).

This chapter also discusses losses resulting from eddy currents in both linear and nonlinear materials and hysteresis, including development of semi-empirical ways of handling iron losses that are combinations of eddy currents and hysteresis.

8.1 Energy Conversion Process

In a motor the energy conversion process can be thought of in simple terms. In 'steady state', electric power input to the machine is just the sum of electric power inputs to the different phase terminals:

$$P_{\mathrm{c}} = \sum_i v_i i_i$$

Mechanical power is torque multiplied by speed:

$$P_{\mathrm{m}} = T\Omega$$

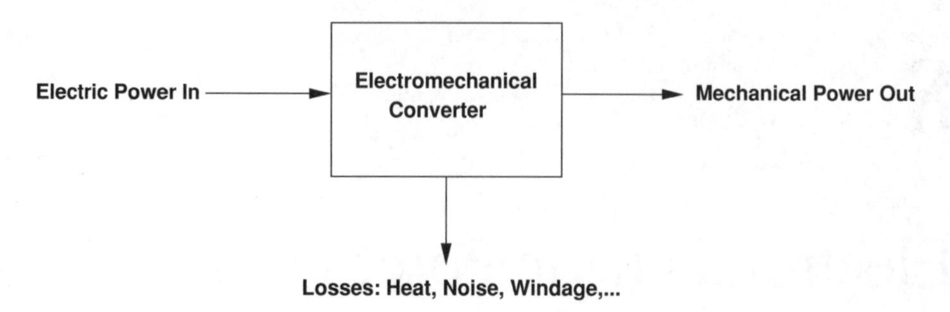

Figure 8.1 Energy conversion process

And the sum of the losses is the difference:

$$P_\mathrm{d} = P_\mathrm{c} - P_\mathrm{m}$$

It will sometimes be convenient to employ the fact that, in most machines, dissipation is small enough to approximate mechanical power with electrical power. In fact, there are many situations in which the loss mechanism is known well enough that it can be idealized away. The 'thermodynamic' arguments for force density take advantage of this and employ a 'conservative' or lossless energy conversion system.

8.1.1 Principle of Virtual Work

In a motor, as in many other types of electromechanical systems, the energy conversion process can be thought of in simple terms. As is illustrated in Figure 8.1, and in accordance with the first low of thermodynamics, electric power into the motor is either converted to mechanical power or dissipated as heat, noise, vibration or fan power. In 'steady state', electric power input to the machine is just the sum of electric power inputs to the different phase terminals. To start, consider some electromechanical system which has two sets of 'terminals', electrical and mechanical, as shown in Figure 8.2. If the system stores energy in magnetic fields, the energy stored depends on the *state* of the system, defined by (in this case) two of the identifiable variables: flux (λ), current (i) and mechanical position (x). In fact, with only a little reflection, you should be able to convince yourself that this state is a single-valued function of two variables and that the energy stored is independent of how the system was brought to this state.

All electromechanical converters have loss mechanisms and so are not themselves conservative. However, the magnetic field system that produces force is, in principle, conservative in

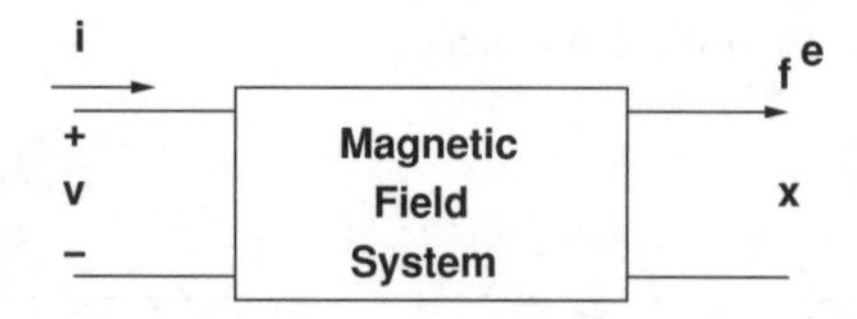

Figure 8.2 Conservative magnetic field system

the sense that its state and stored energy can be described by only two variables. The 'history' of the system is not important.

It is possible to chose the variables in such a way that electrical power *into* this conservative system is:

$$P_c = vi = i\frac{d\lambda}{dt}$$

Similarly, mechanical power *out of* the system is:

$$P_m = f_e\frac{dx}{dt}$$

The difference between these two is the rate of change of energy stored in the system:

$$\frac{dW_m}{dt} = P_e - P_m$$

It is then possible to compute the change in energy required to take the system from one state to another by:

$$W_m(a) - W_m(b) = \int_b^a i d\lambda - F^e dx$$

where the two states of the system are described by $a = (\lambda_a, x_a)$ and $b = (\lambda_b, x_b)$.

If the energy stored in the system is described by two state variables, λ and x, the *total differential* of stored energy is:

$$dW_m = \frac{\partial W_m}{\partial \lambda}d\lambda + \frac{\partial W_m}{\partial x}dx$$

and it is also:

$$dW_m = i d\lambda - f_e dx$$

So there is a direct equivalence between the derivatives and:

$$f_e = -\frac{\partial W_m}{\partial x}$$

In the case of rotary, as opposed to linear, motion, In place of force f_c and displacement x, torque T_c and angular displacement θ are the mechanical variables.

So, in such a rotary system:

$$T_e = -\frac{\partial W_m}{\partial \theta}$$

Generally, it is possible, at least conceptually, to assemble the electromechanical system and to set the mechanical position (X or θ) with zero force if the magnetic excitation of the system is zero (that is, no energy is stored in the system). Then the energy stored in the system is:

$$W_{\mathrm{m}} = \int_0^{\lambda} i \mathrm{d}\lambda$$

The principle of virtual work generalizes in the case of multiple electrical terminals and/or multiple mechanical terminals. For example, a situation with multiple electrical terminals will have:

$$\mathrm{d}W_{\mathrm{m}} = \sum_k i_k \mathrm{d}\lambda_k - f_{\mathrm{e}} \mathrm{d}x$$

In many cases one might consider a system which is electrically *linear*, in which case inductance is a function only of the mechanical position x.

$$\lambda(x) = L(x)i$$

In this case, assuming that the energy integral is carried out from $\lambda = 0$ (so that the part of the integral carried out over x is zero),

$$W_{\mathrm{m}} = \int_0^{\lambda} \frac{1}{L(x)} \lambda \mathrm{d}\lambda = \frac{1}{2} \frac{\lambda^2}{L(x)}$$

This makes

$$f_{\mathrm{e}} = -\frac{1}{2} \lambda^2 \frac{\partial}{\partial x} \frac{1}{L(x)}$$

Note that this is numerically equivalent to

$$f_{\mathrm{e}} = \frac{1}{2} i^2 \frac{\partial}{\partial x} L(x)$$

This is true *only* in the case of a linear system. Note that substituting $L(x)|_{i=\lambda}$ too early in the derivation produces erroneous results: in the case of a linear system it is a sign error, but in the case of a nonlinear system it is just wrong.

8.1.1.1 Example: Lifting Magnet

Consider the example shown in Figure 8.3. This represents, in cartoon form, something that might be a lifting magnet. The shaded volumes are highly permeable material ($\mu \rightarrow \infty$). They

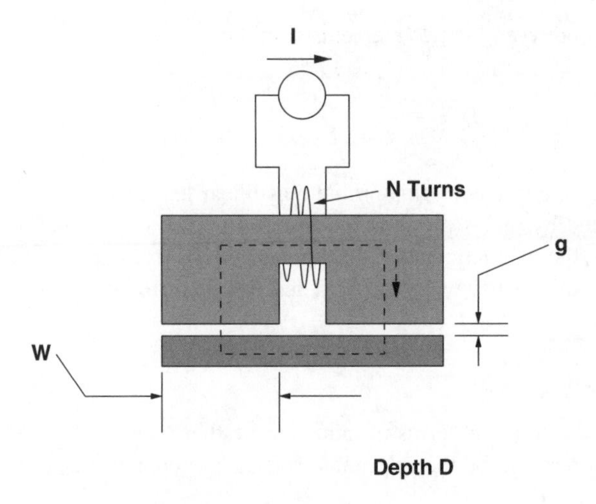

Figure 8.3 Lifting magnet example

are uniform in shape in the direction perpendicular to the view with dimension D. Wrapped around the upper body is a coil with N turns. That coil is driven with a DC (constant) current of magnitude I. To find flux, consider Ampere's Law around the loop indicated as a dotted line.

$$\oint \vec{H} \cdot d\vec{\ell} = NI$$

If magnetic field $\vec{H}$ is taken to be down in the right-hand gap and up in the left-hand gap, and if the gap dimension g is relatively small so the field is uniform over the gap area, and since there is no magnetic field $\vec{H}$ in the magnetic material the field in the gap is:

$$H_g = \frac{NI}{2g}$$

Note that, since magnetic flux has no divergence, the flux leaving the right-hand pole of the upper magnetic body must be matched by flux entering that body in the left-hand pole. Total flux is flux density times area:

$$\Phi = \mu_0 WD \frac{NI}{2g}$$

and since flux linked by the coil is just $N\Phi$, the inductance of the system is:

$$L(g) = \mu_0 \frac{N^2 WD}{2g}$$

The force acting between the two elements would then be:

$$f_e = -\frac{\lambda^2}{2}\frac{\partial}{\partial g}\frac{1}{L(g)} = \frac{\mu_0 I^2 N^2 W D}{2g^2}$$

It should be clear that viewing this problem as linear leads to major limitations: the force cannot, as suggested by this expression, go to infinity as $g \rightarrow 0$. What is happening here is that, at some point, the magnetic material saturates and the force approaches a constant value. This will be seen more clearly when the field description of forces is considered.

8.1.2 Coenergy

Systems are often described in terms of inductance rather than its reciprocal, so that current, rather than flux, appears to be the relevant variable. It is convenient to derive a new energy variable, called *co-energy*, by:

$$W'_m = \sum_k \lambda_k i_k - W_m$$

and in this case it is straightforward to show that the energy differential is (for a single mechanical variable) simply:

$$dW'_m = \sum_k \lambda_k di_k + f_e dx$$

so that force produced is:

$$f_e = \frac{\partial W'_m}{\partial x}$$

8.1.2.1 Example: Coenergy Force problem

Consider the made-up geometry of Figure 8.4. This resembles some magnetic solenoid actuators that are used to close switches, valves, etc. A bar of magnetic material can slide into a gap in a magnetic circuit, excited by a coil of N turns. As in the previous example, both elements have uniform depth D into the paper. In this problem, inductance is readily estimated to be:

$$L = \mu_0 N^2 D\left(\frac{W-x}{g} + \frac{x}{g-h}\right)$$

Co-energy is

$$W'_m = \frac{1}{2}L(x)I^2$$

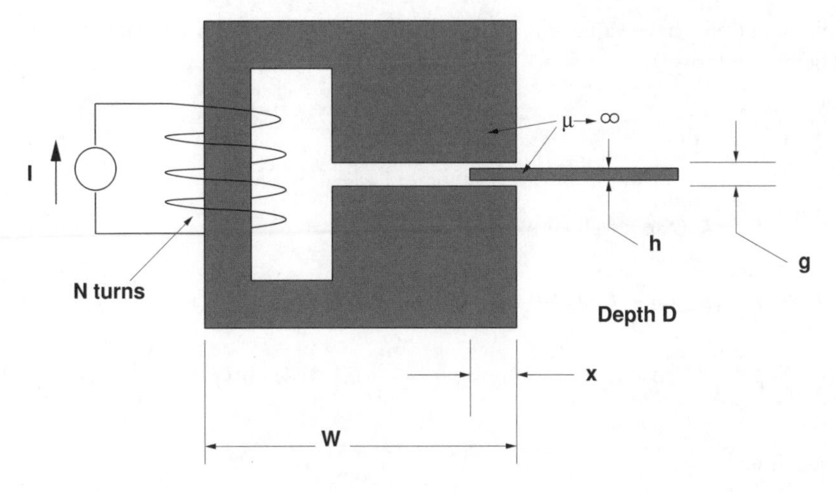

Figure 8.4 Magnetic actuator

And then force is:

$$f_e = \frac{I^2}{2}\frac{\partial L}{\partial x} = \mu_0 N^2 I^2 D\left(\frac{1}{g-h} - \frac{1}{g}\right)$$

In this case, force is uniform with x as long as the bar is engaging the gap: $0 < x < W$.

8.1.2.2 Electric Machine Model

Consider a simple electric machine example in which there is a single winding on a rotor (call it the *field* winding) and a polyphase armature. Suppose the rotor is round so that the flux linkages may be described as:

$$\lambda_a = L_a i_a + L_{ab} i_b + L_{ab} i_c + M\cos(p\theta) i_f$$
$$\lambda_b = L_{ab} i_a + L_a i_b + L_{ab} i_c + M\cos\left(p\theta - \frac{2\pi}{3}\right) i_f$$
$$\lambda_c = L_{ab} i_a + L_{ab} i_b + L_a i_c + M\cos\left(p\theta + \frac{2\pi}{3}\right) i_f$$
$$\lambda_f = M\cos(p\theta) i_a + M\cos\left(p\theta - \frac{2\pi}{3}\right) i_b + M\cos\left(p\theta + \frac{2\pi}{3}\right) i_c + L_f i_f$$

This system can be simply described in terms of coenergy. With multiple excitation it is important to exercise some care in taking the coenergy integral (to ensure that it is taken over a valid path in the multidimensional space). In this case there are actually five dimensions, but only four are important since one can position the rotor with all currents at zero so there is no contribution to coenergy from setting rotor position. Suppose the rotor is at some angle θ and that the four currents have values i_{a0}, i_{b0}, i_{c0} and i_{f0}. One of many correct path integrals to take would be:

$$
\begin{aligned}
W'_m = & \int_0^{i_{a0}} L_a i_a \mathrm{d}i_a \\
& + \int_0^{i_{b0}} (L_{ab} i_{a0} + L_a i_b)\, \mathrm{d}i_b \\
& + \int_0^{i_{c0}} (L_{ab} i_{a0} + L_{ab} i_{b0} + L_a i_c)\, \mathrm{d}i_c \\
& + \int_0^{i_{f0}} \left(M\cos(p\theta) i_{a0} + M\cos(p\theta - \frac{2\pi}{3}) i_{b0} + M\cos(p\theta + \frac{2\pi}{3}) i_{c0} + L_f i_f \right) \mathrm{d}i_f
\end{aligned}
$$

The result is:

$$
\begin{aligned}
W'_m = & \frac{1}{2} L_a \left(i_{a0}^2 + i_{b0}^2 + i_{c0}^2 \right) + L_{ab} \left(i_{a0} i_{b0} + i_{a0} i_{c0} + i_{c0} i_{b0} \right) \\
& + M i_{f0} \left(i_{a0} \cos(p\theta) + i_{b0} \cos(p\theta - \frac{2\pi}{3}) + i_{c0} \cos(p\theta + \frac{2\pi}{3}) \right) + \frac{1}{2} L_f i_{f0}^2
\end{aligned}
$$

If there is no variation of the stator inductances with rotor position θ, (which would be the case if the rotor were perfectly round), the terms that involve L_a and L_{ab} contribute zero so that torque is given by:

$$
T_e = \frac{\partial W'_m}{\partial \theta} = -pM i_{f0} \left(i_{a0} \sin(p\theta) + i_{b0} \sin(p\theta - \frac{2\pi}{3}) + i_{c0} \sin(p\theta + \frac{2\pi}{3}) \right)
$$

8.2 Continuum Energy Flow

At this point, it is instructive to think of electromagnetic energy flow as described by *Poynting's Theorem:*

$$
\vec{S} = \vec{E} \times \vec{H}
$$

Energy flow $\vec{S}$, called *Poynting's Vector*, describes electromagnetic power in terms of electric and magnetic fields. It is power density: power per unit area, with units in the SI system of units of watts per square meter.

To calculate electromagnetic power *into* some volume of space, for example the volume occupied by the rotor of a motor, Poynting's Vector may be integrated over the surface of that volume, and then using the divergence theorem:

$$
P = -\oiint \vec{S} \cdot \vec{n} \mathrm{d}a = -\int_{\mathrm{vol}} \nabla \cdot \vec{S} \mathrm{d}v
$$

The divergence of the Poynting Vector is, using a vector identity, Ampere's Law and Faraday's Law:

$$\nabla \cdot \vec{S} = \nabla \cdot \left(\vec{E} \times \vec{H}\right) = \vec{H} \cdot \nabla \times \vec{E} - \vec{E} \cdot \nabla \times \vec{H}$$
$$= -\vec{H} \cdot \frac{\partial \vec{B}}{\partial t} - \vec{E} \cdot \vec{J}$$

The power crossing into a region of space is then:

$$P = \int_{\mathrm{vol}} \left(\vec{E} \cdot \vec{J} + \vec{H} \cdot \frac{\partial \vec{B}}{\partial t}\right) \mathrm{d}v$$

In the absence of material motion, interpretation of the two terms in this equation is fairly simple. The first term describes dissipation:

$$\vec{E} \cdot \vec{J} = |\vec{E}|^2 \sigma = |\vec{J}|^2 \rho$$

The second term is interpreted as rate of change of magnetic stored energy. In the absence of hysteresis it is:

$$\frac{\partial W_{\mathrm{m}}}{\partial t} = \vec{H} \cdot \frac{\partial \vec{B}}{\partial t}$$

Note that in the case of free space,

$$\vec{H} \cdot \frac{\partial \vec{B}}{\partial t} = \mu_0 \vec{H} \cdot \frac{\partial \vec{H}}{\partial t} = \frac{\partial}{\partial t}\left(\frac{1}{2}\mu_0 |\vec{H}|^2\right)$$

which is straightforwardly interpreted as rate of change of magnetic stored energy density:

$$W_{\mathrm{m}} = \frac{1}{2}\mu_0 |H|^2$$

Some materials exhibit hysteretic behavior, in which stored energy is not a single valued function of either $\vec{B}$ or $\vec{H}$. This is a most important case that will be considered later.

8.2.1 Material Motion

In the presence of material motion $\vec{v}$, electric field $\vec{E}'$ in a 'moving' frame is related to electric field $\vec{E}$ in a 'stationary' frame and to magnetic field $\vec{B}$ by:

$$\vec{E}' = \vec{E} + \vec{v} \times \vec{B}$$

This is an experimental result obtained by observing charged particles moving in combined electric and magnetic fields. It is a relativistic expression, so that the qualifiers 'moving' and 'stationary' are themselves relative. The electric fields are what would be observed in either

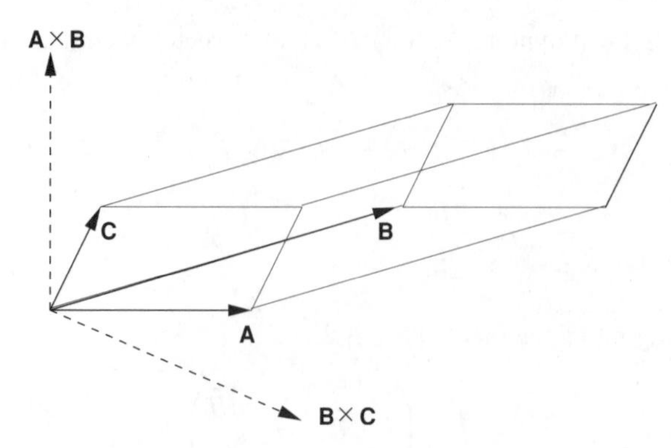

Figure 8.5 Illustration of the scalar triple product, showing that $\vec{A} \times \vec{B} \cdot \vec{C} = \vec{A} \cdot \vec{B} \times \vec{C}$

the 'stationary' or the 'moving' frame. In MQS systems, the magnetic flux density $\vec{B}$ is the same in both frames.

The term relating to current density becomes:

$$\vec{E} \cdot \vec{J} = \left(\vec{E}' - \vec{v} \times \vec{B}\right) \cdot \vec{J}$$

The term $\vec{E}' \cdot \vec{J}$ may be interpreted as dissipation, but the second term bears a little examination. Note that it is in the form of a vector triple (scalar) product, in which the dot and cross products can be interchanged because they describe the volume of a prism in space in which the three vectors describe the prism. This is shown in Figure 8.5.

$$-\vec{v} \times \vec{B} \cdot \vec{J} = -\vec{v} \cdot \vec{B} \times \vec{J} = \vec{v} \cdot \vec{J} \times \vec{B}$$

This is in the form of velocity times force density and represents power conversion from electromagnetic to mechanical form. This is consistent with the Lorentz Force Law (also experimentally observed):

$$\vec{F} = \vec{J} \times \vec{B}$$

This last expression is yet another way of describing energy conversion processes in electric machinery, as the component of apparent electric field produced by material motion through a magnetic field, when reacted against by a current, produces energy conversion to mechanical form rather than dissipation.

8.2.2 Additional Issues in Energy Methods

There are a few more important and interesting issues to consider in the development of forces of electromagnetic origin and their calculation using energy methods. These concern situations that are not simply representable by lumped parameters and situations that involve permanent magnets.

8.2.2.1 Coenergy in Continuous Media

Consider a system with not just a multiplicity of circuits but a continuum of current-carrying paths. In that case the co-energy may be identified as:

$$W'_{\mathrm{m}} = \int_{\mathrm{area}} \int \lambda(\vec{a})\mathrm{d}\vec{J} \cdot \mathrm{d}\vec{a}$$

where that area is chosen to cut all of the current-carrying conductors. This area can be picked to be perpendicular to each of the current filaments since the divergence of current is zero. The flux λ is calculated over a path that coincides with each current filament (such paths exist since current has zero divergence). Then the flux is:

$$\lambda(\vec{a}) = \int \vec{B} \cdot \mathrm{d}\vec{n}$$

If the vector potential $\vec{A}$ for which the magnetic flux density is:

$$\vec{B} = \nabla \times \vec{A}$$

is used, the flux linked by any one of the current filaments is:

$$\lambda(\vec{a}) = \oint \vec{A} \cdot \mathrm{d}\vec{\ell}$$

where $\mathrm{d}\vec{\ell}$ is the path of the current filament. This implies directly that the coenergy is:

$$W'_{\mathrm{m}} = \int_{\mathrm{area}} \int_J \oint \vec{A} \cdot \mathrm{d}\vec{\ell}\mathrm{d}\vec{J} \cdot \mathrm{d}\vec{a}$$

Now: it is possible to make $\mathrm{d}\vec{\ell}$ coincide with $\mathrm{d}\vec{a}$ and be parallel to the current filaments, so that:

$$W'_{\mathrm{m}} = \int_{\mathrm{vol}} \vec{A} \cdot \mathrm{d}\vec{J}\mathrm{d}v$$

8.2.2.2 Permanent Magnets

Permanent magnets are becoming an even more important element in electric machine systems. Often systems with permanent magnets are approached in a relatively *ad hoc* way, made equivalent to a current that produces the same MMF as the magnet itself.

The constitutive relationship for a permanent magnet relates the magnetic flux density $\vec{B}$ to magnetic field $\vec{H}$ and the property of the magnet itself, the *magnetization* $\vec{M}$.

$$\vec{B} = \mu_0 \left(\vec{H} + \vec{M} \right)$$

Now, the effect of the magnetization is to act as if there were a current (called an *amperian current*) with density:

$$\vec{J}^* = \nabla \times \vec{M}$$

This amperian current 'acts' just like ordinary current in making magnetic flux density. Magnetic coenergy is:

$$\mathrm{d}W'_{\mathrm{m}} = \int_{\mathrm{vol}} \vec{A} \cdot \nabla \times \mathrm{d}\vec{M} \mathrm{d}v$$

Next, note the vector identity

$$\nabla \cdot \left(\vec{C} \times \vec{D}\right) = \vec{D} \cdot \left(\nabla \times \vec{C}\right) - \vec{C} \cdot \left(\nabla \times \vec{D}\right)$$

So that:

$$\mathrm{d}W'_{\mathrm{m}} = \int_{\mathrm{vol}} -\nabla \cdot \left(\vec{A} \times \mathrm{d}\vec{M}\right) \mathrm{d}v + \int_{\mathrm{vol}} \left(\nabla \times \vec{A}\right) \cdot \mathrm{d}\vec{M} \mathrm{d}v$$

Then, noting that $\vec{B} = \nabla \times \vec{A}$:

$$W'_{\mathrm{m}} = -\oiint \vec{A} \times \mathrm{d}\vec{M} \mathrm{d}\vec{s} + \int_{\mathrm{vol}} \vec{B} \cdot \mathrm{d}\vec{M} \mathrm{d}v$$

The first of these integrals (closed surface) vanishes if it is taken over a surface just outside the magnet, where $\vec{M}$ is zero. Thus the magnetic coenergy in a system with only a permanent magnet source is

$$W'_{\mathrm{m}} = \int_{\mathrm{vol}} \vec{B} \cdot \mathrm{d}\vec{M} \mathrm{d}v$$

Adding current carrying coils to such a system is done in the obvious way.

8.2.2.3 Energy in the Flux–Current Plane

Following the calculation of energy input to a terminal pair, an important insight into electric machinery behavior may be had by noting that, if power into the terminals is $p = vi = i\frac{\mathrm{d}\lambda}{\mathrm{d}t}$, energy into that terminal pair over some period of time T is:

$$w = \int_0^T i\frac{\mathrm{d}\lambda}{\mathrm{d}t}\mathrm{d}t = \int_0^T i\mathrm{d}\lambda$$

If the voltage and current into that terminal pair is periodic with period T, the energy input per unit cycle is:

$$w_{\mathrm{c}} = \oint i\mathrm{d}\lambda$$

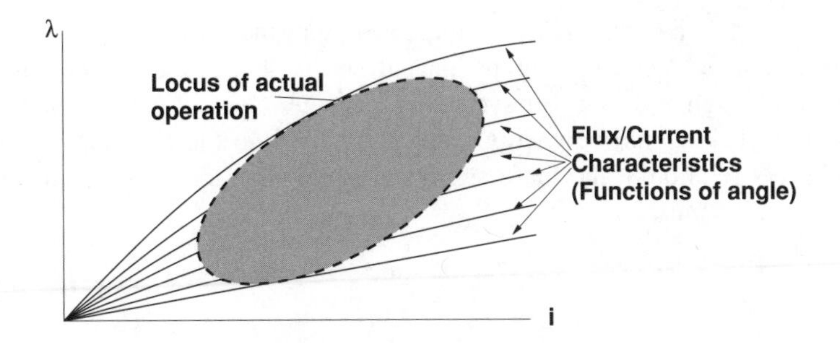

Figure 8.6 Flux-current relationship of a variable reluctance machine

This is simply the area of the locus of the closed curve of flux vs. current. Time average power is simply:

$$< p >= \frac{w_c}{T}$$

Note also that the area of the locus in the flux–current plane could equally validly be taken as co-energy:

$$w_c = \oint i \mathrm{d}\lambda = \oint \lambda \mathrm{d}i$$

This observation can be helpful in understanding phenomena like hysteresis in iron, for which a more thorough explanation is made in Section 8.3.4.

The class of electric machine known as *variable reluctance* machines is not a major focus in this book, but it is relatively easy to understand. Suppose the flux–current relationship for a terminal pair has a mapping $\lambda = \lambda(i, \theta)$. An example, pertinent to the variable reluctance machine is shown in Figure 8.6. For low values of current the machine has an inductance that varies with rotor position. The inductance tends to saturate, and the higher inductance curves saturating more sharply.

The concept of virtual work makes it possible to, in principle, estimate the torque produced by this sort of structure, with co-energy being:

$$W'_m = \int_0^i \lambda(i', \theta) \mathrm{d}i'$$

and then

$$T_e = \frac{\partial W'_m}{\partial \theta}$$

However, if it is noted that, with periodic excitation (and this assumes the period of current excitation is the same as the period of rotational change in inductance), energy input over the period to the terminals will be:

$$w_e = \oint \lambda \mathrm{d}i$$

Shown in Figure 8.6 is a dotted line that might represent the actual locus of operation of a variable reluctance machine; assuming that the current rises when inductance is low and falls when inductance is high, the locus is traversed in a counter-clockwise direction and therefore has positive area. This would be energy input per cycle into the machine. If there are N_s of these cycles per rotation of the shaft, and ignoring resistive losses, windage and any iron losses, average torque must be:

$$< T_e = \frac{N_s w_e}{2\pi}$$

If there are N_a stator poles and N_r rotor poles, the number of cycles per revolution will be $N_s = N_a N_r$.

Of course this result gives only the average torque, and a detailed analysis would yield the fact that torque in systems like this can be very non-uniform. A more thorough explication of this class of machine is beyond the scope of this text.

8.2.3 Electric Machine Description

Actually, this description is of a conventional induction motor. This is a very common type of electric machine and will serve as a reference point. Most other electric machines operate in a similar fashion.

Consider the simplified machine drawing shown in Figure 8.7. Most (but not all!) machines to be considered have essentially this morphology. The rotor of the machine is mounted on a shaft that is supported on some sort of bearing(s). Usually, but not always, the rotor is inside. Figure 8.7 shows a rotor that is round, but this does not need to be the case. rotor conductors are shown in the figure, but sometimes the rotor has permanent magnets either fastened to it or inside, and sometimes (as in variable reluctance machines) it is just a purposely shaped piece of steel. The stator is, in this drawing, on the outside and has windings. With most

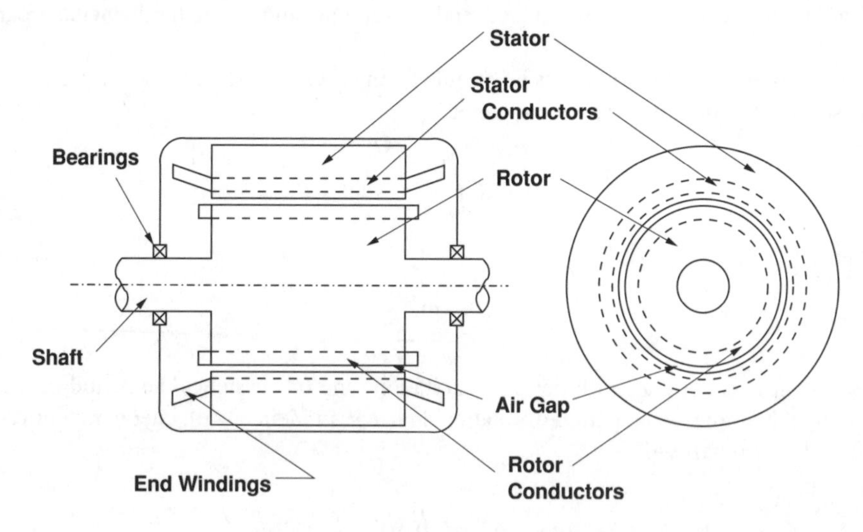

Figure 8.7 Form of electric machine

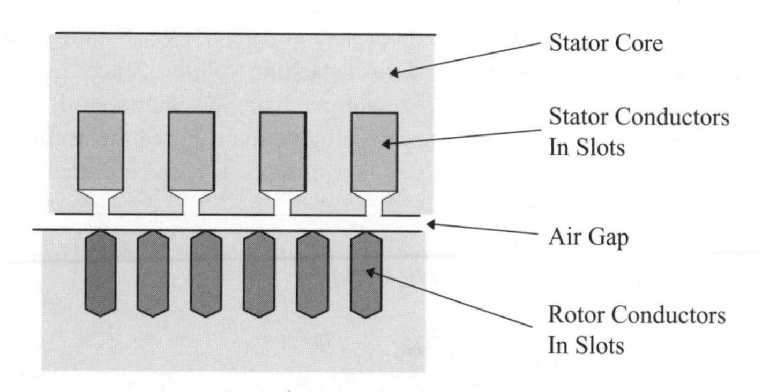

Figure 8.8 Windings in slots

machines, the stator winding is the armature, or electrical power input element. (In DC and universal motors this is reversed, with the armature contained on the rotor. These will be dealt with later.)

In most electrical machines the rotor and the stator are made of highly magnetically permeable materials: steel or magnetic iron. In many common machines such as induction motors the rotor and stator are both made up of stacks of thin sheets of steel, often alloyed with silicon to reduce its electrical conductivity and hence reduce eddy current losses. Punched into those sheets are slots which contain the rotor and stator conductors.

Figure 8.8 is a picture of part of an induction machine distorted so that the air-gap is straightened out (as if the machine had infinite radius). This is actually a convenient way of drawing the machine and often leads to useful methods of analysis. What is shown in this figure is that current carrying conductors are located in slots, rather than being physically *in* the air-gap. As it turns out, the (axial) currents in those slots interact with (radial) magnetic flux crossing the air-gap in a fashion exactly as if those currents were on the surface of the magnetic circuit elements (rotor and stator).

What is important to note for now is that the machine has an air gap g that is relatively small (that is, the gap dimension is much less than the machine radius r). The machine also has a physical length l, parallel to the shaft. The electric machine works by producing a shear stress in the air-gap (with of course side effects such as production of 'back voltage'). It is possible to define the average air-gap shear stress, referred to as τ. Total developed torque is force over the surface area multiplied by moment (rotor radius):

$$T = 2\pi r^2 \ell < \tau >$$

Power transferred by this device is just torque multiplied by speed, which is the same as force multiplied by surface velocity, since surface velocity is $u = r\Omega$:

$$P_{\mathrm{m}} = \Omega T = 2\pi r \ell < \tau > u$$

Noting that active rotor volume is $\pi r^2 \ell$, the ratio of torque to volume is just:

$$\frac{T}{V_{\mathrm{r}}} = 2 < \tau >$$

Determining what can be done in a volume of machine involves two things. First, it is clear that the volume calculated here is not the whole machine volume, since it does not include the stator. The actual estimate of total machine volume from the rotor volume is actually quite complex and detailed. Second, it is necessary to estimate the value of the useful average shear stress. Suppose both the radial flux density B_r and the stator surface current density K_z are sinusoidal flux waves of the form:

$$B_{\mathrm{r}} = \sqrt{2}B_0 \cos\left(p\theta - \omega t\right)$$

$$K_{\mathrm{z}} = \sqrt{2}K_0 \cos\left(p\theta - \omega t\right)$$

Note that this assumes these two quantities are exactly in phase, or oriented to ideally produce torque, so this calculation will yield an optimistic bound. Then the average value of surface traction is:

$$< \tau >= \frac{1}{2\pi} \int_0^{2\pi} B_{\mathrm{r}} K_{\mathrm{z}} \mathrm{d}\theta = B_0 K_0$$

This actually makes some sense in view of the empirically observed Lorentz Force Law.

8.2.4 *Field Description of Electromagnetic Force: The Maxwell Stress Tensor*

The Maxwell stress tensor is a way of describing (and calculating) forces of electromagnetic origin directly from the fields around an object. It leads to a surprisingly simple picture of forces in electric machines and in structures similar to electric machines. The Maxwell stress tensor is consistent with Poynting energy flow, and considering the two methods of analysis can lead to useful insights of electric machine operation.

The development of the previous section is actually enough to describe the forces seen in many machines, but since electric machines have permeable magnetic material and since magnetic fields produce forces on permeable material even in the absence of macroscopic currents it is necessary to observe how force appears on such material. An empirical expression for force density that incorporates both Lorentz Force and forces on magnetic material is:

$$\vec{F} = \vec{J} \times \vec{B} - \frac{1}{2}\left(\vec{H} \cdot \vec{H}\right) \nabla \mu$$

where $\vec{H}$ is the magnetic field intensity and μ is the permeability. This expression expresses the observed fact that permeable magnetic materials are attracted to regions of higher magnetic field.

Noting that current density is the curl of magnetic field intensity:

$$\begin{aligned} \vec{F} &= \left(\nabla \times \vec{H}\right) \times \mu \vec{H} - \frac{1}{2}\left(\vec{H} \cdot \vec{H}\right) \nabla \mu \\ &= \mu \left(\nabla \times \vec{H}\right) \times \vec{H} - \frac{1}{2}\left(\vec{H} \cdot \vec{H}\right) \nabla \mu \end{aligned}$$

And, since:

$$\left(\nabla \times \vec{H}\right) \times \vec{H} = \left(\vec{H} \cdot \nabla\right) \vec{H} - \frac{1}{2}\nabla\left(\vec{H} \cdot \vec{H}\right)$$

force density is:

$$\begin{aligned}\vec{F} &= \mu\left(\vec{H} \cdot \nabla\right)\vec{H} - \frac{1}{2}\mu\nabla\left(\vec{H} \cdot \vec{H}\right) - \frac{1}{2}\left(\vec{H} \cdot \vec{H}\right)\nabla\mu \\ &= \mu\left(\vec{H} \cdot \nabla\right)\vec{H} - \nabla\left(\frac{1}{2}\mu\left(\vec{H} \cdot \vec{H}\right)\right)\end{aligned}$$

This expression can be written by components: the component of force density in the i'th dimension is:

$$F_i = \mu \sum_k \left(H_k \frac{\partial}{\partial x_k}\right) H_i - \frac{\partial}{\partial x_i}\left(\frac{1}{2}\mu \sum_k H_k^2\right)$$

The divergence of magnetic flux density can be written as:

$$\nabla \cdot \vec{B} = \sum_k \frac{\partial}{\partial x_k} \mu H_k = 0$$

and

$$\mu \sum_k \left(H_k \frac{\partial}{\partial x_k}\right) H_i = \sum_k \frac{\partial}{\partial x_k} \mu H_k H_i - H_i \sum_k \frac{\partial}{\partial x_k} \mu H_k$$

Since the last term in that is zero, force density is:

$$F_k = \frac{\partial}{\partial x_i}\left(\mu H_i H_k - \frac{\mu}{2}\delta_{ik} \sum_n H_{\mathrm{n}}^2\right)$$

where the Kronecker delta $\delta_{ik} = 1$ if $i = k$, 0 otherwise.

This force density is in the form of the divergence of a tensor:

$$F_k = \frac{\partial}{\partial x_i} T_{ik}$$

or

$$\vec{F} = \nabla \cdot \mathbf{T}$$

In this case, force on some object that can be surrounded by a closed surface can be found by using the divergence theorem:

$$\vec{f} = \int_{\mathrm{vol}} \vec{F} dv = \int_{\mathrm{vol}} \nabla \cdot \mathbf{T} dv = \oiint \mathbf{T} \cdot \vec{n} da$$

or, surface traction is $\tau_i = \sum_k T_{ik} n_k$, where $\vec{n}$ is the surface normal vector, then the total force in direction i is just:

$$\vec{f} = \oint_s \tau_i \mathrm{d}a = \oint \sum_k T_{ik} n_k \mathrm{d}a$$

The interpretation of all of this is less difficult than the notation suggests. This field description of forces gives a simple picture of surface traction, the force per unit area on a surface. Integrating this traction over the area of a surface surrounding some body yields the electromagnetic force on the body. Note that this works if integration the traction over a surface that is itself in free space but which *surrounds* the body (because no force can be imposed on free space).

Note one more thing about this notation. Sometimes when subscripts are repeated as they are here the summation symbol is omitted. Thus, in that notation

$$\tau_i = \sum_k T_{ik} n_k = T_{ik} n_k$$

In the case of a circular cylinder and if one is interested in torque, this notion of integration of the divergence of the stress tensor does not work directly. However, if the rotor is made up of highly permeable magnetic iron, traction can be computed on the surface of the rotor. The normal vector to the cylinder is just the radial unit vector, and then the circumferential traction must simply be:

$$\tau_\theta = \mu_0 H_r H_\theta$$

Simply integrating this over the surface gives azimuthal force, and then multiplying by radius (moment arm) yields torque. The last step is to note that, if the rotor is made of highly permeable material, the azimuthal magnetic field is equal to surface current density.

8.2.5 Tying the MST and Poynting Approaches Together

Now that the stage is set, consider energy flow and force transfer in a narrow region of space as illustrated by Figure 8.9. This could be the air-gap of a linear machine or the air gap of a rotary machine in the limit of small curvature. The upper and lower surfaces may support currents. Assume that all of the fields, electric and magnetic, are of the form of a traveling wave in the x-direction: $\mathrm{Re}\{e^{j(\omega t - kx)}\}$.

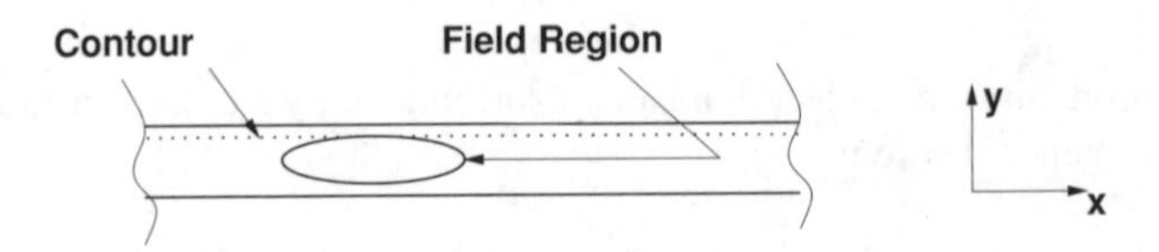

Figure 8.9 Illustrative region of space

If there is no variation in the z-direction (equivalently, the problem is infinitely long in the z-direction), there can be no x-directed currents because the divergence of current is zero: $\nabla \cdot \vec{J} = 0$. In a magnetoquasistatic system this is true of electric field $\vec{E}$ too. Thus one can assume that current is confined to the z-direction and to the two surfaces illustrated in Figure 8.9, and thus the only important fields are:

$$\vec{E} = \vec{i}_z \mathrm{Re}\{\mathrm{E}_z e^{j(\omega t - kx)}\}$$
$$\vec{H} = \vec{i}_x \mathrm{Re}\{\mathrm{H}_x e^{j(\omega t - kx)}\} + \vec{i}_y \mathrm{Re}\{\mathrm{H}_y e^{j(\omega t - kx)}\}$$

Faraday's Law ($\nabla \times \vec{E} = -\frac{\partial \vec{B}}{\partial t}$) establishes the relationship between the electric and magnetic field: the y- component of Faraday's Law is:

$$jk\mathbf{E}_z = -j\omega\mu_0\mathbf{H}_y$$

or

$$\mathbf{E}_z = -\frac{\omega}{k}\mu_0\mathbf{H}_y$$

The phase velocity $u_{ph} = \frac{\omega}{k}$ is a most important quantity. Note that, if one of the surfaces is moving (as it would be in, say, an induction machine), the frequency and hence the apparent phase velocity, will be shifted by the motion.

Energy flow through the surface denoted by the dotted line in Figure 8.9 is the component of Poynting's Vector in the negative y- direction. The relevant component is:

$$S_y = \left(\vec{E} \times \vec{H}\right)_y = E_z H_x = -\frac{\omega}{k}\mu_0 H_y H_x$$

Note that this expression contains the xy component of the Maxwell Stress Tensor $T_{xy} = \mu_0 H_x H_y$ so that power flow downward through the surface is:

$$S = -S_y = \frac{\omega}{k}\mu_0 H_x H_y = u_{ph} T_{xy}$$

The *average* power flow is the same, in this case, for time and for space, and is:

$$< S >= \frac{1}{2}\mathrm{Re}\{\mathbf{E}_z \mathbf{H}_x^*\} = u_{ph}\frac{\mu_0}{2}\mathrm{Re}\{\mathbf{H}_y \mathbf{H}_x^*\}$$

The *surface* impedance is defined as:

$$\mathbf{Z}_s = \frac{\mathbf{E}_z}{-\mathbf{H}_x}$$

which becomes:

$$\mathbf{Z}_s = -\mu_0 u_{ph} \frac{\mathbf{H}_y}{\mathbf{H}_x} = -\mu_0 u_{ph} \boldsymbol{\alpha}$$

where the parameter $\boldsymbol{\alpha}$ is the ratio between y- and x-directed complex field amplitudes. Energy flow through that surface is now:

$$S = -\frac{1}{s}\mathrm{Re}\{\mathbf{E}_z \mathbf{H}_x^*\} = \frac{1}{2}\mathrm{Re}\{|\mathbf{H}_x|^2 \mathbf{Z}_s\}$$

8.2.5.1 Simple Description of a Linear Induction Motor

Consider the geometry described in Figure 8.10. Shown here is *only* the relative motion gap region. This is bounded by two regions of highly permeable material (e.g. iron), comprising the stator and shuttle. On the surface of the stator (the upper region) is a surface current:

$$\vec{K}_s = \vec{i}_z \mathrm{Re}\{\mathbf{K}_{zs} e^{j(\omega t - kx)}\}$$

The shuttle is, in this case, moving in the positive x-direction at some velocity u. It may also be described as an infinitely permeable region with the capability of supporting a surface current with surface conductivity σ_s, so that $K_{zr} = \sigma_s E_z$.

Note that Ampere's Law gives a boundary condition on magnetic field just below the upper surface of this problem: $H_x = K_{zs}$, so that, the ratio between y- and x-directed fields at that location can be established:

$$< T_{xy} >= \frac{\mu_0}{2}\mathrm{Re}\{\mathbf{H}_y \mathbf{H}_x^*\} = \frac{\mu_0}{2}|\mathbf{K}_{zs}|^2 \mathrm{Re}\{\boldsymbol{\alpha}\}$$

Note that the ratio of fields $\mathbf{H}_y/\mathbf{H}_x = \boldsymbol{\alpha}$ is independent of reference frame (it doesn't matter if the fields are observed from the shuttle or the stator), so that the shear stress described by T_{xy} is also frame independent. If the shuttle (lower surface) is moving relative to the upper surface, the velocity of the traveling wave *relative to the shuttle* is:

$$u_s = u_{ph} - u = s\frac{\omega}{k}$$

where the dimensionless *slip* s is the ratio between frequency seen by the shuttle to frequency seen by the stator. This may be used to describe energy flow as described by Poynting's

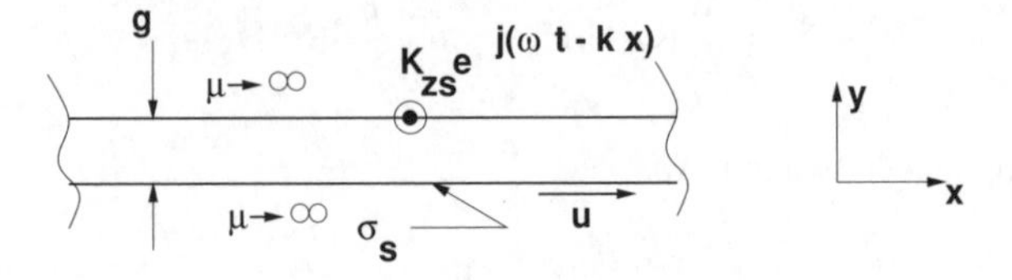

Figure 8.10 Simple description of linear induction motor

Theorem. Energy flow in the stator frame is:

$$\mathbf{S}_{\text{upper}} = u_{\text{ph}} T_{xy}$$

In the frame of the shuttle, however, it is

$$\mathbf{S}_{\text{lower}} = u_{\text{s}} T_{xy} = s\mathbf{S}_{\text{upper}}$$

The interpretation of this is that energy flow out of the upper surface ($\mathbf{S}_{\text{upper}}$) consists of energy *converted* (mechanical power) plus energy dissipated in the shuttle (which is $\mathbf{S}_{\text{lower}}$ here. The difference between these two power flows, calculated using Poynting's Theorem, is power converted from electrical to mechanical form:

$$\mathbf{S}_{\text{converted}} = \mathbf{S}_{\text{upper}}(1-s)$$

To finish the problem, note that surface current in the shuttle is:

$$K_{zr} = E_z'\sigma_{\text{s}} = -u_{\text{s}}\mu_0\sigma_s H_y$$

where the electric field E_z' is measured in the frame of the shuttle.

The magnetic gap g is assumed to be small enough that $kg \ll 1$. Ampere's Law, taken around a contour that crosses the air-gap and has a normal in the z-direction, yields:

$$g\frac{\partial H_x}{\partial x} = K_{zs} + K_{zr}$$

In complex amplitudes, this is:

$$-jkg\mathbf{H}_y = \mathbf{K}_{zs} + \mathbf{K}_{zr} = \mathbf{K}_{zs} - \mu_0 u_{\text{s}}\sigma_{\text{s}}\mathbf{H}_y$$

or, solving for H_y.

$$\mathbf{H}_y = \frac{j\mathbf{K}_{zs}}{kg}\frac{1}{1 + j\mu_0\frac{u_{\text{s}}\sigma_{\text{s}}}{kg}}$$

The quantity $R_e = \frac{\mu_0\sigma_s u_s}{kg}$ is sometimes referred to at the 'magnetic Reynold's number' (see Figure 8.11). It is dimensionless and describes the ratio between period and diffusion time.

Average shear stress is

$$\begin{aligned}
< T_{xy} > &= \frac{\mu_0}{2}\text{Re}\{\mathbf{H}_y\mathbf{H}_x\} \\
&= \frac{\mu_0}{2}\frac{|\mathbf{K}_{zs}|^2}{kg}\text{Re}\{\frac{j}{1 + j\frac{\mu_0 u_{\text{s}}\sigma_{\text{s}}}{kg}}\} \\
&= \frac{\mu_0}{2}\frac{|\mathbf{K}_{zs}|^2}{kg}\frac{\frac{\mu_0 u_{\text{s}}\sigma_{\text{s}}}{kg}}{1 + \left(\frac{\mu_0 u_{\text{s}}\sigma_{\text{s}}}{kg}\right)^2}
\end{aligned}$$

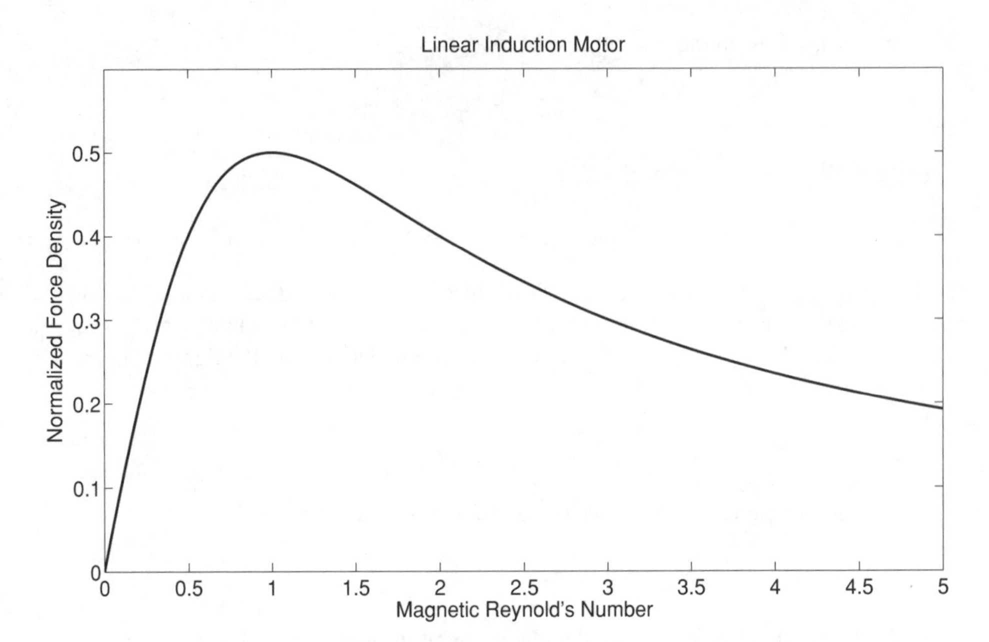

Figure 8.11 Normalized shear vs. magnetic Reynold's number

8.3 Surface Impedance of Uniform Conductors

The objective of this section is to describe the calculation of the surface impedance presented by a layer of conductive material. Two problems are considered here. The first considers a layer of *linear* material backed up by an infinitely permeable surface. This is approximately the situation presented by, for example, surface mounted permanent magnets and is a decent approximation to the conduction mechanism that would be responsible for loss due to asynchronous harmonics in these machines. It is also appropriate for use in estimating losses in solid rotor induction machines and in the poles of turbogenerators. The second problem, which is not worked here, but is simply presented, concerns saturating ferromagnetic material.

8.3.1 Linear Case

The situation and coordinate system are shown in Figure 8.12. The conductive layer is of thickness h and has conductivity σ and permeability μ_0. To keep the mathematical expressions within bounds, rectilinear geometry is assumed. This assumption will present errors which are small to the extent that curvature of the problem is small compared with the wavenumbers encountered. The situation is excited, as it would be in an electric machine, by an axial current sheet of the form $K_z = \text{Re}\left\{\mathbf{K}e^{j(\omega t - kx)}\right\}$

Assume that, in the conductive material that conductivity and permeability are both constant and uniform. Curl of magnetic field is described by Ampere's Law:

$$\nabla \times \vec{H} = \vec{J} = \sigma \vec{E}$$

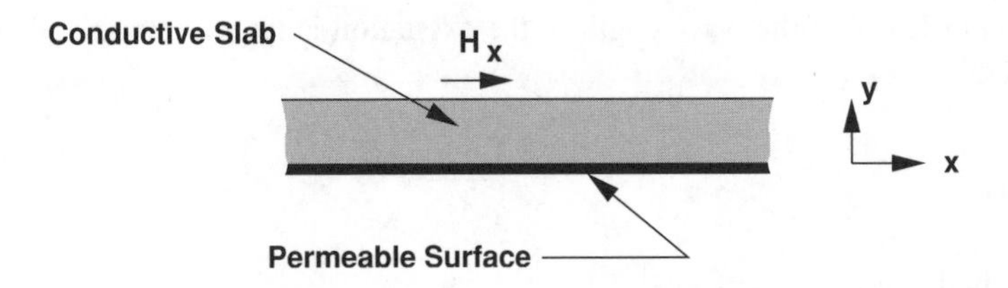

Figure 8.12 Axial view of magnetic field problem

And the curl of electric field is described by Faraday's Law:

$$\nabla \times \vec{E} = -\mu \frac{\partial \vec{H}}{\partial t}$$

Taking the curl of Ampere's Law and substituting:

$$\nabla \times \nabla \times \vec{H} = \sigma \nabla \times \vec{E} = -\mu\sigma \frac{\partial \vec{H}}{\partial t}$$

Then, using the vector identity:

$$\nabla \times \nabla \times \vec{H} = \nabla \left(\nabla \cdot \vec{H} \right) - \nabla^2 \vec{H}$$

If permeability is uniform, $\nabla \cdot \vec{H} = 0$, leaving Bullard's equation: a diffusion equation in $\vec{H}$:

$$\nabla^2 \vec{H} = \mu_0 \sigma \frac{\partial \vec{H}}{\partial t}$$

In view of the boundary condition at the back surface of the material ($\vec{H}_x = 0$), taking that point to be $y = 0$, a general solution for the magnetic field in the material is:

$$H_x = \text{Re}\left\{ A \sinh \gamma y e^{j(\omega t - kx)} \right\}$$
$$H_y = \text{Re}\left\{ j\frac{k}{\gamma} A \cosh \gamma y e^{j(\omega t - kx)} \right\}$$

where the coefficient γ satisfies:

$$\gamma^2 = j\omega\mu_0\sigma + k^2$$

and note that the coefficients above are required to have the specified ratio so that $\vec{H}$ has divergence of zero.

If k is small (that is, if the wavelength of the excitation is large), this spatial coefficient γ becomes

$$\gamma = \frac{1+j}{\delta}$$

where the skin depth is:

$$\delta = \sqrt{\frac{2}{\omega\mu_0\sigma}}$$

To obtain surface impedance, we use Faraday's law:

$$\nabla \times \vec{E} = -\frac{\partial \vec{B}}{\partial t}$$

which gives:

$$\mathbf{E}_z = -\mu_0 \frac{\omega}{k} \mathbf{H}_y$$

Now: the 'surface current' is just

$$\mathbf{K}_s = -\mathbf{H}_x$$

so that the equivalent surface impedance is:

$$\mathbf{Z} = \frac{\mathbf{E}_z}{-\mathbf{H}_x} = j\mu_0 \frac{\omega}{\gamma} \coth \gamma h$$

A pair of limits are interesting here. Assuming that the wavelength is long so that k is negligible, then if γh is *small* (i.e. thin material),

$$\mathbf{Z} \to j\mu_0 \frac{\omega}{\gamma^2 h} = \frac{1}{\sigma h}$$

On the other hand as $\gamma T \to \infty$ and $k \to 0$,

$$\mathbf{Z} \to \frac{1+j}{\sigma\delta}$$

Next it is necessary to transfer this surface impedance across the air-gap of a machine. So, with reference to Figure 8.13, assume a new coordinate system in which the surface of impedance $\mathbf{Z}_s$ is located at $y = 0$, and we wish to determine the impedance $\mathbf{Z} = -\mathbf{E}_z/\mathbf{H}_x$ at $y = g$.

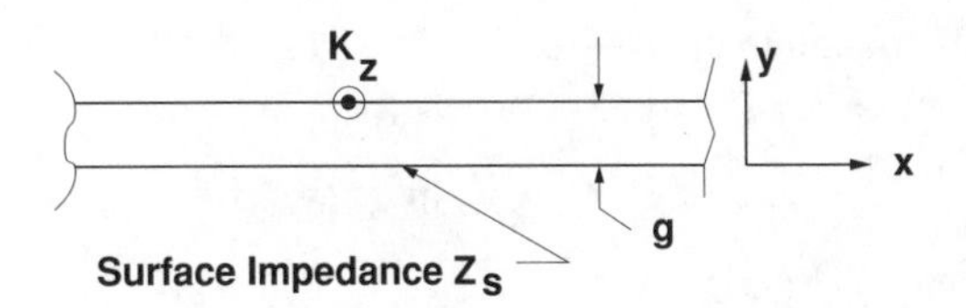

Figure 8.13 Impedance across the air-gap

In the gap there is no current, so magnetic field can be expressed as the gradient of a scalar potential which obeys Laplace's equation:

$$\vec{H} = -\nabla\psi$$

and

$$\nabla^2\psi = 0$$

Ignoring a common factor of $e^{j(\omega t-kx)}$, we can express $\vec{H}$ in the gap as:

$$\mathbf{H}_x = jk\left(\mathbf{\Psi}_+ e^{ky} + \mathbf{\Psi}_- e^{-ky}\right)$$
$$\mathbf{H}_y = -k\left(\mathbf{\Psi}_+ e^{ky} - \mathbf{\Psi}_- e^{-ky}\right)$$

At the surface of the rotor,

$$\mathbf{E}_z = -\mathbf{H}_x\mathbf{Z}_s = -\mu_0\frac{\omega}{k}\mathbf{H}_y = -\omega\mu_0\left(\mathbf{\Psi}_+ - \mathbf{\Psi}_-\right)$$

or

$$-\omega\mu_0\left(\mathbf{\Psi}_+ - \mathbf{\Psi}_-\right) = jk\mathbf{Z}_s\left(\mathbf{\Psi}_+ + \mathbf{\Psi}_-\right)$$

and then, at the surface of the stator,

$$\mathbf{Z} = -\frac{\mathbf{E}_z}{\mathbf{H}_x} = j\mu_0\frac{\omega}{k}\frac{\mathbf{\Psi}_+ e^{kg} - \mathbf{\Psi}_- e^{-kg}}{\mathbf{\Psi}_+ e^{kg} + \mathbf{\Psi}_- e^{-kg}}$$

A bit of manipulation is required to obtain:

$$\mathbf{Z} = j\mu_0\frac{\omega}{k}\left\{\frac{e^{kg}\left(\omega\mu_0 - jk\mathbf{Z}_s\right) - e^{-kg}\left(\omega\mu_0 + jk\mathbf{Z}_s\right)}{e^{kg}\left(\omega\mu_0 - jk\mathbf{Z}_s\right) + e^{-kg}\left(\omega\mu_0 + jk\mathbf{Z}_s\right)}\right\}$$

It is useful to note that, in the limit of $\mathbf{Z}_s \to \infty$, this expression approaches the *gap impedance*

$$\mathbf{Z}_g = j\frac{\omega\mu_0}{k^2 g}$$

and, if the gap is small enough that $kg \to 0$,

$$\mathbf{Z} \to \mathbf{Z}_g || \mathbf{Z}_s$$

8.3.2 Iron

Electric machines employ ferromagnetic materials to carry magnetic flux from and to appropriate places within the machine. Such materials have properties which are interesting, useful and problematical, and the designers of electric machines must deal with this stuff. The purpose of this section is to introduce the most salient properties of the kinds of magnetic materials used in electric machines and to describe approximate ways of dealing with such materials.

The types of materials of interest here exhibit *magnetization*: flux density is something other than $\vec{B} = \mu_0 \vec{H}$. Magnetic materials are often characterized as being *hard* or *soft*. Hard materials are those in which the magnetization tends to be permanent, while soft materials are used in magnetic circuits of electric machines and transformers. While the language and analytical methods used in dealing with these two types of material are closely related, the applications to which they are put are not.

8.3.3 Magnetization

It is possible to relate, in all materials, magnetic flux density to magnetic field intensity with a constitutive relationship of the form:

$$\vec{B} = \mu_0 \left(\vec{H} + \vec{M} \right)$$

where magnetic field intensity H and magnetization M are the two important properties. In linear magnetic material magnetization is a simple linear function of magnetic field:

$$\vec{M} = \chi_{\mathrm{m}} \vec{H}$$

so that the flux density is also a linear function:

$$\vec{B} = \mu_0 (1 + \chi_{\mathrm{m}}) \vec{H}$$

Note that in the most general case the magnetic susceptibility χ_{m} might be a tensor, leading to flux density being not co-linear with magnetic field intensity, or it might be complex and frequency dependent, leading to phase shift and lossy behavior. But such a relationship would still be linear. Generally this sort of complexity does not have a major effect on electric machines.

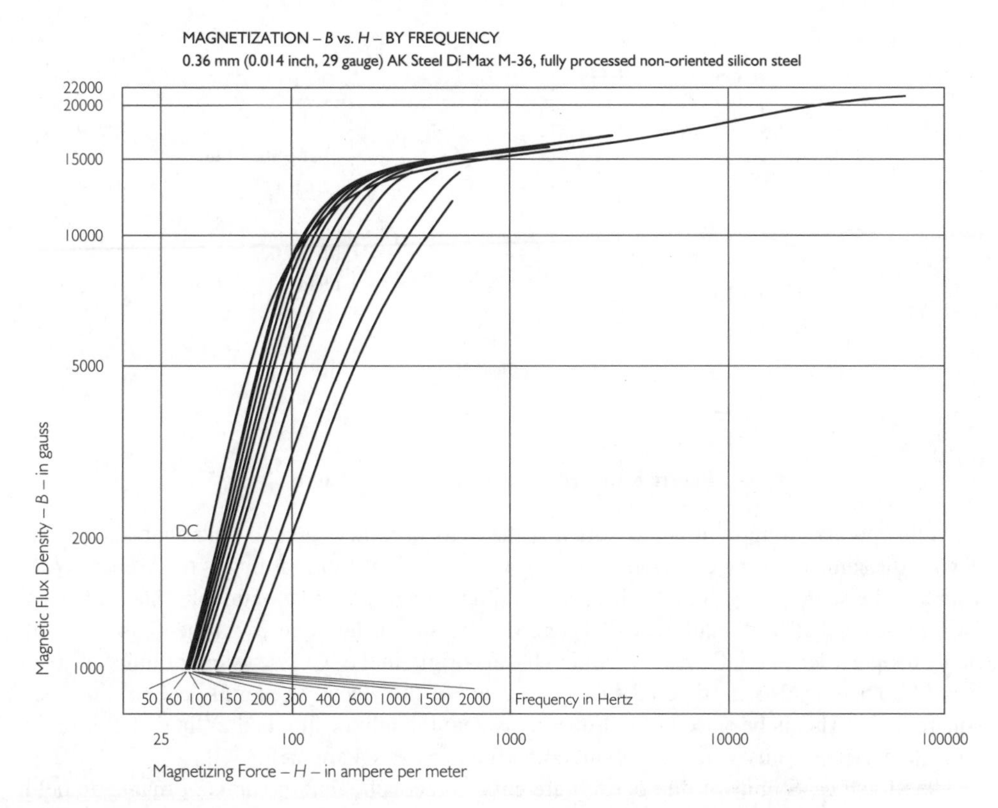

Figure 8.14 Saturation curve for one type of commercial silicon iron. Adapted from *Lamination steels,* 3rd edn, Steve Sprague, editor, ©2007. Reproduced by permission of the Electric Motor Education and Research Foundation

8.3.4 Saturation and Hysteresis

In useful magnetic materials this analytically benign relationship is not correct and there is a need for more general view. The microscopic picture is not dealt with here, except to note that the magnetization is due to the alignment of groups of magnetic dipoles, the groups often called *domains*. There are only so many magnetic dipoles available in any given material, so that once the flux density is high enough the material is said to *saturate*, and the relationship between magnetic flux density and magnetic field intensity is nonlinear.

Shown in Figure 8.14, for example, is a 'saturation curve' for a magnetic sheet steel that is sometimes used in electric machinery sprague (2007)[1]. Note the magnetic field intensity is on a logarithmic scale. If this were plotted on linear coordinates the saturation would appear to be quite abrupt. It is also important to recognize that data such as this is typical, and before actually designing a machine it is important to verify actual properties of the materials to be used.

[1] Chart adapted for this publication by Steve Sprague, based on the original chart in *Lamination steels*, 3rd editon Sprague, S. (ed.) a CD-ROM published by EMERF (Electric Motor Education and Research Foundation).

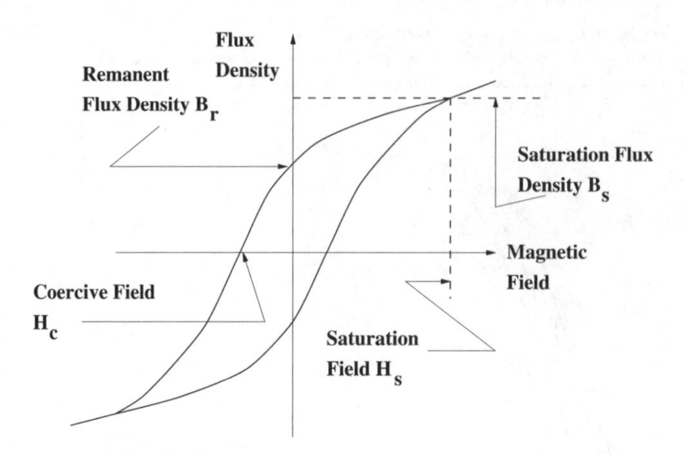

Figure 8.15 Hysteresis curve nomenclature

At this point it is appropriate to note that the units used in magnetic field analysis are not always the same nor even consistent. In almost all systems the unit of flux is the Weber (Wb), which is the same as a volt-second. In the commonly accepted international system of units (sometimes called 'Rationalized MKS' or 'SI', the unit of flux density is the Tesla (T), but many people refer to the Gauss (G), which has its origin in the 'CGS' system of units. 10,000 G = 1 T. There is also an Imperial system measure of flux density generally called 'lines per square inch'. This is because in the Imperial system the unit of flux is the 'line'. 10^8 lines is equal to a Weber. Thus, a Tesla is about 64.5 kilolines per square inch.

The SI and CGS units of flux density are easy to reconcile, but the units of magnetic field are a bit harder. In SI, H is expressed in A/m (or ampere-turns per meter). The permeability of free space is $\mu_0 = 4\pi \times 10^{-7}$H/m. So the magnetic field corresponding to one Tesla in free space is $H = \frac{1}{\mu_0} \approx 795,775$ A/m. In the CGS system of units the unit of magnetic field is the Oersted and the permeability of free space is taken to be one Gauss per Oersted, So 79.5775 A/m is 1 Oe.

In most useful magnetic materials the magnetic domains tend to be somewhat 'sticky', and a more-than-incremental magnetic field is required to get them to move. This leads to the property called 'hysteresis', both useful and problematical in many magnetic systems.

Hysteresis loops take many forms; a generalized picture of one is shown in Figure 8.15. Salient features of the hysteresis curve are the remanent magnetization B_r and the coercive field H_c. Note that the actual loop that will be traced out is a function of field amplitude and history. Thus there are many other 'minor loops' that might be traced out by the B-H characteristic of a piece of material, depending on just what the fields and fluxes have done and are doing.

Hysteresis is important for two reasons. First, it represents the mechanism for 'trapping' magnetic flux in a piece of material to form a permanent magnet. Second, hysteresis is a loss mechanism. To show this, consider some arbitrary chunk of material for which MMF and flux are:

$$F = NI = \int \vec{H} \cdot d\vec{\ell}$$

$$\Phi = \int \frac{V}{N} dt = \iint_{\text{Area}} \vec{B} \cdot d\vec{A}$$

Energy input to the chunk of material over some period of time is

$$w = \int VI\mathrm{d}t = \int F\mathrm{d}\Phi = \int_t \int \vec{H} \cdot \mathrm{d}\vec{\ell} \int \int \mathrm{d}\vec{B} \cdot \mathrm{d}\vec{A}\ \mathrm{d}t$$

Now, imagine carrying out the second (double) integral over a continuous set of surfaces which are perpendicular to the magnetic field H. (This is possible!) The energy becomes:

$$w = \int_t \int_{\text{volume}} \vec{H} \cdot \mathrm{d}\vec{B}\mathrm{dvol}\ \mathrm{d}t$$

and, done over a complete cycle of some input waveform, that is:

$$w = \int_{\text{volume}} W_{\mathrm{m}}\mathrm{dvol}$$

where the loss per unit volume at a point in the material is:

$$W_{\mathrm{m}} = \oint_t \vec{H} \cdot \mathrm{d}\vec{B}$$

That last expression simply expresses the area of the hysteresis loop for the particular cycle. Note the product of magnetic field intensity and magnetic flux density has units of Joules per cubic meter:

$$1A/\mathrm{m} \times 1Wb/\mathrm{m}^2 = 1J/\mathrm{m}^3$$

Generally, for most electric machine applications magnetic materials are characterized as 'soft', having a narrow hysteresis loop (and therefore a low hysteresis loss). At the other end of the spectrum are 'hard' magnetic materials that are used to make permanent magnets. The terminology comes from steel, in which soft, annealed steel material tends to have narrow loops and hardened steel tends to have wider loops. However, permanent magnet technology has advanced to the point where the coercive forces possible in even cheap ceramic magnets far exceed those of the hardest steels.

8.3.5 Conduction, Eddy Currents and Laminations

Steel, being a metal, is an electrical conductor. Thus when time varying magnetic fields pass through it they cause eddy currents to flow, and of course those currents produce dissipation. In fact, for almost all applications involving 'soft' iron, eddy currents are the dominant source of loss. To reduce the eddy current loss, magnetic circuits of transformers and electric machines are almost invariably laminated, or made up of relatively thin sheets of steel. To further reduce losses the steel is alloyed with elements (often silicon) which reduce, or 'poison' the electrical conductivity.

There are several approaches to estimating the loss due to eddy currents in steel sheets and in the surface of solid iron, and it is worthwhile to look at a few of them. It should be noted that this is a difficult problem, since the behavior of the material itself is hard to characterize.

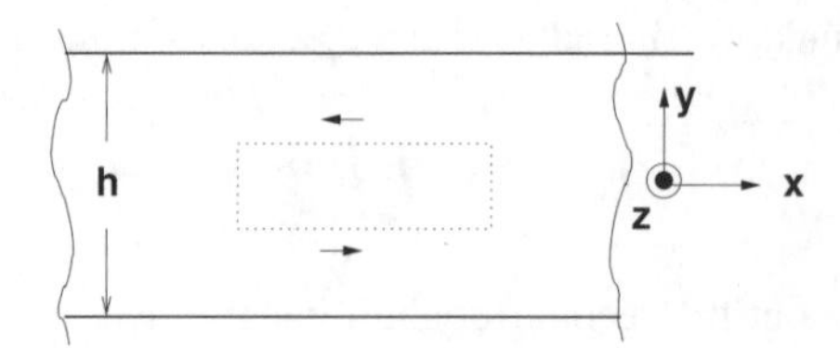

Figure 8.16 Lamination section for loss calculation

8.3.5.1 Complete Penetration Case

Consider the problem of a stack of laminations. In particular, consider one sheet in the stack represented in Figure 8.16. It has thickness h and conductivity σ. Assume that the 'skin depth' is much greater than the sheet thickness so that magnetic field penetrates the sheet completely. Further, assume that the applied magnetic flux density is parallel to the surface of the sheets:

$$\vec{B} = \vec{i}_z \mathrm{Re}\{\sqrt{2} B_0 e^{j\omega t}\}$$

Now use Faraday's law to determine the electric field and therefore current density in the sheet. If the problem is uniform in the x- and z-directions.

$$\frac{\partial \mathbf{E}_x}{\partial y} = -j\omega_0 B_0$$

Note also that, unless there is some net transport current in the x-direction, E must be anti-symmetric about the center of the sheet. Thus if the origin of y is in the center of the sheet, electric field and current are:

$$\mathbf{E}_x = -j\omega B_0 y$$
$$\mathbf{J}_x = -j\omega B_0 \sigma y$$

Local power dissipated is

$$P(y) = \omega^2 B_0^2 \sigma y^2 = \frac{|J|^2}{\sigma}$$

To find average power dissipated, integrate over the thickness of the lamination:

$$< P >= \frac{2}{h} \int_0^{\frac{h}{2}} P(y) \mathrm{d}y = \frac{2}{h} \omega^2 B_0^2 \sigma \int_0^{\frac{h}{2}} y^2 \mathrm{d}y = \frac{1}{12} \omega^2 B_0^2 h^2 \sigma$$

Pay attention to the orders of the various terms here: average power dissipated in a volume of sheets of material is proportional to the square of flux density and to the square of frequency. It is also proportional to the square of the lamination thickness.

As an aside, consider a simple magnetic circuit made of this material, with some length ℓ and area A so that volume of material is ℓA. Flux linked by a coil of N turns would be:

$$\Lambda = N\Phi = NAB_0$$

and voltage is of course just $V = j\omega\Lambda$. Total power dissipated in this core would be:

$$P_c = A\ell\frac{1}{12}\omega^2 B_0^2 h^2 \sigma = \frac{V^2}{R_c}$$

where the equivalent core resistance is now

$$R_c = 12\frac{A}{\ell}\frac{N^2}{\sigma h^2}$$

8.3.6 Eddy Currents in Saturating Iron

The objective in this section is to establish the surface impedance of a conductive, ferromagnetic surface, although we consider only the one-dimensional problem ($k \rightarrow 0$). The problem was worked out by Maclean (1954) and Agarwal (1959). They assumed that the magnetic field at the surface of the flat slab of material was sinusoidal in time and of high enough amplitude to saturate the material. This is true if the material has high permeability and the magnetic field is strong. What happens is that the impressed magnetic field saturates a region of material near the surface, leading to a magnetic flux density parallel to the surface. The depth of the region affected changes with time, and there is a separating surface (in the flat problem this is a plane) that moves away from the top surface in response to the change in the magnetic field. An electric field is developed to move the surface, and that magnetic field drives eddy currents in the material.

Assume that the material has a perfectly rectangular magnetization curve as shown in Figure 8.17, so that flux density in the x-direction is:

$$B_x = B_0 \text{sign}(H_x)$$

The flux per unit width (in the z- direction) is:

$$\Phi = \int_0^{-\infty} B_x \mathrm{d}y$$

and Faraday's law becomes:

$$E_z = \frac{\partial \Phi}{\partial t}$$

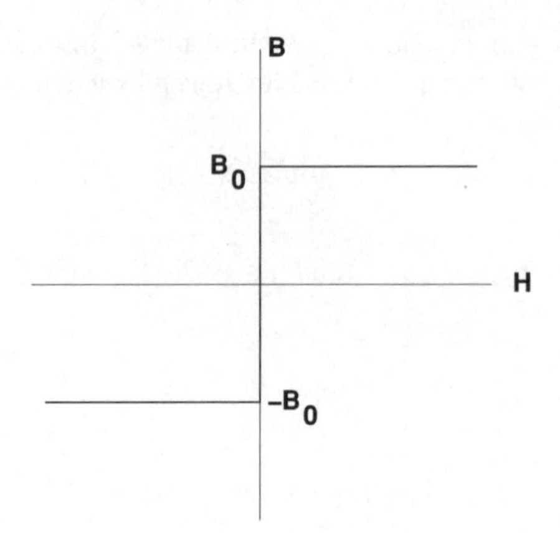

Figure 8.17 Idealized saturating characteristic

while Ampere's law in conjunction with Ohm's law is:

$$\frac{\partial H_x}{\partial y} = \sigma E_z$$

McLean suggested a solution to this set in which there is a 'separating surface' at depth ζ below the surface, as shown in Figure 8.18. At any given time:

$$H_x = H_s(t)\left(1 + \frac{y}{\zeta}\right)$$

$$J_z = \sigma E_z = \frac{H_s}{\zeta}$$

That is, in the region between the separating surface and the top of the material, electric field E_z is uniform and magnetic field H_x is a linear function of depth, falling from its impressed value at the surface to zero at the separating surface. Electric field is produced by the rate of

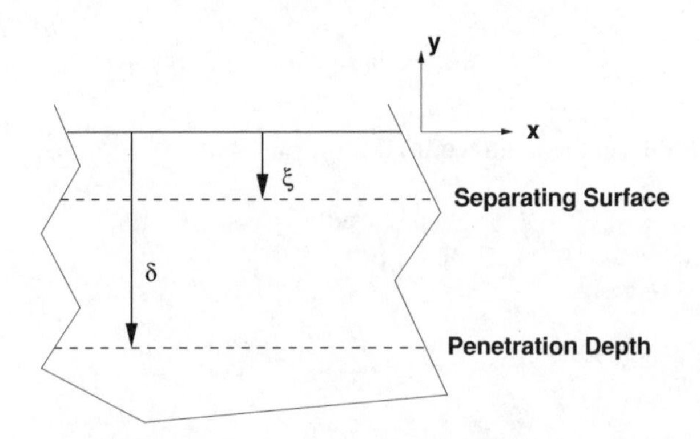

Figure 8.18 Separating surface and penetration depth

change of flux which is:

$$E_z = \frac{\partial \Phi}{\partial t} = 2B_x \frac{\partial \zeta}{\partial t}$$

Eliminating E

$$2\zeta \frac{\partial \zeta}{\partial t} = \frac{H_s}{\sigma B_x}$$

and then, if the impressed magnetic field is sinusoidal, this becomes:

$$\frac{d\zeta^2}{dt} = \frac{H_0}{\sigma B_0} |\sin \omega t|$$

This is easy to solve, assuming that $\zeta = 0$ at $t = 0$,

$$\zeta = \sqrt{\frac{2H_0}{\omega\sigma B_0}} \sin \frac{\omega t}{2}$$

The separating surface always moves in the downward direction (as is shown in Figure 8.18), so at in each half cycle a new surface is created: the old one just stops moving at a maximum position which becomes the penetration depth:

$$\delta = \sqrt{\frac{2H_0}{\omega\sigma B_0}}$$

This penetration depth is analogous to the ‘skin depth’ of the linear theory. However, it is an absolute penetration depth. Below that depth there are no field variations at all.

The resulting electric field is:

$$\mathrm{E}_z = \frac{2\mathrm{H}_0}{\sigma\delta} \cos \frac{\omega t}{2} \qquad 0 < \omega t < \pi$$

This may be Fourier analyzed: noting that if the impressed magnetic field is sinusoidal, only the time fundamental component of electric field is important, leading to:

$$E_z = \frac{8}{3\pi} \frac{H_0}{\sigma\delta} (\cos \omega t + 2 \sin \omega t + \ldots)$$

Complex surface impedance is the ratio between the complex amplitude of electric and magnetic field, which becomes:

$$\mathbf{Z}_s = \frac{\mathbf{E}_z}{\mathbf{H}_x} = \frac{8}{3\pi} \frac{1}{\sigma\delta} (2 + j)$$

Thus, in practical applications, this surface can be treated much in the same fashion as linear conductive surfaces, by establishing a skin depth and assuming that current flows within that skin depth of the surface. The resistance is modified by the factor of $\frac{16}{3\pi}$ and the 'power factor' of this surface is about 89% (as opposed to a linear surface where the 'power factor' is $\frac{1}{\sqrt{2}}$ or about 71%.

Agarwal suggests using a value for B_0 of about 75% of the saturation flux density of the steel.

8.4 Semi-Empirical Method of Handling Iron Loss

Neither of the models described so far are fully satisfactory in describing the behavior of laminated iron, because losses are a combination of eddy current and hysteresis losses. The rather simple model employed for eddy currents is imprecise because of its assumption of abrupt saturation. The hysteresis model, while precise, would require an empirical determination of the size of the hysteresis loops anyway. So resort is often made to empirical loss data. Manufacturers of lamination steel sheets publish data, usually in the form of curves, for many of their products. Here are a few ways of looking at the data.

A low frequency flux density vs. magnetic field ('saturation') curve was shown in Figure 8.14. In *some* machine applications either the 'total' inductance (ratio of flux to MMF) or 'incremental' inductance (slope of the flux to MMF curve) is required. In the limit of low frequency these numbers may be useful.

For designing electric machines, however, a second way of looking at steel may be more useful. This is to measure the real and reactive power as a function of magnetic flux density and (sometimes) frequency. In principle, this data is immediately useful. In any well-designed electric machine the flux density in the core is distributed fairly uniformly and is not strongly affected by eddy currents in the core. Under such circumstances one can determine the flux density in each part of the core. With that information one can go to the published empirical data for real and reactive power and determine core loss and reactive power requirements.

Figures 8.19 and 8.20 show core loss and 'apparent' power per unit mass as a function of (RMS) induction for 29 gage, fully processed M-36 steel (Sprague, 2007). The use of this data is quite straightforward. If the flux density in a machine is estimated for each part of the machine and the mass of steel calculated, then with the help of this chart a total core loss and apparent power can be estimated[2]. Then the effect of the core may be approximated with a pair of elements in parallel with the terminals, with:

$$R_c = \frac{q|V|^2}{P}$$

$$X_c = \frac{q|V|^2}{Q}$$

$$Q = \sqrt{P_a^2 - P^2}$$

[2] This data is representative of actual steel, but it is important to test and verify any material before using it in a product.

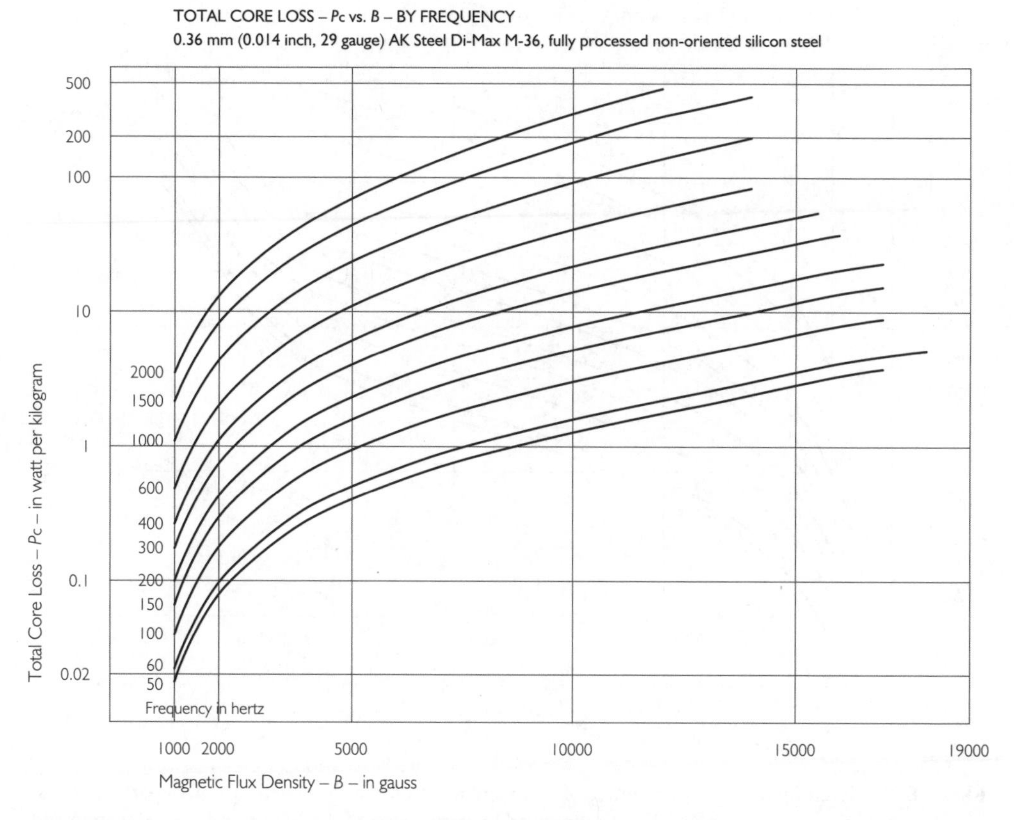

Figure 8.19 Core loss for one grade of commercial sheet steel. Adapted from *Lamination Steels*, 3rd edn, Steve Sprague, editor. ©2007. Reproduced by permission of the Electric Motor Education and Research Foundation

Where q is the number of machine phases and V is *phase* voltage. Note that this picture is, strictly speaking, only valid for the voltage and frequency for which the flux density was calculated. But it will be approximately true for small excursions in either voltage or frequency and therefore useful for estimating voltage drop due to exciting current and such matters. In design program applications these parameters can be recalculated repeatedly if necessary.

'Looking up' this data is a bit awkward for design studies, so it is often convenient to do a 'curve fit' to the published data. There are a large number of possible ways of doing this. One method that has been found to work reasonably well for silicon iron is an 'exponential fit':

$$P \approx P_0 \left(\frac{B}{B_0}\right)^{\epsilon_B} \left(\frac{f}{f_0}\right)^{\epsilon_F}$$

This fit is appropriate if the data appears on a log-log plot to lie in approximately straight lines. Figure 8.21 shows such a fit for another grade of steel sheet. Table 8.1 gives some exemplar parameters for commonly used steel sheets.

For 'apparent power' or 'exciting power' (they mean the same), the same sort of method can be used. It appears, however, that the simple exponential fit that works well for real power

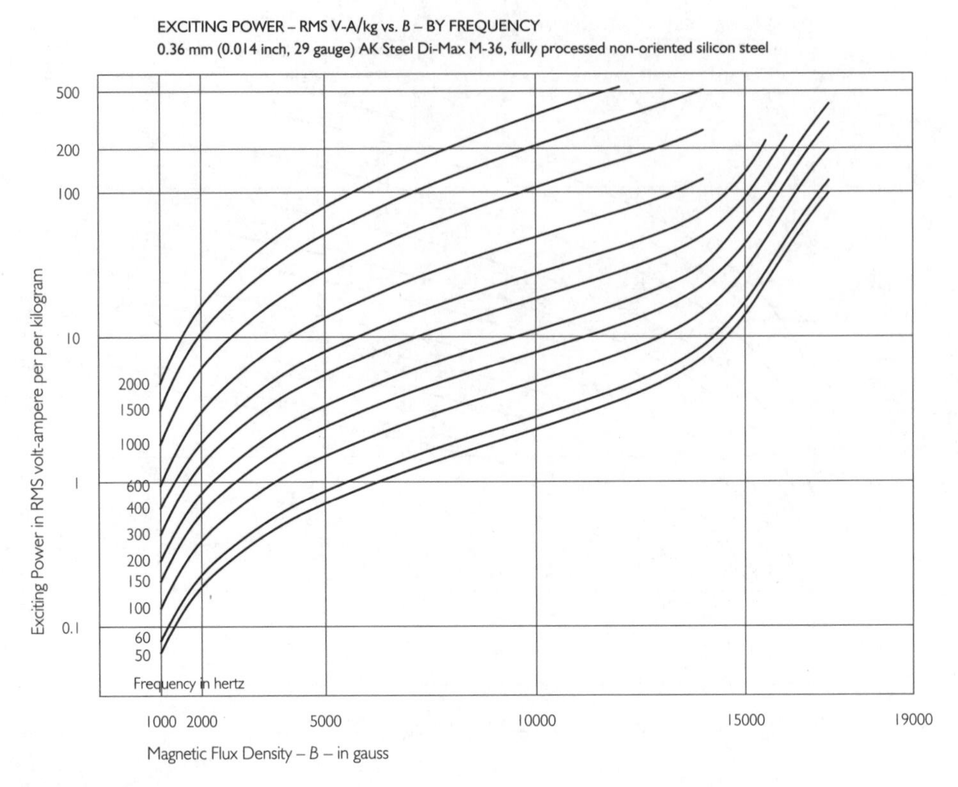

Figure 8.20 Exciting power for one grade of commercial sheet steel. Adapted from *Lamination Steels*, 3rd edn, Steve Sprague, editor. ©2007. Reproduced by permission of the Electric Motor Education and Research Foundation

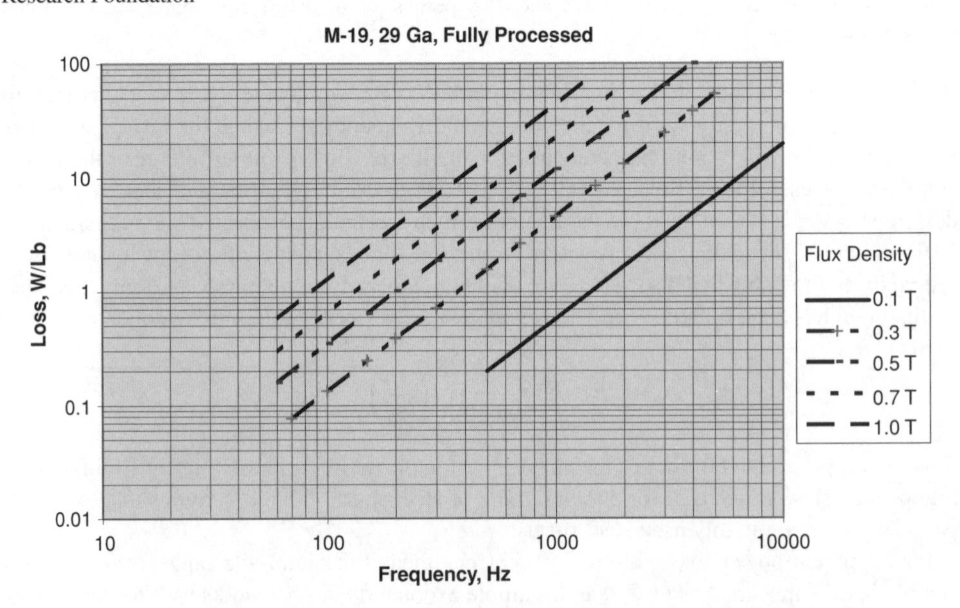

Figure 8.21 Steel sheet core loss fit vs. flux density and frequency

Table 8.1 Exponential fit parameters for two steel sheets

		29 Ga, fully processed	
		M-19	M-36
Base flux density	B_0	1 T	1 T
Base frequency	f_0	60 Hz	60 Hz
Base power (w/lb)	P_0	0.59	0.67
Flux exponent	ϵ_B	1.88	1.86
Frequency exponent	ϵ_F	1.53	1.48
Base apparent power 1	VA_0	1.08	1.33
Base apparent power 2	VA_1	0.0144	0.0119
Flux exponent	ϵ_0	1.70	2.01
Flux exponent	ϵ_1	16.1	17.2

Data from manufacturer's data sheets

is inadequate, at least if relatively high flux densities are to be used. This is because, as the steel saturates, the reactive component of exciting current rises rapidly. The author has had some success with a 'double exponential' fit:

$$VA \approx VA_0 \left(\frac{B}{B_0}\right)^{\epsilon_0} + VA_1 \left(\frac{B}{B_0}\right)^{\epsilon_1}$$

To first order the reactive component of exciting current will be linear in frequency.

8.5 Problems

1. An induction motor is rated at 1 kW at 3,000 r.p.m. It can achieve a shear stress of 4 kPa. If the rotor of that machine has an aspect ratio $\frac{L}{D} = 1$, what are the rotor diameter and length (which are the same!)?
2. A magnetic structure with a single air-gap is shown in cross-section Figure 8.22. The gap is 1 mm and has a surface area of 4 cm^2. The coil has 100 turns and is carrying a current of 10 A. What is the force on the air-gap?

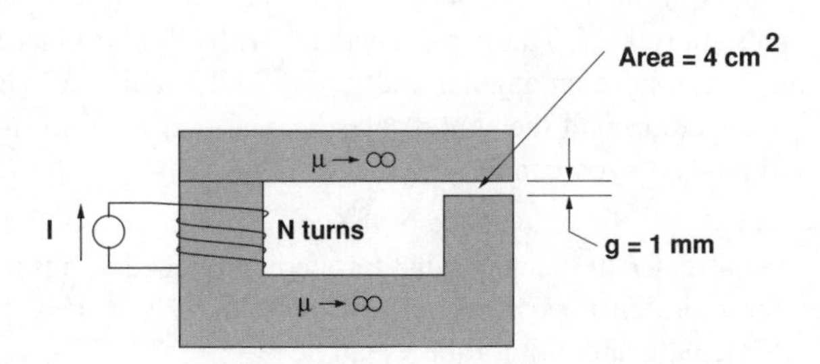

Figure 8.22 Magnetic circuit

3. A linear actuator is shown in Figure 8.23. This is cylindrical in shape: a circular rod is engaged with a circular cylindrical structure and a coil, as shown. The gap between the

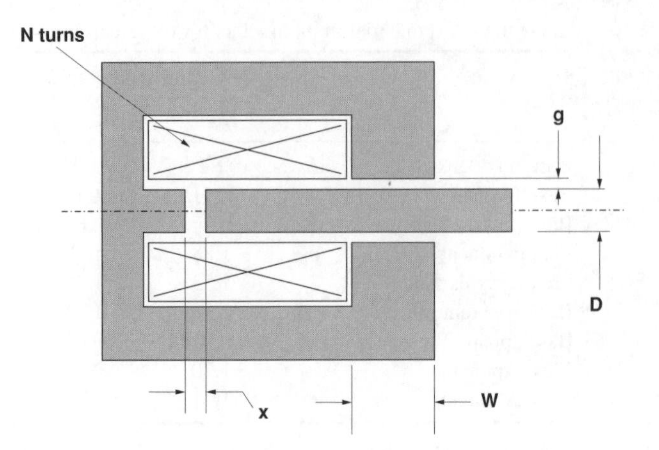

Figure 8.23 Magnetic circuit

rod and the outer magnetic circuit is 0.5 mm. The rod is 1 cm in diameter. The dimension W is 2 cm. Both the rod and the outer magnetic circuit are of very high permeability. The gap between the end of the rod and the surface of the magnetic circuit is shown as dimension 'x', and can vary between 0 and 5 mm.

(a) If the coil has 100 turns and is carrying 10 A, what is the force on the rod?
(b) If the coil has 100 turns, how much current is required to make the flux density across the gap equal to 1.8 T?

4. A rotary structure has inductance that is a function of an angle. That inductance is:

$$L = L_0 + L_2 \cos 2\theta$$

(a) If this structure has a current I in the coil, what torque is produced?
(b) Suppose the thing is operated with $I = I_0 \cos \omega t$ and the rotor is turning so that $\theta = \omega t + \delta$. What is torque as a function of time and that angle δ?

5. A primitive rotary actuator is shown in Figure 8.24. A highly permeable salient rotor can turn within a highly permeable magnetic circuit. The rotor can be thought of as a circular rod with its sides shaved off. The stator has poles with circular inner surfaces. The poles of the rotor and stator have an angular width of θ_0 and a radius R. The gap dimension is g, The coils wrapped around the stator poles have a total of N turns. The structure has length (in the dimension you cannot see) L.

(a) Estimate and sketch the inductance of the coil as a function of the angle θ.
(b) If there is a current I in the coil, what torque is produced as a function of angle?
(c) Now use these dimensions: $R = 2\,\text{cm}$, $g = 0.5\,\text{mm}$, $N = 100$, $L = 10\,\text{cm}$, $\theta_0 = \frac{\pi}{6}$, $I = 10A$. Calculate and plot torque vs. angle.

6. Figure 8.25 shows a 'cartoon' view of a 'rail gun'. This has a depth D in the dimension you can't see (into the paper). The rails are spaced apart by width w and the railgun has barrel length L. The current source establishes a surface current which is uniform

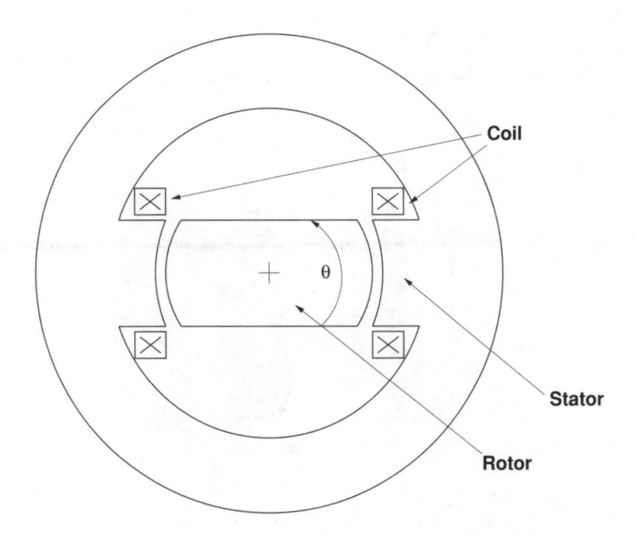

Figure 8.24 Rotary structure

in that direction. Current flows through the dark shaded block of material, which is the projectile. Magnetic fields produce force, which pushes the projectile to the right. Assume that friction, windage and resistive losses are all negligible. Assume that current is a step:

$$I(t) = I_0 u_{-1}(t)$$

(a) What is the force on the block, in terms of the dimensions D, w and current I_0?
(b) If the block is moving to the right with velocity u, what is the voltage across the current source?
(c) Find the velocity and position of the block, assuming that it has mass M, starts at position $x = 0$ and with velocity $u = 0$.
(d) Find power converted into mechanical motion and power out of the current source.
(e) Defining 'efficiency' as the ratio of energy converted into mechanical motion to energy from the current source, what is the efficiency of the rail gun? Assume that energy input stops when the block leaves the end of the barrel and that energy stored in magnetic field is somehow allowed to dissipate through a mechanism that someone else will design.

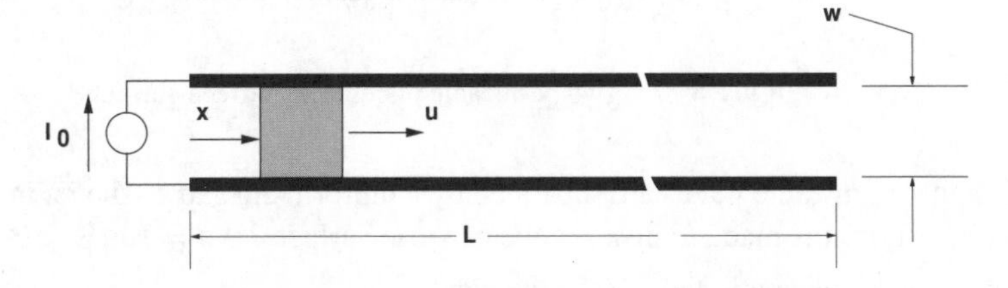

Figure 8.25 Cartoon view of idealized rail gun

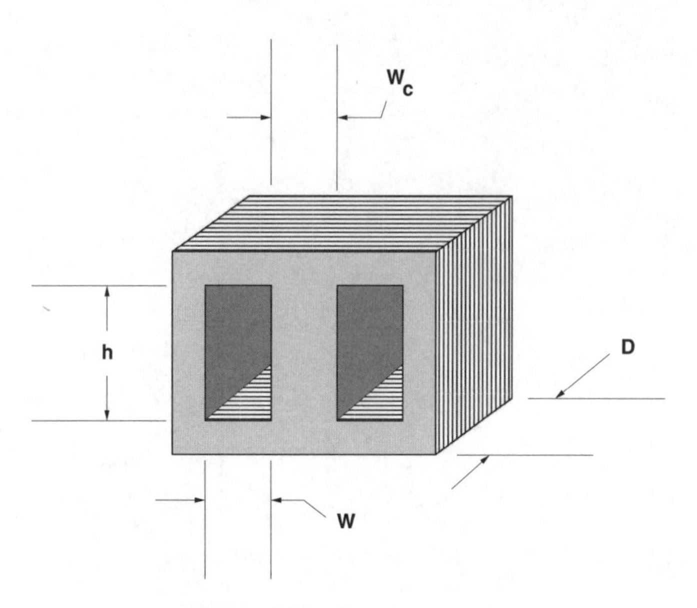

Figure 8.26 Transformer core

7. A transformer core is shown in cartoon form in Figure 8.26. This is made up of, perhaps implausibly, 29 gage M-36 laminations: the same sort of material described by Figures 8.19 and 8.20. Assume a density of 7,700 kg/m^3. Dimensions as called out in the figure are:

Core central column width	W_c	250 mm
Window width	W	250 mm
Window height	h	500 mm
Depth	D	250 mm

The primary winding of this core is wound with 96 turns. It is connected to a 60 Hz voltage source of 2,400 V, RMS.

(a) How much current does the winding draw?
(b) How much power is dissipated in the core?

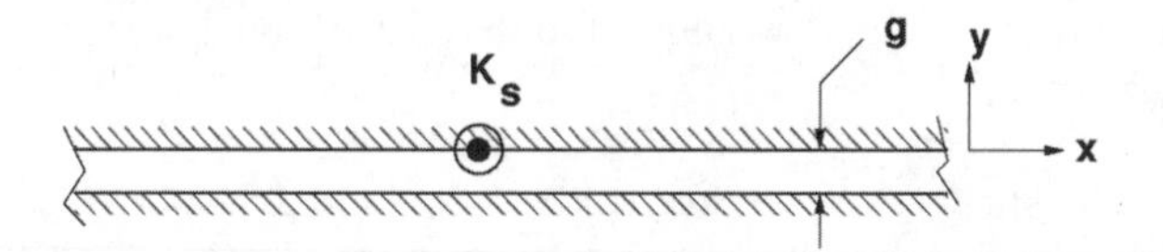

Figure 8.27 Two permeable plates with surface current

8. Two highly permeable plates are separated by a uniform air-gap as shown in Figure 8.27. A surface current is made to flow on one of those surfaces: $K_S = \text{Re}\{\mathbf{K}_{zs} e^{j(\omega t - kx)}\}$.

(a) What is the magnetic field H_y in the gap?
(b) Use the Maxwell Stress Tensor to get the force on the lower plate.

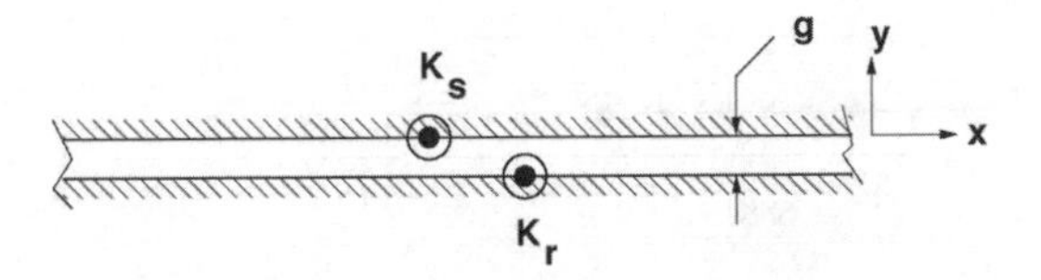

Figure 8.28 Two permeable plates with surface currents

9. Two highly permeable plates are separated by a uniform air-gap as shown in Figure 8.28. Surface currents are made to flow on both of those surfaces:

$$K_S = \mathrm{Re}\{\mathbf{K}_{zs}\mathrm{e}^{j(\omega t - kx)}\}$$
$$K_R = \mathrm{Re}\{\mathbf{K}_{zr}\mathrm{e}^{j(\omega t - k(x - x_0))}\}$$

(a) What is the magnetic field H_y in the gap? Assume both coefficients $\mathbf{K}_{zs}$ and $\mathbf{K}_{zr}$ are real.
(b) Use the Maxwell Stress Tensor go get average force on the lower plate as a function of x_0.

10. Figure 8.29 shows a solid conductor in a slot. Assume the material surrounding the slot is both highly permeable and laminated so that it cannot conduct current in the direction perpendicular to the paper. The conductor is made of copper with an electrical conductivity of $\sigma = 5.81 \times 10^7$ S/m. The width of the conductor is $W = 1$ cm.

(a) What is the resistance per unit length for DC current if the depth $D = 5$ cm?
(b) What is the resistance per unit length for 60 Hz current if the depth is very large?
(c) What is the reactance per unit length for 60 Hz current if the depth is very large?
(d) What is the resistance per unit length for 60 Hz current if the depth is $D = 5$ cm?
(e) What is the reactance per unit length for 60 Hz current if the depth is $D = 5$ cm?
(f) Calculate, compare and plot the resistance per unit length for two cases: one is very large depth and the other is for $D = 5$ cm over a frequency range from $1 < f < 1000$ Hz.
(g) Calculate, compare and plot the reactance per unit length for two cases: one is very large depth and the other is for $D = 5$ cm over a frequency range from $1 < f < 1000$ Hz.

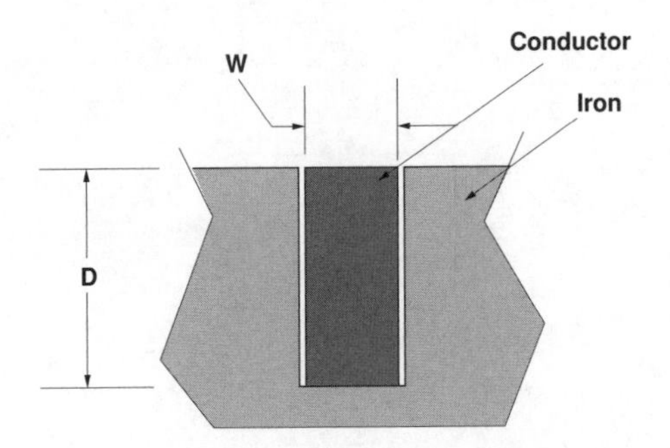

Figure 8.29 Conductor in slot

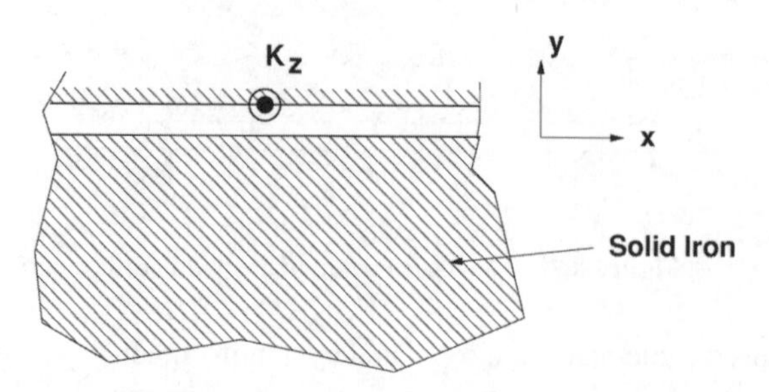

Figure 8.30 Iron surface with magnetic field parallel

11. Figure 8.30 shows the situation. An axial current $K_z = \mathrm{Re}\{\mathrm{K_s} e^{j\omega t}\}$ flows on the surface of an assumed permeable surface. Current must return in the region near the surface of an iron conductor (the lower shaded volume in the figure). Use the Mclean/Agarwal theory presented in Section 8.3.6, with an assumed saturation flux density of $B_0 = 1$ T.

(a) Assume the conductivity of the iron is $\sigma = 6 \times 10^6$ S/m. What is the loss per unit area in the iron if the surface current density $|K_s| = 10,000$ A/m (peak)at 60 Hz?

(b) Estimate and plot the loss per unit area for 60 Hz currents between 10,000 and 100,000 A/m (peak). Use the same conductivity of $\sigma = 6 \times 10^6$ S/m.

12. Using the picture of Figure 8.30 and the situation of the previous problem, but with the conductor of aluminum with a conductivity of $\sigma = 3 \times 10^7$ S/m

(a) what is the loss per unit area for current density $|K_s| = 10,000$ A/m (peak)at 60 Hz?

(b) Estimate and plot the loss per unit area for 60 Hz currents between 10,000 and 100,000 A/m (peak).

References

Agarwal, P.D. (1959) Eddy-current losses in solid and laminated iron. *Trans. AIEE*, 78, 169–171.

MacLean, W. (1954) Theory of strong electromagnetic waves in massive iron. *Journal of Applied Physics*, 25(10).

Melcher, JR. and Woodson H.H. (1968) *Electromechanical Dynamics* (3 volumes). Chichester, John Wiley & Sons, Ltd.

Sprague, S. (ed.) (2007) *Lamination Steels*, 3rd edition. South Dartmouth, MA: The Electric Motor Education and Research Foundation. CD-ROM. Second printing.

9

Synchronous Machines

Virtually all generators in power plants are *synchronous machines*, justifying a substantial amount of attention to this class of electric machine. Synchronous machines are named for the property that they rotate in synchronism with the electrical system to which they are connected. Synchronous generators come in a very wide range of ratings, up to, as of this writing, nearly 2 GVA.

Synchronous machines can have either wound field or permanent magnet excitation. Most machines used as generators and many synchronous motors have field windings to permit control of voltage and reactive power; this is the type of machine we deal with in this chapter. Permanent magnet synchronous machines, controlled by power electronic converters, are sometimes used as high performance motors. This class of machine will be dealt with in Chapter 15.

The basic operation of a synchronous machine is quite simple. The field winding (or permanent magnet array) produces magnetic flux that varies, roughly sinusoidally, around the periphery of the rotor. The number of cycles of variation of this flux density is referred to as the number of *pole pairs* of the rotor. The armature of the machine, usually mounted on the stator, consists of a number of phase windings (typically three), wound in such a way as to link the flux from the rotor winding. As the rotor turns, the variation of flux from the rotor induces a voltage in the armature windings.

There are two types of wound field synchronous machines used as power plant generators. Turbogenerators, machines driven by steam or gas turbines, have relatively high shaft speed (3,600 or 1,800 r.p.m. in 60 Hz systems, 3,000 or 1,500 r.p.m. in 50 Hz systems). Such machines generally have field windings mounted in slots machined into a cylinder of magnetic steel. These are called 'round rotor' machines because the inductances of the phase windings of the armature do not vary much with rotor position. Generators with lower shaft speeds, such as those for use with hydraulic turbines or reciprocating engines, usually have field windings disposed around discrete magnetic poles mounted on a rotor body. These 'salient poles' cause the armature inductances to vary with rotor position. Performance of machines of these two classes is similar but salient pole machines require a slightly different analysis. The discussion starts with round rotor machines.

Electric Power Principles: Sources, Conversion, Distribution and Use James L. Kirtley

9.1 Round Rotor Machines: Basics

Figure 9.1 shows the structure of a round rotor synchronous machine. The field winding is situated in slots in the magnetic rotor which turns at some speed. Figure 9.1 actually shows a *two-pole* synchronous machine in which the rotational speed of the rotor is the same as the electrical frequency. It is possible, however, to build synchronous machines with four, six, etc., poles, and the analysis to follow takes this into account. The number of pole pairs is designated as p. The stator of the machine carries an armature winding, typically with three phases, wound in slots.

The three armature windings, or coils, are situated so that they link flux produced by currents in the rotor, or field winding. Such a machine may be described electrically by its self and mutual inductances. In a three phase machine the three armature coils are located symmetrically, with 120° (electrical) separating them. Each of the three armature windings will have a the same self inductance, L_a. There will be a mutual inductance between the armature windings, L_{ab}, and if the windings are located with a uniform 120° separation, those mutual inductances will all be the same. The field winding will have a self-inductance L_f. The mutual inductances between the field winding and the armature phase windings will depend on rotor position ϕ and will be, to a very good approximation:

$$M_{af} = M \cos(p\phi) \tag{9.1}$$

$$M_{bf} = M \cos\left(p\phi - \frac{2\pi}{3}\right) \tag{9.2}$$

$$M_{af} = M \cos\left(p\phi + \frac{2\pi}{3}\right) \tag{9.3}$$

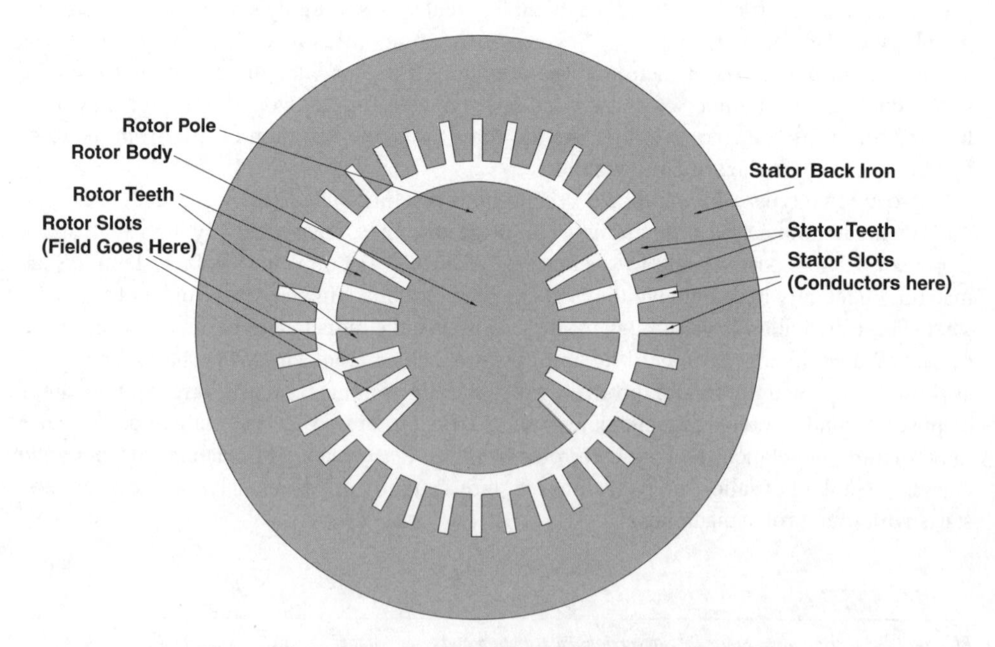

Figure 9.1 Round rotor synchronous machine: magnetic parts

Then the current-flux relationship for the whole machine is:

$$\begin{bmatrix} \lambda_a \\ \lambda_b \\ \lambda_c \\ \lambda_f \end{bmatrix} = \begin{bmatrix} L_a & L_{ab} & L_{ab} & M\cos p\phi \\ L_{ab} & L_a & L_{ab} & M\cos\left(p\phi - \frac{2\pi}{3}\right) \\ L_{ab} & L_{ab} & L_a & M\cos\left(p\phi + \frac{2\pi}{3}\right) \\ M\cos p\phi & M\cos\left(p\phi - \frac{2\pi}{3}\right) & M\cos\left(p\phi + \frac{2\pi}{3}\right) & L_f \end{bmatrix} \begin{bmatrix} i_a \\ i_b \\ i_c \\ i_f \end{bmatrix} \tag{9.4}$$

Using the Principle of Virtual Work method for finding forces of electromagnetic origin, torque is found to be:

$$T^{em} = -pMi_a i_f \sin p\phi - pMi_b i_f \sin\left(p\phi - \frac{2\pi}{3}\right) - pMi_c i_f \sin\left(p\phi + \frac{2\pi}{3}\right)$$

9.1.1 Operation with a Balanced Current Source

Suppose the machine is operated in this fashion: the rotor turns at a constant angular velocity $p\phi = \omega t + \delta_i$; the field current is held constant $i_f = I_f$ and the three stator currents are sinusoids in time, with the same amplitude and with phases that differ by 120°. (This is a *balanced* current set). Note that the rotational speed of the machine matches the electrical frequency of the currents: $p\Omega = \omega$, where Ω is the mechanical rotational speed of the rotor and ω is the electrical frequency. This is why it is a *synchronous* machine.

$$i_a = I\cos(\omega t)$$
$$i_b = I\cos\left(\omega t - \frac{2\pi}{3}\right)$$
$$i_c = I\cos\left(\omega t + \frac{2\pi}{3}\right)$$

Straightforward (but tedious) manipulation yields an expression for torque:

$$T = -\frac{3}{2}pMII_f \sin\delta_i$$

Operated in this way, with balanced currents and with the mechanical speed consistent with the electrical frequency ($p\Omega = \omega$), the machine exhibits a *constant* torque. The phase angle δ_i is the *current* torque angle. It is important to use some caution, as there is more than one torque angle.

Note that this machine can produce either positive or negative torque, depending on the sign of the angle δ_i. If $-\pi < \delta_i < 0$, the sine is negative and torque is positive: the machine is a motor. On the other hand, if $0 < \delta_i < \pi$, torque is negative: the machine is a generator.

9.1.2 Operation with a Voltage Source

Next, look at the machine from the electrical terminals. Flux linked by Phase a will be:

$$\lambda_a = L_a i_a + L_{ab} i_b + L_{ab} i_c + MI_f \cos p\phi$$

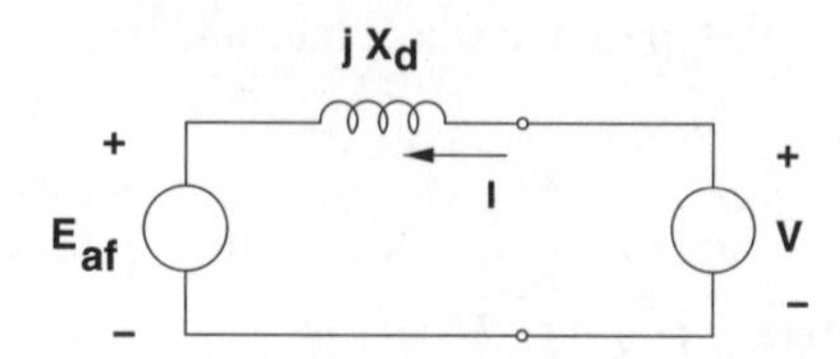

Figure 9.2 Round rotor synchronous machine equivalent circuit, connected to a voltage source

Noting that the sum of phase currents is, under balanced conditions, zero and that the mutual phase–phase inductances are equal, this simplifies to:

$$\lambda_{\mathrm{a}} = (L_{\mathrm{a}} - L_{\mathrm{ab}})\, i_{\mathrm{a}} + M I_{\mathrm{f}} \cos p\phi = L_d i_{\mathrm{a}} + M I_{\mathrm{f}} \cos p\phi$$

where the notation L_{d} denotes synchronous inductance.

If the machine is turning at a speed consistent with the electrical frequency, it is operating synchronously, and it is possible to employ complex notation in the sinusoidal steady state. Then, note:

$$i_{\mathrm{a}} = I \cos(\omega t + \theta_{\mathrm{i}}) = \mathrm{Re}\left\{ I \mathrm{e}^{j(\omega t + \theta_{\mathrm{i}})} \right\}$$

The complex amplitude of flux is:

$$\lambda_{\mathrm{a}} = \mathrm{Re}\left\{ \mathbf{\Lambda}_{\mathrm{a}} \mathrm{e}^{j\omega t} \right\}$$

and the following complex notation for currents is also used:

$$\mathbf{I} = I \mathrm{e}^{j\theta_{\mathrm{i}}}$$
$$\mathbf{I}_{\mathrm{f}} = I_{\mathrm{f}} \mathrm{e}^{j\theta_{\mathrm{m}}}$$

The angle θ_m accounts for the position of the rotor at the time origin.

Terminal voltage of this system is:

$$v_{\mathrm{a}} = \frac{\mathrm{d}\lambda_{\mathrm{a}}}{\mathrm{d}t} = \mathrm{Re}\left\{ j\omega \mathbf{\Lambda}_{\mathrm{a}} \mathrm{e}^{j\omega t} \right\}$$

This system is described by the equivalent circuit shown in Figure 9.2.
where the internal voltage is:

$$\mathbf{E}_{\mathrm{af}} = j\omega M I_{\mathrm{f}} \mathrm{e}^{j\theta_{\mathrm{m}}}$$

If that is connected to a voltage source (i.e. if V is fixed), terminal current is:

$$\mathbf{I} = \frac{V - E_{\mathrm{af}} \mathrm{e}^{j\delta}}{j X_d}$$

where $X_d = \omega L_d$ is the *synchronous reactance*. The angle δ is the angle of the internal voltage relative to the terminal voltage and is referred to as the 'voltage torque angle', or simply as the 'torque angle'.

Then real and reactive power (in phase a) are:

$$P + jQ = \frac{1}{2}\mathbf{VI}^* = \frac{1}{2}\mathbf{V}\left(\frac{\mathbf{V} - E_{af}e^{j\delta}}{jX_d}\right)^* = \frac{1}{2}\frac{|V|^2}{-jX_d} - \frac{1}{2}\frac{VE_{af}e^{j\delta}}{-jX_d}$$

This makes real and reactive power for a single phase:

$$P_a = -\frac{1}{2}\frac{VE_{af}}{X_d}\sin\delta$$

$$Q_a = \frac{1}{2}\frac{V^2}{X_d} - \frac{1}{2}\frac{VE_{af}}{X_d}\cos\delta$$

Considering all three phases, real power is

$$P = -\frac{3}{2}\frac{VE_{af}}{X_d}\sin\delta$$

Now, look at actual operation of these machines, which can serve either as motors or as generators.

Writing Kirchhoff's Voltage Law around the loop that includes the internal voltage, reactance and terminal voltage of the synchronous machine, vector diagrams that describe operation as a motor and as a generator are generated. These vector diagrams are shown in Figures 9.3 and 9.4, respectively.

Operation as a generator is not much different from operation as a motor, but it is common to make notations with the terminal current given the opposite ('generator') sign. Note that, as with the current source case, the difference between motor and generator operation is the sign of the torque angle δ. Note also that the angle δ is *not* the same as the current torque angle δ_i. Note also that reactive power of the machine is affected by field current. Generally, increasing field current, which increases internal voltage, tends to reduce the amount of reactive power

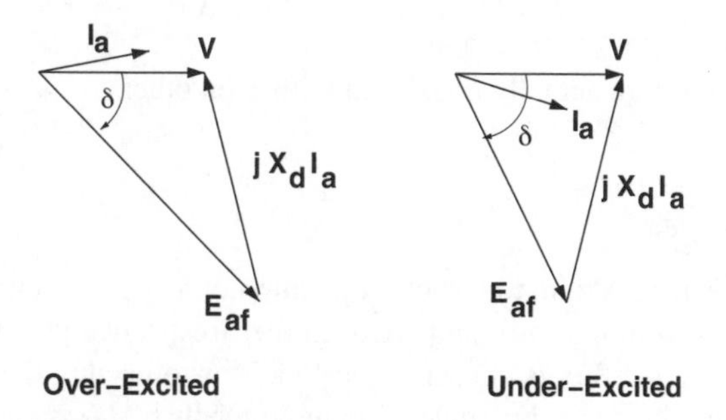

Figure 9.3 Motor operation, under- and overexcited

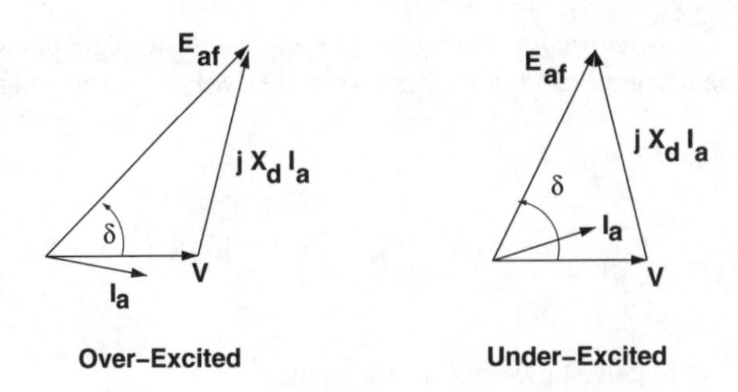

Figure 9.4 Generator operation, under- and overexcited

drawn by the synchronous machine or to increase the amount of reactive power produced by the machine.

9.2 Reconciliation of Models

Power and/or torque characteristics of the synchronous machine may be calculated from two points of view. First, currents in the rotor and stator are used in an expression for torque vs. a power angle:

$$T = -\frac{3}{2} pMII_{\mathrm{f}} \sin \delta_{\mathrm{i}}$$

From a circuit point of view, it is possible to derive an expression for power in terms of internal and terminal voltage:

$$P = -\frac{3}{2} \frac{VE_{\mathrm{af}}}{X_d} \cos \delta$$

and of course since power is torque times speed, this implies that:

$$T = -\frac{3}{2} \frac{VE_{\mathrm{af}}}{\Omega X_{\mathrm{d}}} \sin \delta = -\frac{3}{2} \frac{pVE_{\mathrm{af}}}{\omega X_{\mathrm{d}}} \sin \delta$$

These two expressions are actually consistent with each other.

9.2.1 *Torque Angles*

Figure 9.5 shows a vector diagram that shows operation of a synchronous motor. It represents the MMF's and fluxes from the rotor and stator in their respective positions in *space* during normal operation. Terminal flux is chosen to be 'real', or to occupy the horizontal position. In motor operation the rotor lags by angle δ, so the rotor flux MI_{f} is shown in that position, rotated by $-\delta$. Stator current is also shown, and the torque angle between it and the rotor, δ_i is

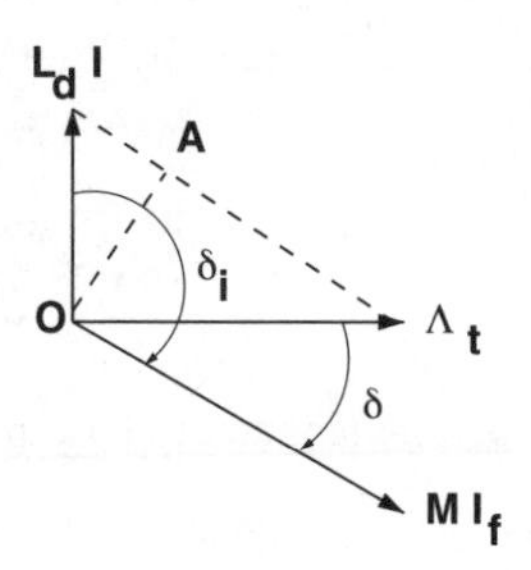

Figure 9.5 Synchronous machine phasor addition

also shown. Now, note that the dotted line OA, drawn perpendicular to a line drawn between the stator flux $L_d I$ and terminal flux Λ_t, has length:

$$|OA| = L_d I \sin \delta_i = \Lambda_t \sin \delta$$

Then, noting that terminal voltage $V = \omega\Lambda_t$, $E_a = \omega M I_f$ and $X_d = \omega L_d$, straightforward substitution yields:

$$\frac{3}{2}\frac{pVE_{af}}{\omega X_d} \sin \delta = \frac{3}{2} pMII_f \sin \delta_i$$

So the current- and voltage based pictures *do* give the same result for torque.

9.3 Per-Unit Systems

At this point, a short detour is taken to apply per-unit systems to synchronous machines. This is a notational device that, in addition to being convenient, will also be conceptually helpful. As has already been seen, the basic notion is quite simple: for most variables a base quantity is established and then, by dividing the variable by the base a 'per-unit', or normalized version of that variable is derived. Generally one would want to tie the base quantity to some aspect of normal operation. For example, the base voltage and current correspond with machine rating might be used. If that is the case, then power base becomes:

$$P_B = 3V_B I_B$$

and, in similar fashion, the impedance base is:

$$Z_B = \frac{V_B}{I_B}$$

Now, a little caution is required here. Voltage base is defined as line–neutral and current base as line current (both RMS). That is not necessary. In a three-phase system one could very well have defined base voltage to have been line–line and base current to be current in a *delta*-connected element:

$$V_{B\Delta} = \sqrt{3}V_B \qquad I_{B\Delta} = \frac{I_B}{\sqrt{3}}$$

In that case the base power would be unchanged but base impedance would differ by a factor of three:

$$P_B = V_{B\Delta} I_{B\Delta} \qquad Z_{B\Delta} = 3Z_B$$

However, if one were consistent with actual impedances (note that a *delta* connection of elements of impedance $3Z$ is equivalent to a *wye* connection of Z), the per-unit impedances of a given system are not dependent on the particular connection. In fact one of the major advantages of using a per-unit system is that per-unit values are uniquely determined, while ordinary variables can be line–line, line–neutral, RMS, peak, etc., for a large number of variations.

Base quantities are usually given as line–line voltage and base power. So that:

$$I_B = \frac{P_B}{\sqrt{3}V_{B\Delta}} \qquad Z_B = \frac{V_B}{I_B} = \frac{1}{3}\frac{V_{B\Delta}}{I_{B\Delta}} = \frac{V_{B\Delta}^2}{P_B}$$

In this text we will usually write per-unit variables as lower-case versions of the ordinary variables:

$$v = \frac{V}{V_B} \qquad p = \frac{P}{P_B} \qquad \text{etc.}$$

Thus, written in per-unit notation, real and reactive power for a synchronous machine operating in steady state are:

$$p = -\frac{v e_{af}}{x_d}\sin\delta \qquad q = \frac{v^2}{x_d} - \frac{v e_{af}}{x_d}\cos\delta$$

These are, of course, in *motor* reference coordinates, and represent real and reactive power *into* the terminals of the machine.

9.4 Normal Operation

The synchronous machine may be used, essentially interchangeably, as a motor and as a generator. Note that this type of machine produces time-average torque only when it is running at synchronous speed. This is not, of course, a problem for a generator that is started by its prime mover (e.g. a steam turbine), but it is an issue for motors that must be started from rest. Many synchronous motors are started as induction machines on their damper cages (sometimes called starting cages). And of course with power electronic drives the machine can often be considered to be 'in synchronism' even down to zero speed.

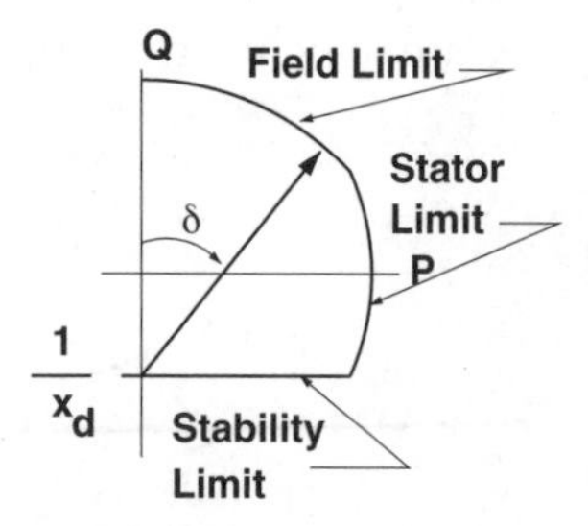

Figure 9.6 Synchronous generator capability diagram

As either a motor or as a generator, the synchronous machine can either produce or consume reactive power. In normal operation real power is dictated by the load (if a motor) or the prime mover (if a generator), and reactive power is determined by the real power and by field current.

9.4.1 Capability Diagram

Figure 9.6 shows one way of representing the *capability* of a synchronous machine. This picture represents operation as a generator, so the signs of p and q are reversed from motor notation, but all of the other elements of operation are as would ordinarily be expected. If p and q (calculated in the normal way) are plotted against each other, the construction shown is generated.

Starting at a location $p = 0$, $q = -v^2/x_d$, then the locus of p and q is what would be obtained by swinging a vector of length ve_{af}/x_d over an angle δ. This is called a *capability chart* because it is an easy way of visualizing what the synchronous machine (in this case generator) is capable of doing. There are three easily noted limits to capability. The upper limit is a circle (the one traced out by that vector) which is referred to as *field* capability. The second limit is a circle that describes constant $|p + jq|$. This is, of course, related to the magnitude of armature current and so this limit is called *armature* capability. The final limit is related to machine stability, since the torque angle cannot go beyond 90°. In actuality there are often other limits that can be represented on this type of a chart. For example, large synchronous generators typically have a problem with heating of the stator iron if operated in highly underexcited conditions (q strongly negative), so that one will often see another limit that prevents the operation of the machine near its stability limit. In very large machines with more than one cooling state (e.g. different values of cooling hydrogen pressure) there may be multiple curves for some or all of the limits.

9.4.2 Vee Curve

Another way of describing the limitations of a synchronous machine is embodied in the *vee curve*. An example is shown in Figure 9.7. This is a cross-plot of magnitude of armature current with field current. Note that the field and armature current limits are straightforward (and are the right-hand and upper boundaries, respectively, of the chart). The machine stability limit is what terminates each of the curves at the upper left-hand edge. Note that each

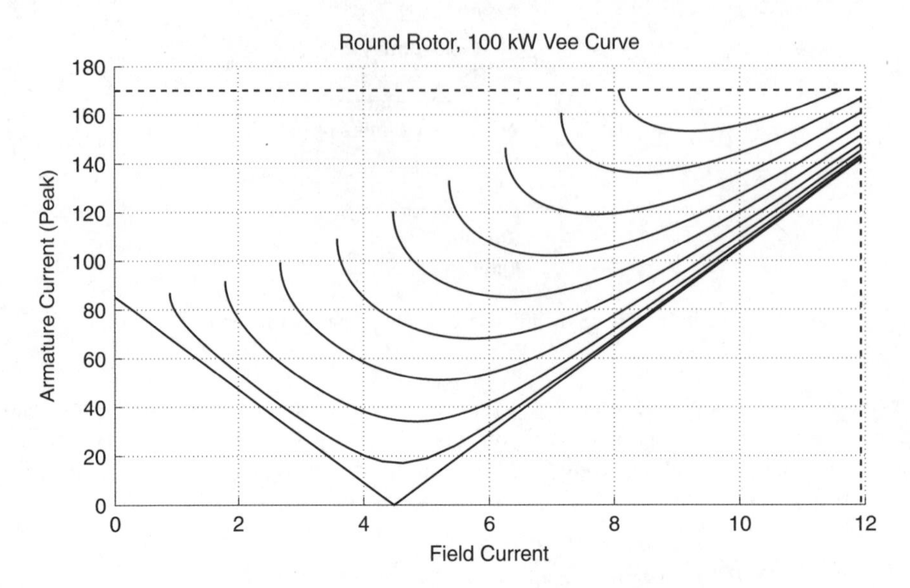

Figure 9.7 Synchronous machine vee curve

curve has a minimum at unity power factor. In fact, there is yet another cross-plot possible, called a *compounding curve*, in which field current is plotted against real power for fixed power factor.

9.5 Salient Pole Machines: Two-Reaction Theory

So far, what has been described are referred to as 'round rotor' machines, in which stator reactance is not dependent on rotor position. This is a pretty good approximation for large turbine generators and many smaller two-pole machines, but it is not a good approximation for many synchronous motors nor for slower speed generators. For many such applications it is more cost effective to wind the field conductors around steel bodies (called poles) which are then fastened onto the rotor body with bolts or dovetail joints. These produce magnetic anisotropies into the machine which affect its operation. The theory that follows is an introduction to two-reaction theory and consequently for the rotating field transformations that form the basis for most modern dynamic analyses.

Figure 9.8 shows a very schematic picture of the salient pole machine, intended primarily to show how to frame this analysis. As with the round rotor machine the stator winding is located in slots in the surface of a highly permeable stator core annulus. The field winding is wound around steel pole pieces. The stator current sheet is separated into two components: one aligned with, and one in quadrature to the field. Remember that these two current components are themselves (linear) combinations of the stator phase currents. A formal statement of the transformation between phase currents and the *d*- and *q-axis* components will appear in Section 9.7. The direct axis is aligned with the field winding, while the quadrature axis leads

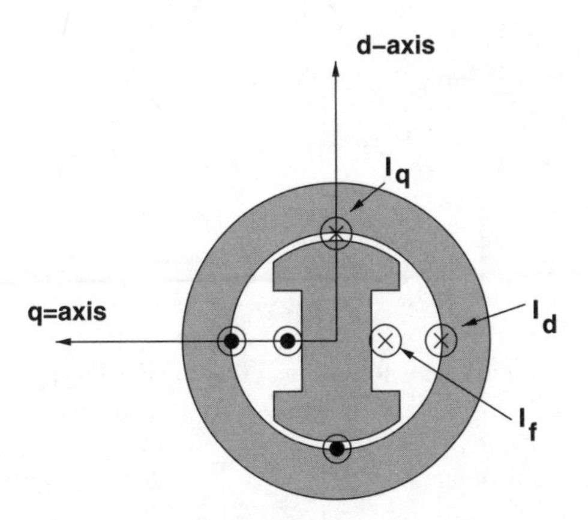

Figure 9.8 Cartoon of a salient pole synchronous machine

the direct by 90°. Then, if ϕ is the angle between the direct axis and the axis of phase a, flux linking phase a is:

$$\lambda_a = \lambda_d \cos\phi - \lambda_q \sin\phi$$

Then, in steady state operation, if $V_a = \frac{d\lambda_a}{dt}$ and $\phi = \omega t + \delta$,

$$V_a = -\omega\lambda_d \sin\phi - \omega\lambda_q \cos\phi$$

which results in the definition of direct and quadrature axis voltages:

$$V_d = -\omega\lambda_q$$
$$V_q = \omega\lambda_d$$

one might think of the 'voltage' vector as leading the 'flux' vector by 90°.

Now, if the machine is magnetically linear, those fluxes are given by:

$$\lambda_d = L_d I_d + M I_f$$
$$\lambda_q = L_q I_q$$

Note that, in general, $L_d \neq L_q$. In wound-field synchronous machines, usually $L_d > L_q$. The reverse is true for most salient (buried magnet) permanent magnet machines.

Referring to Figure 9.9, one can resolve terminal voltage into these components:

$$V_d = V \sin\delta$$
$$V_q = V \cos\delta$$

or:

$$V_d = -\omega\lambda_q = -\omega L_q I_q = V \sin\delta$$
$$V_q = \omega\lambda_d = \omega L_d I_d + \omega M I_f = V \cos\delta$$

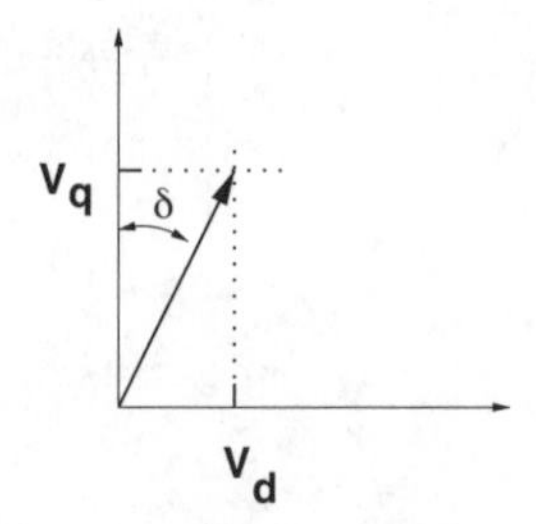

Figure 9.9 Resolution of terminal voltage

which are easily inverted to produce:

$$I_d = \frac{V\cos\delta - E_{\text{af}}}{X_d}$$
$$I_q = -\frac{V\sin\delta}{X_q}$$

where

$$X_d = \omega L_d \qquad X_q = \omega L_q \qquad E_{\text{af}} = \omega M I_{\text{f}}$$

The derivation is, at this point, in ordinary variables and each of these variables is peak amplitude. Then, in a complex frame of reference:

$$\text{V} = V_d + jV_q$$
$$\text{I} = I_d + jI_q$$

complex power is:

$$P + jQ = \frac{3}{2}\text{VI}^* = \frac{3}{2}\left\{\left(V_d I_d + V_q I_q\right) + j\left(V_q I_d - V_d I_q\right)\right\}$$

or:

$$P = -\frac{3}{2}\left(\frac{VE_{\text{af}}}{X_d}\sin\delta + \frac{V^2}{2}\left(\frac{1}{X_q} - \frac{1}{X_d}\right)\sin 2\delta\right)$$
$$Q = \frac{3}{2}\left(\frac{V^2}{2}\left(\frac{1}{X_d} + \frac{1}{X_q}\right) - \frac{V^2}{2}\left(\frac{1}{X_q} - \frac{1}{X_d}\right)\cos 2\delta - \frac{VE_{af}}{X_d}\cos\delta\right)$$

A phasor diagram for a salient pole machine is shown in Figure 9.10. This is a little different from the equivalent picture for a round-rotor machine, in that stator current has been separated into its *d*- and *q-axis* components, and the voltage drops associated with those components

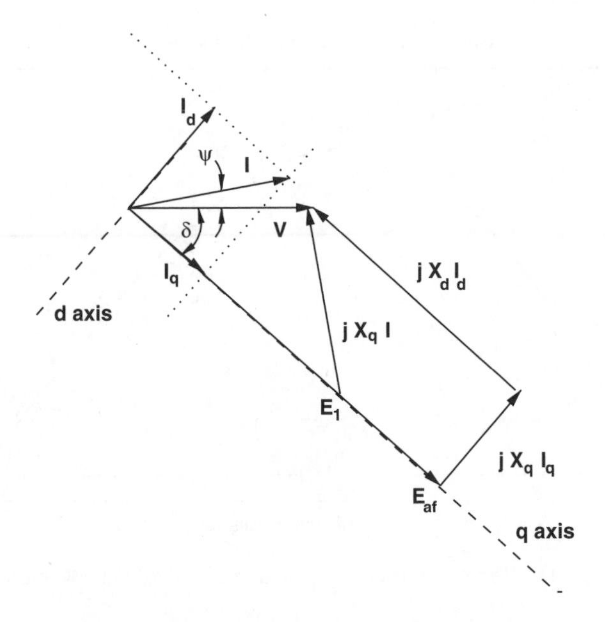

Figure 9.10 Phasor diagram: salient pole motor

have been drawn separately. It is interesting and helpful to recognize that the internal voltage E_{af} can be expressed as:

$$E_{\text{af}} = E_1 + \left(X_d - X_q\right) I_d$$

where the voltage E_1 is on the quadrature axis. In fact, E_1 would be the internal voltage of a round rotor machine with reactance X_q and the same stator current and terminal voltage. Then the operating point is found fairly easily:

$$\delta = -\tan^{-1}\left(\frac{X_q I \cos\psi}{V + X_q I \sin\psi}\right)$$

$$E_1 = \sqrt{\left(V + X_q I \sin\psi\right)^2 + \left(X_q I \cos\psi\right)^2}$$

A comparison of torque-angle curves for a pair of machines, one with a round, one with a salient rotor is shown in Figure 9.11. It is not too difficult to see why power systems analysts often neglect saliency in doing things like transient stability calculations.

9.6 Synchronous Machine Dynamics

Because torque produced by a synchronous machine is dependent on rotor phase angle, which is really the rotor position with respect to the rotation of the voltage source driving the machine, that torque is very much like a nonlinear spring. This, combined with the inertia of the rotor, might be expected to result in second order dynamics similar to a simple spring-mass system.

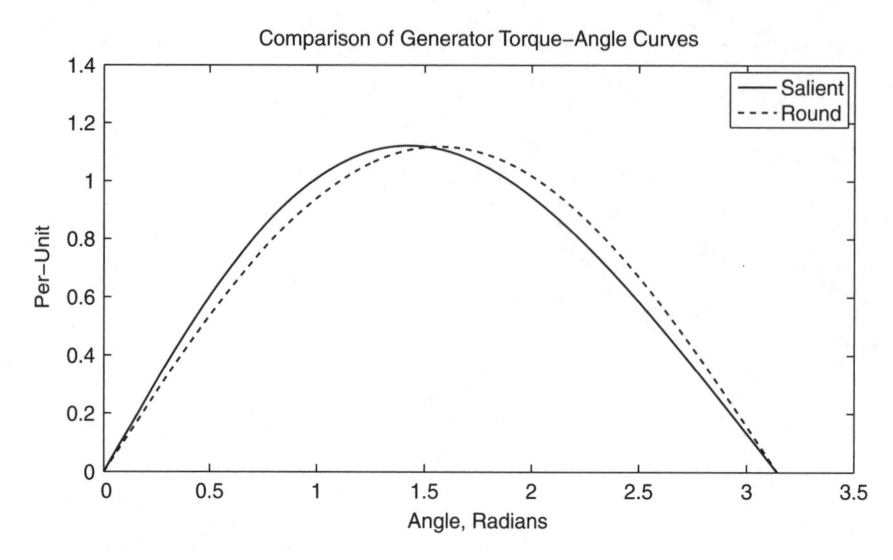

Figure 9.11 Torque-angle curves: round rotor and salient pole machines

Consider, for example, a generator driven by a steam turbine. If the turbine produces a constant torque T^m and the combined generator and turbine rotors have a total inertia constant J, one would expect the system to be described by the following ordinary differential equation, in which V and E_{af} are RMS amplitudes:

$$J\frac{\mathrm{d}^2(\frac{\delta}{p})}{\mathrm{d}t^2} = T^m - 3\frac{p}{\omega}\frac{VE_{af}}{X_d}\sin\delta$$

Note that the physical angle of motion is $\frac{\delta}{p}$. If we assume that the speed of the machine does not vary much from its steady state value *and* that any deviations on the angle δ are small, this system may be linearized to a more easily handled form. Noting also that in the steady state, prime mover torque must be matched by electrical torque:

$$T^m = 3\frac{p}{\omega_0}\frac{VE_{af}}{X_d}\sin\delta_0$$

electrical torque may be linearized by noting that, for small variations in δ, assuming $\delta = \delta_0 + \delta_1$

$$T^e \approx 3\frac{p}{\omega_0}\frac{VE_{af}}{X_d}\sin\delta_0 + \left(3\frac{p}{\omega_0}\frac{VE_{af}}{X_d}\cos\delta_0\right)\delta_1$$

Then the linearized ordinary differential equation of small angle displacement becomes:

$$\frac{\mathrm{d}^2\delta_1}{\mathrm{d}t^2} = \left(\frac{3}{J}\frac{p^2}{\omega_0}\frac{VE_{af}}{X_d}\cos\delta_0\right)\delta_1$$

This would be expected to produce sinusoidal oscillations with a natural frequency:

$$\Omega \approx \sqrt{\frac{3}{J}\frac{p^2}{\omega_0}\frac{V E_{\mathrm{af}}}{X_d}\cos\delta_0} \tag{9.5}$$

These small-scale 'swing' oscillations are a bit tricky to observe but can be seen using a line synchronized strobe lamp. It should be noted that the observed frequency is actually higher than what is predicted by Equation 9.5, indicating that the effective reactance of the machine is smaller than the 'synchronous' value X_d. This leads to the need for a more thorough description of synchronous machine dynamics.

9.7 Synchronous Machine Dynamic Model

To allow a more thorough examination of the dynamic properties of synchronous machines, A dynamic model must be built. To do this, focus on those elements of the machine that store energy in electromagnetic or kinetic form and look at the time evolution of energy stored in those elements. This model formulation can be integrated by any of a number of procedures (this is called 'simulation'), or in linearized form can be solved directly.

9.7.1 Electromagnetic Model

The elementary electromagnetic model is shown in Figure 9.12. The armature of the machine (usually on the stator) consists of three phase coils (A, B, C), located symmetrically, 120° apart. The rotor of the machine has two coils: the field winding is driven by, typically, a voltage source. In quadrature to that is a 'damper'. This may be a winding but is more typically

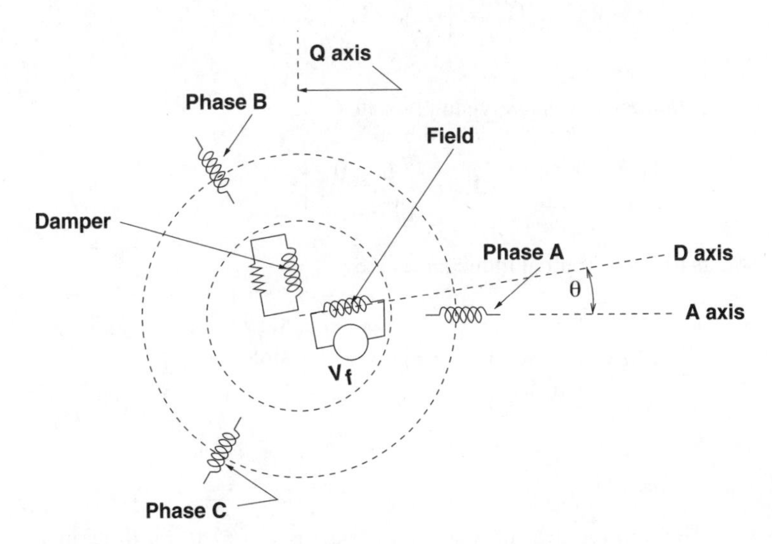

Figure 9.12 Electromagnetic winding model of synchronous machine

some sort of a cage structure, or perhaps even just conductive parts of the rotor structure. It should be noted that this is a rather simplified model for the machine: many investigators consider 'damper' coils on the direct axis, linking the field. This simplified model captures the important dynamic properties of the synchronous machine system and does not unduly complicate the calculations.

As it's name implies, the rotor can turn with respect to the stator, and the mutual inductances between rotor and stator vary sinusoidally with the rotational angle. The flux linkages and currents are:

$$\begin{bmatrix} \boldsymbol{\lambda}_{\mathrm{ph}} \\ \boldsymbol{\lambda}_{\mathrm{R}} \end{bmatrix} = \begin{bmatrix} \mathbf{L}_{\mathrm{ph}} & \mathbf{M} \\ \mathbf{M}^T & \mathbf{L}_{\mathrm{R}} \end{bmatrix} \begin{bmatrix} \mathbf{I}_{\mathrm{ph}} \\ \mathbf{I}_{\mathrm{R}} \end{bmatrix} \tag{9.6}$$

where *phase* and *rotor* fluxes (and, similarly, currents) are:

$$\boldsymbol{\lambda}_{\mathrm{ph}} = \begin{bmatrix} \lambda_{\mathrm{a}} \\ \lambda_{\mathrm{b}} \\ \lambda_{\mathrm{c}} \end{bmatrix} \qquad \mathbf{I}_{\mathrm{ph}} = \begin{bmatrix} I_{\mathrm{a}} \\ I_{\mathrm{b}} \\ I_{\mathrm{c}} \end{bmatrix}$$

$$\boldsymbol{\lambda}_{\mathrm{R}} = \begin{bmatrix} \lambda_{\mathrm{f}} \\ \lambda_{\mathrm{k}} \end{bmatrix} \qquad \mathbf{I}_{\mathrm{R}} = \begin{bmatrix} I_{\mathrm{f}} \\ I_{\mathrm{k}} \end{bmatrix}$$

The subscripts here are a, b and c, referring to the three phases of the armature winding, and f, and k, referring to the field winding and the quadrature axis damper structure, respectively.

There are three inductance submatrices. The first of these describes armature winding inductances:

$$\mathbf{L}_{\mathrm{ph}} = \begin{bmatrix} L_{\mathrm{a}} & L_{\mathrm{ab}} & L_{\mathrm{ac}} \\ L_{\mathrm{ab}} & L_{\mathrm{b}} & L_{\mathrm{bc}} \\ L_{\mathrm{ac}} & L_{\mathrm{bc}} & L_{\mathrm{c}} \end{bmatrix}$$

The rotor inductances are relatively simply stated:

$$\mathbf{L}_{\mathrm{R}} = \begin{bmatrix} L_{\mathrm{f}} & 0 \\ 0 & L_{\mathrm{k}} \end{bmatrix}$$

And the stator- to- rotor mutual inductances are:

$$\mathbf{M} = \begin{bmatrix} M\cos\theta & -L_{\mathrm{ak}}\sin\theta \\ M\cos\left(\theta - \frac{2\pi}{3}\right) & -L_{\mathrm{ak}}\sin\left(\theta - \frac{2\pi}{3}\right) \\ M\cos\left(\theta + \frac{2\pi}{3}\right) & -L_{\mathrm{ak}}\sin\left(\theta + \frac{2\pi}{3}\right) \end{bmatrix}$$

9.7.2 Park's Equations

The first step in the development of a suitable model is to *transform* the armature winding variables to a coordinate system in which the rotor is stationary. This is done because the stator of the machine has a very regular form and can be represented easily even if it is moving with respect to the coordinate system in which analysis is done. The rotor is not as symmetric and cannot be so easily represented. We identify equivalent armature windings in the *direct* and *quadrature* axes. The *direct axis* armature winding is the equivalent of one of the phase windings, but aligned directly with the field. The *quadrature* winding is situated so that its axis *leads* the field winding by 90 *electrical* degrees. The transformation used to map the armature currents, fluxes and so forth onto the *direct* and *quadrature* axes is the celebrated *Park's Transformation*, named after Robert H. Park, an early investigator into transient behavior in synchronous machines. The mapping takes the form:

$$\begin{bmatrix} u_d \\ u_q \\ u_0 \end{bmatrix} = \mathbf{u}_{\mathrm{dq}} = \mathbf{T}\mathbf{u}_{\mathrm{ph}} = \mathbf{T}\begin{bmatrix} u_{\mathrm{a}} \\ u_{\mathrm{b}} \\ u_{\mathrm{c}} \end{bmatrix}$$

Where the transformation and its inverse are:

$$\mathbf{T} = \frac{2}{3}\begin{bmatrix} \cos\theta & \cos\left(\theta - \frac{2\pi}{3}\right) & \cos\left(\theta + \frac{2\pi}{3}\right) \\ -\sin\theta & -\sin\left(\theta - \frac{2\pi}{3}\right) & -\sin\left(\theta + \frac{2\pi}{3}\right) \\ \frac{1}{2} & \frac{1}{2} & \frac{1}{2} \end{bmatrix} \tag{9.7}$$

$$\mathbf{T}^{-1} = \begin{bmatrix} \cos\theta & -\sin\theta & 1 \\ \cos\left(\theta - \frac{2\pi}{3}\right) & -\sin\left(\theta - \frac{2\pi}{3}\right) & 1 \\ \cos\left(\theta + \frac{2\pi}{3}\right) & -\sin\left(\theta + \frac{2\pi}{3}\right) & 1 \end{bmatrix} \tag{9.8}$$

The angle θ in the transformation is the *electrical* angle of the rotor, which is p times the physical angle. This transformation maps *balanced* sets of phase currents into *constant* currents in the *d-q* frame. That is, if rotor angle is $\theta = \omega t + \theta_0$, and phase currents are:

$$I_{\mathrm{a}} = I\cos\omega t \qquad I_{\mathrm{b}} = I\cos\left(\omega t - \frac{2\pi}{3}\right) \qquad I_{\mathrm{c}} = I\cos\left(\omega t + \frac{2\pi}{3}\right)$$

Then the transformed set of currents is:

$$\begin{aligned} I_d &= I\cos\theta_0 \\ I_q &= -I\sin\theta_0 \\ I_0 &= 0 \end{aligned}$$

Now apply this transformation to Equation 9.6 to express fluxes and currents in the armature in the *d-q* reference frame. To do this, extract the top line in Equation 9.6:

$$\boldsymbol{\lambda}_{\mathrm{ph}} = \mathbf{L}_{\mathrm{ph}}\mathbf{I}_{\mathrm{ph}} + \mathbf{M}\mathbf{I}_{\mathrm{R}}$$

The transformed flux is obtained by premultiplying this whole expression by the transformation matrix. Phase current may be obtained from *d-q* current by multiplying by the inverse of the transformation matrix. Thus:

$$\boldsymbol{\lambda}_{dq} = \mathbf{T}\mathbf{L}_{\mathrm{ph}}\mathbf{T}^{-1}\mathbf{I}_{dq} + \mathbf{T}\mathbf{M}\mathbf{I}_{\mathrm{R}}$$

The same process carried out for the lower line of Equation 9.6 yields:

$$\boldsymbol{\lambda}_{\mathrm{R}} = \mathbf{M}^{T}\mathbf{T}^{-1}\mathbf{I}_{dq} + \mathbf{L}_{\mathrm{R}}\mathbf{I}_{\mathrm{R}}$$

Thus the fully transformed version of Equation 9.6 is:

$$\begin{bmatrix} \boldsymbol{\lambda}_{dq} \\ \boldsymbol{\lambda}_{\mathrm{R}} \end{bmatrix} = \begin{bmatrix} \mathbf{L}_{dq} & \mathbf{L}_{\mathrm{C}} \\ \frac{3}{2}\mathbf{L}_{\mathrm{C}}^{T} & \mathbf{L}_{\mathrm{R}} \end{bmatrix} \begin{bmatrix} \mathbf{I}_{\mathrm{dq}} \\ \mathbf{I}_{\mathrm{R}} \end{bmatrix} \tag{9.9}$$

The inductance submatrices of Equation 9.9 are of particularly simple form. (Please note that a substantial amount of algebra has been left out here!)

$$\mathbf{L}_{dq} = \begin{bmatrix} L_d & 0 & 0 \\ 0 & L_q & 0 \\ 0 & 0 & L_0 \end{bmatrix} \tag{9.10}$$

$$\mathbf{L}_{\mathrm{C}} = \begin{bmatrix} M & 0 \\ 0 & L_{\mathrm{ak}} \\ 0 & 0 \end{bmatrix} \tag{9.11}$$

Note that Equations 9.9 through 9.11 express three *separate* sets of apparently independent flux/current relationships. These may be re-cast into the following form:

$$\begin{bmatrix} \lambda_d \\ \lambda_{\mathrm{f}} \end{bmatrix} = \begin{bmatrix} L_d & M \\ \frac{3}{2}M & L_{\mathrm{f}} \end{bmatrix} \begin{bmatrix} I_d \\ I_{\mathrm{f}} \end{bmatrix} \tag{9.12}$$

$$\begin{bmatrix} \lambda_q \\ \lambda_{\mathrm{k}} \end{bmatrix} = \begin{bmatrix} L_q & L_{\mathrm{ak}} \\ \frac{3}{2}L_{\mathrm{ak}} & L_k \end{bmatrix} \begin{bmatrix} I_q \\ I_k \end{bmatrix} \tag{9.13}$$

$$\lambda_0 = L_0 I_0 \tag{9.14}$$

It should be noted that there are some mathematically specific restrictions on the forms of the machine reactances to ensure the simple forms of Equations 9.12 through 9.14. These restrictions, which we will not worry about or even state here, are almost always at least approximately satisfied even in salient pole machines.

Next, armature voltage is given by:

$$\mathbf{V}_{\mathrm{ph}} = \frac{\mathrm{d}}{\mathrm{d}t}\boldsymbol{\lambda}_{\mathrm{ph}} + \mathbf{R}_{\mathrm{ph}}\mathbf{I}_{\mathrm{ph}} = \frac{\mathrm{d}}{\mathrm{d}t}\mathbf{T}^{-1}\boldsymbol{\lambda}_{dq} + \mathbf{R}_{\mathrm{ph}}\mathbf{T}^{-1}\mathbf{I}_{\mathrm{dq}}$$

and that the *transformed* armature voltage must be:

$$\begin{aligned}\mathbf{V}_{dq} &= \mathbf{T}\mathbf{V}_{\mathrm{ph}}\\ &= \mathbf{T}\frac{\mathrm{d}}{\mathrm{d}t}(\mathbf{T}^{-1}\boldsymbol{\lambda}_{dq}) + \mathbf{T}\mathbf{R}_{\mathrm{ph}}\mathbf{T}^{-1}\mathbf{I}_{dq}\\ &= \frac{\mathrm{d}}{\mathrm{d}t}\boldsymbol{\lambda}_{dq} + (\mathbf{T}\frac{\mathrm{d}}{\mathrm{d}t}\mathbf{T}^{-1})\boldsymbol{\lambda}_{dq} + \mathbf{R}_{dq}\mathbf{I}_{dq}\end{aligned}$$

Note the resistance matrices are stated in a formal way. Since the phase resistances are independent of each other:

$$\mathbf{R}_{\mathrm{ph}} = \begin{bmatrix} R_{\mathrm{a}} & 0 & 0 \\ 0 & R_{\mathrm{a}} & 0 \\ 0 & 0 & R_{\mathrm{a}} \end{bmatrix}$$

And when we multiply the diagonal matrix by the transform and its inverse:

$$\mathbf{R}_{dq} = \begin{bmatrix} R_{\mathrm{a}} & 0 & 0 \\ 0 & R_{\mathrm{a}} & 0 \\ 0 & 0 & R_{\mathrm{a}} \end{bmatrix}$$

Straightforward manipulation reduces the second term of the voltage equation, resulting in:

$$\mathbf{T}\frac{\mathrm{d}}{\mathrm{d}t}\mathbf{T}^{-1} = \begin{bmatrix} 0 & -\frac{\mathrm{d}\theta}{\mathrm{d}t} & 0 \\ \frac{\mathrm{d}\theta}{\mathrm{d}t} & 0 & 0 \\ 0 & 0 & 0 \end{bmatrix}$$

This expresses the *speed voltage* that arises from a coordinate transformation. The important stator voltage/flux relationships are:

$$\begin{aligned}V_d &= \frac{\mathrm{d}\lambda_d}{\mathrm{d}t} - \omega\lambda_q + R_{\mathrm{a}}I_d\\ V_q &= \frac{\mathrm{d}\lambda_q}{\mathrm{d}t} + \omega\lambda_d + R_{\mathrm{a}}I_q\end{aligned}$$

where:

$$\omega = \frac{\mathrm{d}\theta}{\mathrm{d}t}$$

9.7.3 Power and Torque

Instantaneous *power* is given by:

$$P = V_a I_a + V_b I_b + V_c I_c \tag{9.15}$$

Using the transformations Equations 9.7 and 9.8, this can be shown to be:

$$P = \frac{3}{2} V_d I_d + \frac{3}{2} V_q I_q + 3 V_0 I_0$$

which, in turn, is:

$$P = \omega \frac{3}{2}(\lambda_d I_q - \lambda_q I_d) + \frac{3}{2}\left(\frac{d\lambda_d}{dt} I_d - \frac{d\lambda_q}{dt} I_q\right) + 3\frac{d\lambda_0}{dt} I_0 + \frac{3}{2} R_a \left(I_d^2 + I_q^2 + 2I_0^2\right) \tag{9.16}$$

Equation 9.16 has three principal parts. From the right is *dissipation* in the armature resistance, then a set of terms that relate to energy stored in magnetic fields, or more precisely, rate of change of energy stored in magnetic fields. The leftmost term, which is proportional to rotational speed, must be energy conversion.

Noting that *electrical* speed ω and shaft speed Ω are related by $\omega = p\Omega$ and that (9.16) describes electrical terminal power as the sum of shaft power and rate of change of stored energy, one may deduce that torque is given by:

$$T = \frac{3}{2} p(\lambda_d I_q - \lambda_q I_d) \tag{9.17}$$

9.7.4 Per-Unit Normalization

The next step in this analysis is to normalize the voltage, current, flux, etc. into a *per-unit* system. This is done in a fashion very similar to what has already been done, but now it is necessary to include the rotor quantities. Important normalizing *base* parameters are, for the stator:

- Base power

$$P_B = \frac{3}{2} V_B I_B$$

- Base impedance

$$Z_B = \frac{V_B}{I_B}$$

- Base flux

$$\lambda_B = \frac{V_B}{\omega_0}$$

- Base torque

$$T_B = \frac{p}{\omega_0} P_B$$

Note that base *voltage* and *current* are expressed as *peak* quantities. Base voltage is taken on a phase basis (line to neutral for a *wye*-connected machine), and base current is similarly taken on a phase basis, (line current for a *wye*-connected machine).

Normalized, or *per-unit* quantities are derived by dividing the *ordinary* variable (with units) by the corresponding *base*. For example, per-unit flux is:

$$\psi = \frac{\lambda}{\lambda_B} = \frac{\omega_0 \lambda}{V_B} \tag{9.18}$$

In this derivation, per-unit quantities will usually be designated by lower case letters. Two notable exceptions are flux, where the letter ψ is used for the per-unit quantity and λ is the ordinary variable, and torque, where the upper case T is used for both per-unit and ordinary quantities. Ordinary variable torque is generally superscripted (T^e, T^m), while per-unit torque is subscripted (T_e, T_m), so that $T_e = T^e / T_B$.

Note that there will be *base* quantities for voltage, current and frequency for each of the different coils represented in the model. While it is reasonable to expect that the *frequency* base will be the same for all coils in a problem, the *voltage* and *current* bases may be different. The per-unit equivalent of (9.12) is:

$$\begin{bmatrix} \psi_d \\ \psi_f \end{bmatrix} = \begin{bmatrix} \frac{\omega_0 I_{dB}}{V_{dB}} L_d & \frac{\omega_0 I_{fB}}{V_{dB}} M \\ \frac{\omega_0 I_{dB}}{V_{fB}} \frac{3}{2} M & \frac{\omega_0 I_{fB}}{V_{fB}} L_f \end{bmatrix} \begin{bmatrix} i_d \\ i_f \end{bmatrix} \tag{9.19}$$

The quadrature axis expression is similar:

$$\begin{bmatrix} \psi_q \\ \psi_k \end{bmatrix} = \begin{bmatrix} \frac{\omega_0 I_{qB}}{V_{qB}} L_q & \frac{\omega_0 I_{kB}}{V_{qB}} L_{kq} \\ \frac{\omega_0 I_{qB}}{V_{kB}} \frac{3}{2} L_{kq} & \frac{\omega_0 I_{kB}}{V_{kB}} L_k \end{bmatrix} \begin{bmatrix} i_q \\ i_k \end{bmatrix}$$

where $i = I / I_B$ denotes *per-unit*, or normalized current.

Note that Equation 9.19 may be written in simple form:

$$\begin{bmatrix} \psi_d \\ \psi_f \end{bmatrix} = \begin{bmatrix} x_d & x_{ad} \\ x_{ad} & x_f \end{bmatrix} \begin{bmatrix} i_d \\ i_f \end{bmatrix} \tag{9.20}$$

Note that Equation 9.20 *assumes* reciprocity in the normalized system. To wit, the following expressions are implied:

$$x_d = \omega_0 \frac{I_{dB}}{V_{dB}} L_d$$

$$x_f = \omega_0 \frac{I_{fB}}{V_{fB}} L_f$$

$$x_{ad} = \omega_0 \frac{I_{\text{fB}}}{V_{d\text{B}}} M$$
$$= \frac{3}{2} \omega_0 \frac{I_{d\text{B}}}{V_{\text{fB}}} M$$

This implies:

$$\frac{3}{2} V_{d\text{B}} I_{d\text{B}} = V_{\text{fB}} I_{\text{fB}} \tag{9.21}$$

That is, the base power for the rotor is the same as for the stator.

The per-unit flux/current relationship for the q- axis is:

$$\begin{bmatrix} \psi_q \\ \psi_{\text{k}} \end{bmatrix} = \begin{bmatrix} x_q & x_{aq} \\ x_{aq} & x_k \end{bmatrix} \begin{bmatrix} i_q \\ i_k \end{bmatrix} \tag{9.22}$$

The voltage equations, including speed voltage terms, (9.15) and (9.15), are:

$$V_d = \frac{\mathrm{d}\lambda_d}{\mathrm{d}t} - \omega\lambda_q + R_{\text{a}} I_d$$
$$V_q = \omega\lambda_d + \frac{\mathrm{d}\lambda_q}{\mathrm{d}t} + R_{\text{a}} I_q$$

The *per-unit* equivalents of these are:

$$v_d = \frac{1}{\omega_0} \frac{d\psi_d}{\mathrm{d}t} - \frac{\omega}{\omega_0} \psi_q + r_{\text{a}} i_d$$
$$v_q = \frac{\omega}{\omega_0} \psi_d + \frac{1}{\omega_0} \frac{\mathrm{d}\psi_q}{\mathrm{d}t} + r_{\text{a}} i_q$$

Where the per-unit armature resistance is just $r_{\text{a}} = \frac{R_{\text{a}}}{Z_{\text{B}}}$

Note that neither the field nor damper circuits in this model have *speed voltage* terms, nor does the zero axis circuit, so their voltage expressions are exactly what we might expect:

$$v_{\text{f}} = \frac{1}{\omega_0} \frac{\mathrm{d}\psi_{\text{f}}}{\mathrm{d}t} + r_{\text{f}} i_{\text{f}}$$
$$v_k = \frac{1}{\omega_0} \frac{\mathrm{d}\psi_{\text{k}q}}{\mathrm{d}t} + r_k i_k$$
$$v_0 = \frac{1}{\omega_0} \frac{\mathrm{d}\psi_0}{\mathrm{d}t} + r_{\text{a}} i_0$$

It should be noted that the *damper* winding circuit represents a closed conducting path on the rotor, so the voltage v_{k} is always zero.

Per-unit torque is simply:

$$T_{\text{e}} = \psi_d i_q - \psi_q i_d$$

Often it is necessary to represent the dynamic behavior of the machine, including electromechanical dynamics involving rotor inertia. If J is the rotational inertia constant of the machine system, the rotor dynamics are described by the two ordinary differential equations:

$$\frac{1}{p} J \frac{d\omega}{dt} = T^{e} + T^{m}$$
$$\frac{d\delta}{dt} = \omega - \omega_0$$

where T^{e} and T^{m} represent *electrical* and *mechanical* torques in 'ordinary' variables. The angle δ represents rotor phase angle with respect to some synchronous reference.

It is customary to define an 'inertia constant' which is not dimensionless but which nevertheless fits into the per-unit system of analysis. This is:

$$H \equiv \frac{\text{Rotational kinetic energy at rated speed}}{\text{Base power}}$$

Or:

$$H = \frac{\frac{1}{2} J \left(\frac{\omega_0}{p}\right)^2}{P_B} = \frac{J \omega_0}{2 p T_B}$$

Then the per-unit equivalent to Equation 9.23 is:

$$\frac{2H}{\omega_0} \frac{d\omega}{dt} = T_c + T_m$$

where now we use T_e and T_m to represent *per-unit* torques.

9.7.5 *Equivalent Circuits*

The flux-current relationships implied by Equations 9.20 and 9.22 may be represented by the equivalent circuits of Figures 9.13 and 9.14, if the 'leakage' reactances are:

$$x_{al} = x_d - x_{ad}$$
$$= x_q - x_{aq}$$

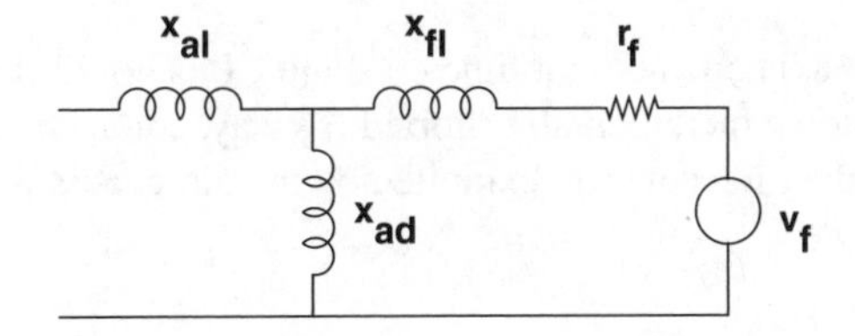

Figure 9.13 *d*-axis per-unit equivalent circuit

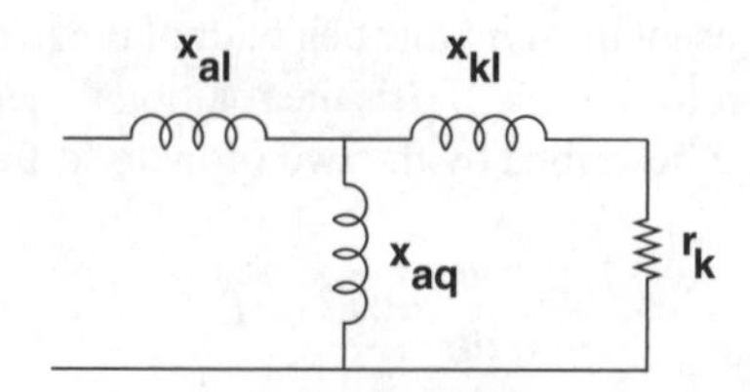

Figure 9.14 Q-Axis per-unit equivalent circuit

$$x_{\mathrm{fl}} = x_{\mathrm{f}} - x_{ad}$$
$$x_{\mathrm{kl}} = x_{\mathrm{k}} - x_{ad}$$

Note that not all of these elements are really needed: one or more of the leakage inductances could be further normalized away while still preserving the dynamic behavior of the machine. However, we have taken the normal practice of assuming the same armature leakage element x_{al} on both axes (something that may make physical sense). This element may be picked arbitrarily, and is often taken to be the per-unit equivalent of the actual armature leakage reactance.

9.7.6 *Transient Reactances and Time Constants*

Examining Figures 9.13 and 9.14, one can see how, during transient swings, the apparent reactance (inductance) of the machine is less than the synchronous reactance. For sufficiently high frequencies, one can see that the magnetizing branch of the equivalent circuit is shorted out by the leakage reactance and equivalent resistance. Note that, for frequencies that are high enough, or for transients that are short enough, the resistance is not consequential. The transient reactances are thus:

$$x_d' = x_{\mathrm{al}} + x_{\mathrm{a}d}||x_{\mathrm{fl}}$$
$$x_q' = x_{\mathrm{al}} + x_{\mathrm{a}q}||x_{\mathrm{kl}}$$

Associated with the dynamical properties of the direct- and quadrature axes are the time constants. If the armature winding were open circuited, for example, the direct axis would exhibit a time constant:

$$T_{d\mathrm{o}}' = \frac{x_{\mathrm{f}}}{\omega_0 r_{\mathrm{f}}}$$

This is called the 'direct axis open circuit time constant'. It is not difficult to show that, if the armature winding were to be incrementally shorted (by, say, connection to a voltage source in normal operation), that the time constant exhibited by the direct axis would be the 'direct axis short circuit time constant':

$$T_d' = \frac{x_d'}{x_d} T_{d\mathrm{o}}'$$

The quadrature axis is similar:

$$T'_{qo} = \frac{x_{\mathrm{k}}}{\omega_0 r_{\mathrm{k}}}$$
$$T'_q = \frac{x'_q}{x_q} T'_{qo}$$

9.8 Statement of Simulation Model

Now a simulation model may be written. This model may be used in an actual simulation routine, as is demonstrated below, or it might be linearized and solved directly using classical techniques.

The state variables are the two stator fluxes ψ_d, ψ_q, the 'damper' flux ψ_{k}, field flux ψ_{f}, and rotor speed ω and torque angle δ. The most straightforward way of stating the model employs currents as auxiliary variables, and these are:

$$\begin{bmatrix} i_d \\ i_{\mathrm{f}} \end{bmatrix} = \begin{bmatrix} x_d & x_{\mathrm{ad}} \\ x_{\mathrm{ad}} & x_{\mathrm{f}} \end{bmatrix}^{-1} \begin{bmatrix} \psi_d \\ \psi_{\mathrm{f}} \end{bmatrix}$$

$$\begin{bmatrix} i_q \\ i_{\mathrm{k}q} \end{bmatrix} = \begin{bmatrix} x_q & x_{\mathrm{a}q} \\ x_{\mathrm{a}q} & x_{\mathrm{k}q} \end{bmatrix}^{-1} \begin{bmatrix} \psi_q \\ \psi_{\mathrm{k}q} \end{bmatrix}$$

These are simply:

$$i_d = \frac{x_{\mathrm{f}}\psi_d - x_{\mathrm{ad}}\psi_{\mathrm{f}}}{x_d x_{\mathrm{f}} - x_{\mathrm{ad}}^2}$$
$$i_{\mathrm{f}} = \frac{x_d\psi_{\mathrm{f}} - x_{\mathrm{ad}}\psi_d}{x_d x_{\mathrm{f}} - x_{\mathrm{ad}}^2}$$
$$i_q = \frac{x_k\psi_q - x_{\mathrm{a}q}\psi_k}{x_q x_k - x_{\mathrm{a}q}^2}$$
$$i_k = \frac{x_q\psi_k - x_{\mathrm{a}q}\psi_q}{x_q x_k - x_{\mathrm{a}q}^2}$$

Then the state equations are:

$$\frac{\mathrm{d}\psi_{\mathrm{d}}}{\mathrm{d}t} = \omega_0 v_d + \omega\psi_q - \omega_0 r_{\mathrm{a}} i_d$$
$$\frac{\mathrm{d}\psi_q}{\mathrm{d}t} = \omega_0 v_q - \omega\psi_d - \omega_0 r_{\mathrm{a}} i_q$$
$$\frac{\mathrm{d}\psi_{\mathrm{k}}}{\mathrm{d}t} = -\omega_0 r_{\mathrm{k}} i_{\mathrm{k}}$$

$$\frac{d\psi_f}{dt} = -\omega_0 r_f i_f$$
$$\frac{d\omega}{dt} = \frac{\omega_0}{2H}(T_e + T_m)$$
$$\frac{d\delta}{dt} = \omega - \omega_0$$

and, of course,

$$T_e = \psi_d i_q - \psi_q i_d$$

9.8.1 Example: Transient Stability

These models may be used for a variety of problems. This is a situation of substantial importance. With a generator operating normally there is a balance between prime mover torque (positive) and generator torque (negative). If something happens to short-circuit the output (e.g. a lightning strike on the transmission line), power output of the generator and hence torque falls to nearly zero. The generator will accelerate until the fault is cleared and generator torque can be restored. If that is not done quickly, the phase angle of the generator will become large enough that torque cannot be effectively restored and the system will be unstable: the generator must be quickly shut down to avoid overspeed and consequent damage to the system.

Appended to this chapter in Section 9.9 is a MATLAB source file that does a simulation of a scenario in which the terminal voltage actually goes to zero for a period of time. There are really two steps to this process. The first is to determine the initial conditions for the simulation, and those are found using the steady-state models for the synchronous machine that have been developed in this chapter. The next step is the simulation itself.

It is normal to verify that initial conditions are correct by simulating the system for a short period of time with currents and voltages at their normal levels. In this case, field voltage and terminal voltage are set to their normal levels. The output should be (and is, in this case) steady. Then the terminal voltage is set to zero for a period of time called the 'clearing time'. Finally, terminal voltage is restored and the system is simulated some more. The 'critical clearing time' is the maximum value of clearing time for which the system comes back to equilibrium. Figure 9.15 shows the simulated value of torque angle δ for a value of clearing time very nearly critical (195 ms).

9.8.2 Equal Area Transient Stability Criterion

An approximate method for assessing transient stability starts by assuming that torque is a function only of phase angle and not of rotor speed. This assumption is an approximation to reality and so gives only approximate results. Because it ignores dissipation in the rotor circuits due to relative motion between the rotor and stator flux it tends to be conservative (to underestimate transient stability). The method employs an 'energy well' approach, by calculating changes in rotational kinetic energy.

Figure 9.15 Simulation of a near-critical swing

The differential equation that describes acceleration and thus angle is:

$$\frac{\mathrm{d}^2\delta}{\mathrm{d}t^2} = \frac{\omega_0}{2H}\left(T_\mathrm{c}(\delta) + T_\mathrm{m}\right)$$

Multiplying the whole expression by $\frac{d\delta}{dt}$ and integrating over time, and taking into account:

$$\frac{\mathrm{d}\delta}{\mathrm{d}t}\frac{\mathrm{d}^2\delta}{\mathrm{d}t^2} = \frac{\mathrm{d}}{\mathrm{d}t}\frac{1}{2}\left(\frac{\mathrm{d}\delta}{\mathrm{d}t}\right)^2$$

$$\int_{t_1}^{t_2}\frac{\mathrm{d}}{\mathrm{d}t}\frac{1}{2}\left(\frac{\mathrm{d}\delta}{\mathrm{d}t}\right)^2\mathrm{d}t = \int_{t_1}^{t_2}\frac{\mathrm{d}\delta}{\mathrm{d}t}\frac{\omega_0}{2H}\left(T_\mathrm{c}(\delta) + T_\mathrm{m}\right)\mathrm{d}t \tag{9.23}$$

Now, Equation 9.23 simply expresses the change in the square of rotor speed (in the synchronous frame) in terms of the area under the torque vs. angle curve:

$$\frac{1}{2}\left(\frac{\mathrm{d}\delta}{\mathrm{d}t}\right)^2|_{t_1} - \frac{1}{2}\left(\frac{\mathrm{d}\delta}{\mathrm{d}t}\right)^2|_{t_2} = \int_{t_1}^{t_2}\frac{\omega_0}{2H}\left(T_\mathrm{c}(\delta) + T_\mathrm{m}\right)d\delta$$

Consider the situation as it is illustrated in Figure 9.16. Initially, the machine is operating at its stable, steady-state angle in which the generating torque (inverted in this picture for clarity) just matches the mechanical prime mover torque. When the fault occurs the electrical torque becomes very small (assumed here to be zero), so the machine accelerates under the influence of the prime mover. If the fault is removed when the rotor is at a position called the

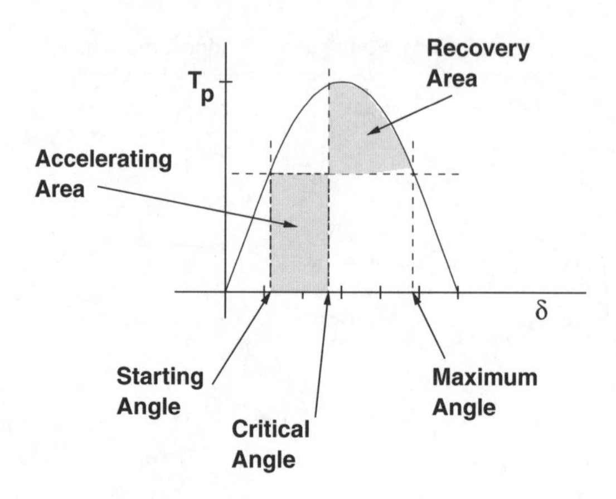

Figure 9.16 Equal area criterion for transient stability

'clearing angle', the generating torque becomes the value shown by the original torque-angle curve. This will be higher than the prime mover torque, slowing the machine down. If the area between the generating and prime mover torque from the clearing angle to the maximum angle is at least as great as the area under the mechanical torque from the steady state angle to the clearing angle, the rotor angle δ will stop increasing and the machine will recover (meaning it will be 'transiently stable'.

The 'critical clearing angle' is that angle for which the area above the mechanical torque and to the right of the clearing angle is just equal to the accelerating area under the mechanical torque between the initial and critical angle. If we note T_{m} as mechanical torque, T_{p} as peak generating torque and if the torque angle is sinusoidal so that $T_{\mathrm{m}} = T_{\mathrm{p}} \sin \delta_0$, that critical angle is:

$$\delta_{\mathrm{c}} = \cos^{-1}\left(\frac{T_{\mathrm{m}}}{T_{\mathrm{p}}}(\pi - 2\delta_0) - \cos \delta_0\right)$$

Since, under acceleration with constant torque T_{m} produces an angle that is quadratic in time, the critical clearing angle can be converted to a critical clearing time:

$$\delta_{\mathrm{c}} = \delta_0 + \frac{\omega_0}{2H}\frac{t_{\mathrm{c}}^2}{2}$$

So the critical clearing time is simply:

$$t_{\mathrm{c}} = \sqrt{\frac{4H}{\omega_0}(\delta_{\mathrm{c}} - \delta_0)}$$

Because the machine produces a typically higher peak torque during rapid transients, it is important to estimate that peak torque for equal-area estimates. Figure 9.17 shows how one might estimate the voltage behind transient reactance for such a situation. For the transient stability case simulated in the previous section of this chapter, equal area results in an estimate

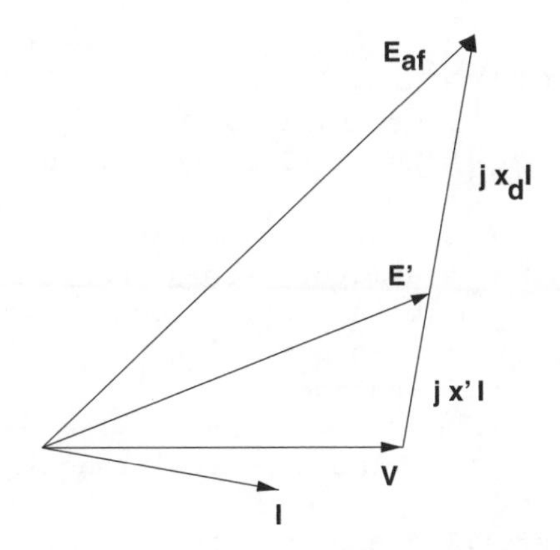

Figure 9.17 Voltage behind transient reactance

for critical clearing time of about 162 ms. This is actually quite conservative as simulation estimates about 195 ms. That is because torque is not really dependent only on angle: the action of the damper takes energy out of the swing.

9.9 Appendix: Transient Stability Code

There are two files here: the main simulation code and a subroutine that calculates the first derivatives of all of the state variables as required by MATLAB. This simulation uses the MATLAB routine ODE23().

```
% simulation of transient stability incident
% copyright 2009  James L Kirtley Jr
% this is done in three steps:
% first, a short time simulation to ensure that initial conditions
% are correct (simulation should be stationary)
% second, terminal voltage is set to zero until the clearing time
% third, terminal voltage is restored and the simulation is run out

global xd xq xad xaq xal xf xk ra rf rk omz vf H V TM

% put those times here for convenience
T_i = .1;                   % initial time to confirm initial
                            % conditions
T_c = .195;                 % clearing time
T_f = 8;                    % simulate to this time

% a little data
xd = 2.0;                   % d- axis synchronous reactance
xdp = .4;                   % transient reactance
xq = 2.0;                   % q- axis synchronous reactance
xqp = .4;                   % transient reactance
omz = 2*pi*60;              % here in the USA
```

```
Tdop = 4.0;                % transient, open circuit time constant
Tqop = 0.1;                % same for q- axis
V0 = 1.0;                   % terminal voltage
I = 1.0;                   % terminal current magnitude
psi = 0;                   % power factor
H = 2.0;                   % inertia constant
ra = 0.0;                  % armature resistance
TM = V0*I*cos(psi);        % mechanical torque
xal = 0.1;                 % need something to use for armature
                           % leakage

xad = xd-xal;              % values of model reactances
xaq = xq-xal;
xfl = (xdp-xal)/(xd-xdp);% field leakage
xkl = (xqp-xal)/(xq-xqp);% damper leakage
xf = xad+xfl;              % total field winding reactance
xk = xaq+xkl;              % total damper winding reactance
rf  = xf/(omz*Tdop);       % field resistance
rk = xk/(omz*Tqop);        % damper resistance

% need to get initial conditions
E_1 = V0 + xq*I*sin(psi) + j*xq*I*cos(psi);   % establishes
                                      % angle of q axis
delt0 = angle(E_1);                   % initial torque angle
id0 = I*sin(delt0 + psi);             % direct-axis current
E_af = abs(E_1) + (xd-xq)*id0;        % required internal voltage
psid0 = V0*cos(delt0);                % initial d- axis flux
psiq0 = -V0*sin(delt0);               % initial q- axis flux
i_f0 = E_af/xad;                      % initial field current
vf = rf*i_f0;                         % field voltage: hold constant
psik0 = psiq0;                        % damper starts with q axis flux
psiad0 = psid0 + xal*id0;             % mag branch flux (d- axis)
psif0 = psiad0 + xfl*i_f0;            % initial field flux
iq0 = I*cos(delt0+psi);               % quadrature axis current
Trq_0 = psid0*iq0-psiq0*id0;          % indicated initial torque

% a bit of summary output
fprintf('Transient Stability Analysis\n')
fprintf('Initial Conditions:\n')
fprintf('Torque Angle delta = %g\n', delt0);
fprintf('Direct Axis Flux psid = %g\n', psid0)
fprintf('Quadrature Axis Flux psiq = %g\n', psiq0)
fprintf('Direct Axis Current I_d = %g\n', id0)
fprintf('Quadrature Axis Current I_q = %g\n', iq0)
fprintf('Torque = %g\n', Trq_0);
fprintf('Required Internal Voltage E_{af} = %g\n', E_af)
fprintf('Field Flux psif = %g\n', psif0)

stz = [psid0 psiq0 psif0 psik0 omz delt0]; % initial state of
                                           % the system
```

```
% first, simulate for a short time to see if initial conditions
% are right
dt = .001;             % establish a time step
t0 = 0:dt:T_i;         % this is the first time period
V = V0;                % should simulate as steady operation
[ti, sti] = ode23('ds', t0, stz);  % this step does the simulation

t0 = T_i+dt:dt:T_i+T_c;               % fault period
V=0;                                  % machine is shorted
                                      % initial conditions should
                                      % not change
[tf, stf] = ode23('ds', t0, stz); % this step does the simulation

psid_0 = stf(length(tf), 1);  % initial conditions for recovery
psiq_0 = stf(length(tf), 2);  % period are those at end of fault
psif_0 = stf(length(tf), 3);  % period
psik_0 = stf(length(tf), 4);
om_0 = stf(length(tf), 5);
delt_0 = stf(length(tf), 6);

stz = [psid_0 psiq_0 psif_0 psik_0 om_0 delt_0]; % start next state
t_0 = T_c+T_i+dt:dt:T_c+T_i+T_f;     % time vector for simulation
V=V0;                                % voltage is restored
[tr, str] = ode23('ds', t_0, stz);   % this step does the simulation
t = [ti; tf; tr]';                   % total simulation time
st = [sti; stf ; str];               % system state for whole period
psid = st(:,1);                      % and these are the actual
psiq = st(:,2);                      % states
psif = st(:,3);
psik = st(:,4);
om = st(:,5);
delt = st(:,6);

titstr = sprintf('Transient Simulation: Clearing Time = %g',T_c);
figure(1)                            % and now plot the output
plot(t, delt)
title(titstr)
ylabel('Torque Angle, radians')
xlabel('seconds')
grid on

%%%%%%%%%%%%%%%%%%%%%%%%%%

function ss = ds(t, st)
% simulation routine: machine open

global xd xq xad xaq xal xf xk ra rf rk omz vf H V TM

% unpack variables
psid = st(1);
```

```
psiq = st(2);
psif = st(3);
psik = st(4);
om = st(5);
delt = st(6);

vd = V*sin(delt);
vq = V*cos(delt);

id = (xf*psid-xad*psif)/(xd*xf-xad^2);
iq = (xk*psiq-xaq*psik)/(xq*xk-xaq^2);
i_f = (psif*xd-psid*xad)/(xd*xf-xad^2);
ik = (psik*xq-psiq*xaq)/(xq*xk-xaq^2);

dpsid = om*psiq + omz*vd - omz*ra*id;
dpsiq = -om*psid + omz*vq - omz*ra*iq;
dpsif = omz*vf - omz*rf*i_f;
dpsik = -omz*rk*ik;
dom = (omz/(2*H)) * (TM + (psid*iq - psiq*id));
ddelt = om - omz;

ss = [dpsid dpsiq dpsif dpsik dom ddelt]';
```

9.10 Appendix: Winding Inductance Calculation

The purpose of this section is to show how the inductances of windings in round-rotor machines with narrow air gaps may be calculated. We deal only with the idealized air-gap magnetic fields, and do not consider slot, end winding, peripheral or skew reactances. We do, however, consider the space harmonics of winding magneto-motive force (MMF).

To start, consider the MMF of a full-pitch, concentrated winding. Assuming that the winding has a total of N turns over p pole-pairs, the MMF is:

$$F = \sum_{\substack{n=1 \\ n \text{ odd}}}^{\infty} \frac{4}{n\pi}\frac{NI}{2p} \sin np\phi$$

This leads directly to magnetic flux density in the air-gap:

$$B_{\mathrm{r}} = \sum_{\substack{n=1 \\ n \text{ odd}}}^{\infty} \frac{\mu_0}{g}\frac{4}{n\pi}\frac{NI}{2p} \sin np\phi$$

Note that a real winding, which will most likely not be full-pitched and concentrated, will have a *winding factor* which is the product of pitch and breadth factors, to be discussed later.

Now, suppose that there is a polyphase winding, consisting of more than one phase (we will use three phases), driven with one of two types of current. The first of these is *balanced*, current:

$$
\begin{aligned}
I_{\mathrm{a}} &= I\cos(\omega t) \\
I_{\mathrm{b}} &= I\cos\left(\omega t - \frac{2\pi}{3}\right) \\
I_{\mathrm{c}} &= I\cos\left(\omega t + \frac{2\pi}{3}\right)
\end{aligned} \tag{9.24}
$$

Conversely, there are *zero sequence* currents, for which:

$$I_{\mathrm{a}} = I_{\mathrm{b}} = I_{\mathrm{c}} = I\cos\omega t$$

Then it is possible to express magnetic flux density for the two distinct cases. For the *balanced* case:

$$B_{\mathrm{r}} = \sum_{n=1}^{\infty} B_{\mathrm{r}n}\sin(np\phi \mp \omega t)$$

where

- The upper sign holds for $n = 1, 7, \ldots$
- The lower sign holds for $n = 5, 11, \ldots$
- all other terms are zero

and

$$B_{\mathrm{r}n} = \frac{3}{2}\frac{\mu_0}{g}\frac{4}{n\pi}\frac{NI}{2p} \tag{9.25}$$

The zero-sequence case is simpler: it is non-zero only for the *triplen* harmonics:

$$B_{\mathrm{r}} = \sum_{\mathrm{n}=3,9,\ldots}^{\infty} \frac{\mu_0}{g}\frac{4}{n\pi}\frac{NI}{2p}\frac{3}{2}(\sin(np\phi - \omega t) + \sin(np\phi + \omega t)) \tag{9.26}$$

Next, consider the flux from the field winding, on the rotor: that will have the same form as the flux produced by a single armature winding, but will be referred to the rotor position:

$$B_{\mathrm{rf}} = \sum_{\substack{n=1 \\ n\,\mathrm{odd}}}^{\infty} \frac{\mu_0}{g}\frac{4}{n\pi}\frac{N_{\mathrm{f}}I_{\mathrm{f}}}{2p}\sin np\phi'$$

which is, substituting $\phi' = \phi - \frac{\omega t}{p}$,

$$B_{\mathrm{rf}} = \sum_{\substack{n=1 \\ n \text{ odd}}}^{\infty} \frac{\mu_0}{g} \frac{4}{n\pi} \frac{N_{\mathrm{f}} I_{\mathrm{f}}}{2p} \sin n(p\phi - \omega t)$$

The next step here is to find the flux linked if we have some air- gap flux density of the form:

$$B_{\mathrm{r}} = \sum_{n=1}^{\infty} B_{\mathrm{r}n} \sin(np\phi \pm \omega t)$$

It is possible to calculate flux linked by a single- turn, full- pitched winding by:

$$\Phi = \int_0^{\frac{\pi}{p}} B_{\mathrm{r}} Rl d\phi$$

and this is:

$$\Phi = 2Rl \sum_{n=1}^{\infty} \frac{B_{\mathrm{rn}}}{np} \cos(\omega t)$$

This allows computation of self-and mutual-inductances, since winding flux is:

$$\lambda = N\Phi$$

The end of this is a set of expressions for various inductances. It should be noted that, in the real world, most windings are not full- pitched nor concentrated. Fortunately, these shortcomings can be accommodated by the use of *winding factors*.

The simplest and perhaps best definition of a winding factor is the ratio of flux linked by an actual winding to flux that would have been linked by a full-pitch, concentrated winding with the same number of turns. That is:

$$k_{\mathrm{w}} = \frac{\lambda_{\mathrm{actual}}}{\lambda_{\mathrm{full-pitch}}}$$

It is relatively easy to show, using reciprocity arguments, that the winding factors are also the ratio of effective MMF produced by an actual winding to the MMF that would have been produced by the same winding were it to be full-pitched and concentrated. The argument goes as follows: mutual inductance between any pair of windings is reciprocal. That is, if the windings are designated *one* and *two*, the mutual inductance is flux induced in winding *one* by current in winding *two*, and it is also flux induced in winding *two* by current in winding *one*. Since each winding has a winding factor that influences its linking flux, and since the mutual inductance must be reciprocal, the same winding factor must influence the MMF produced by the winding.

The winding factors are often expressed for each space harmonic, although sometimes when a winding factor is referred to without reference to a harmonic number, what is meant is the space factor for the space fundamental.

Two winding factors are commonly specified for ordinary, regular windings. These are usually called *pitch* and *breadth* factors, reflecting the fact that often windings are not *full-pitched*, which means that individual turns do not span a full π electrical radians and that the windings occupy a range or breadth of slots within a phase belt. The breadth factors are ratios of flux linked by a given winding to the flux that would be linked by that winding were it full-pitched and concentrated. These two winding factors are discussed in a little more detail below. What is interesting to note, although we do not prove it here, is that the winding factor of any given winding is the *product* of the pitch and breadth factors:

$$k_{\mathrm{w}} = k_{\mathrm{p}} k_{\mathrm{b}}$$

With winding factors as defined here and in the sections below, it is possible to define winding inductances. For example, the *synchronous* inductance of a winding will be the apparent inductance of one phase when the polyphase winding is driven by a *balanced* set of currents. This is, approximately:

$$L_d = \sum_{n=1,5,7,\ldots}^{\infty} \frac{3}{2}\frac{4}{\pi}\frac{\mu_0 N^2 R l k_{\mathrm{w}n}^2}{p^2 g n^2}$$

This expression is approximate because it ignores the asynchronous interactions between higher order harmonics and the rotor of the machine.

Zero-sequence inductance is the ratio of flux to current if a winding is excited by zero sequence currents:

$$L_0 = \sum_{n=3,9,\ldots}^{\infty} 3\frac{4}{\pi}\frac{\mu_0 N^2 R l k_{\mathrm{w}n}^2}{p^2 g n^2}$$

Mutual inductance, as between a *field* winding (f) and an *armature* winding (a), is:

$$L_{\mathrm{af}}(\theta) = \sum_{\substack{n=1 \\ n \text{ odd}}}^{\infty} \frac{4}{\pi}\frac{\mu_0 N_{\mathrm{f}} N_{\mathrm{a}} k_{\mathrm{f}n} k_{\mathrm{a}n} R l}{p^2 g n^2}\cos(np\theta)$$

Now we turn our attention to computing the winding factors for simple, regular winding patterns. We do not prove but only state that the winding factor can, for regular winding patterns, be expressed as the product of a *pitch* factor and a *breadth* factor, each of which can be estimated separately.

9.10.1 Pitch Factor

Pitch factor is found by considering the flux linked by a less than full-pitched winding. Consider the situation in which radial magnetic flux density is:

$$B_r = B_n \sin(np\phi - \omega t)$$

A winding with pitch α will link flux:

$$\lambda = Nl \int_{\frac{\pi}{2p}-\frac{\alpha}{2p}}^{\frac{\pi}{2p}+\frac{\alpha}{2p}} B_n \sin(np\phi - \omega t) R d\phi$$

Pitch α refers to the angular displacement between sides of the coil, expressed in *electrical* radians. For a full-pitch coil, $\alpha = \pi$.

The flux linked is:

$$\lambda = \frac{2NlRB_n}{np} \sin\left(\frac{n\alpha}{2}\right) \sin\left(\frac{n\pi}{2}\right)$$

The *pitch* factor is seen to be:

$$k_{pn} = \sin\frac{n\alpha}{2} \sin\frac{n\pi}{2}$$

Since $\sin\frac{n\pi}{2} = \pm 1$, in situations in which only the magnitude of an inductance component is important, the pitch factor may be taken to be $k_{pn} = |\sin\frac{n\alpha}{2}|$,

9.10.2 Breadth Factor

Now for *breadth* factor. This describes the fact that a winding may consist of a number of coils, each linking flux slightly out of phase with the others. A regular winding will have a number (say m) coil elements, separated by *electrical* angle γ.

A full-pitch coil with one side at angle ξ will, in the presence of sinusoidal magnetic flux density, link flux:

$$\lambda = Nl \int_{\frac{\xi}{p}}^{\frac{\pi}{p}-\frac{\xi}{p}} B_n \sin(np\phi - \omega t) R d\phi$$

This is readily evaluated to be:

$$\lambda = \frac{2NlRB_n}{np} \text{Re}\left(e^{j(\omega t - n\xi)}\right)$$

where complex number notation has been used for convenience in carrying out the rest of this derivation.

Now: if the winding is distributed into m sets of slots and the slots are evenly spaced, the angular position of each slot will be:

$$\xi_{\mathrm{i}} = i\gamma - \frac{m-1}{2}\gamma$$

and the number of turns in each slot will be $\frac{N}{mp}$, so that actual flux linked will be:

$$\lambda = \frac{2NlRB_{\mathrm{n}}}{np}\frac{1}{m}\sum_{\mathrm{i}=0}^{m-1}\mathrm{Re}\left(\mathrm{e}^{j(\omega t - n\xi_{\mathrm{i}})}\right)$$

The *breadth* factor is then simply:

$$k_{\mathrm{b}} = \frac{1}{m}\sum_{\mathrm{i}=0}^{m-1}\mathrm{e}^{-jn(i\gamma - \frac{m-1}{2}\gamma)}$$

Note that this can be written as:

$$k_{\mathrm{b}} = \frac{\mathrm{e}^{jn\gamma\frac{m-1}{2}}}{m}\sum_{\mathrm{i}=0}^{m}\mathrm{e}^{-jni\gamma}$$

Now, focus on that sum. It is known that any converging geometric sum has a simple sum:

$$\sum_{\mathrm{i}=0}^{\infty}x^{i} = \frac{1}{1-x}$$

and that a truncated sum is:

$$\sum_{\mathrm{i}=0}^{m-1} = \sum_{\mathrm{i}=0}^{\infty} - \sum_{\mathrm{i}=\mathrm{m}}^{\infty}$$

Then the useful sum can be written as:

$$\sum_{\mathrm{i}=0}^{m-1}\mathrm{e}^{-jni\gamma} = \left(1 - \mathrm{e}^{-jnm\gamma}\right)\sum_{\mathrm{i}=0}^{\infty}\mathrm{e}^{-jni\gamma} = \frac{1-\mathrm{e}^{-jnm\gamma}}{1-\mathrm{e}^{-jn\gamma}}$$

Finishing the computation by multiplying by $\mathrm{e}^{jn\gamma\frac{m-1}{2}}/m$, the breadth factor is found:

$$k_{\mathrm{b}n} = \frac{\sin\frac{nm\gamma}{2}}{m\sin\frac{n\gamma}{2}}$$

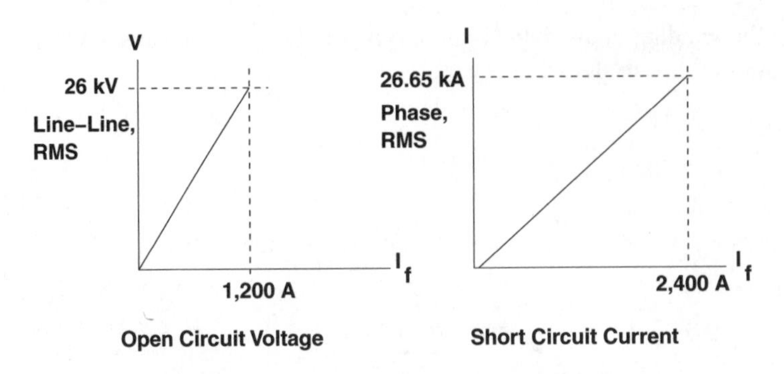

Figure 9.18 Open and short-circuit test data

9.11 Problems

1. Shown in Figure 9.18 are some test data for a large (1,200 MVA) four-pole synchronous generator. It is rated 26 kV, line–line, RMS and about 26.65 kA, RMS (phase current). Under open circuit conditions and at rated speed (1,800 r.p.m.) it achieves rated voltage with a field current of 1,200 A. Under short circuit conditions it achieves rated armature current with a field current of 2,400 A.
   (a) What is the field to armature mutual inductance? (H)
   (b) What is the armature synchronous inductance? (H)

2. A large 2 pole, 60 Hz synchronous machine has a field-phase mutual inductance $M = 56$ mH, phase inductance $L_{\mathrm{a}} = 2.4$ mH and phase-phase mutual inductance $L_{\mathrm{ab}} = -1.2$ mH. When run open circuited it produces a sine wave of voltage.
   (a) If the machine is operated with a field current $I_{\mathrm{f}} = 1000$ A and with a balanced set of currents in the armature with *peak* value of $|I_{\mathrm{a}}| = 31,113$ A, what is torque vs. angle?
   (b) If the machine is operated with the same field current but this time with the armature connected to a balanced voltage source with a phase voltage with a *peak* value of $|V_a| = 21,229$ V, what is torque vs. angle?
   (c) This machine is used as a generator and is producing 1,000 MW at unity power factor.
   - Sketch and label a phasor diagram, showing terminal voltage, terminal current, internal voltage and voltage drop across the synchronous reactance.
   - What are the two torque angles? (That is, phase angle between internal and terminal voltage and phase angle between internal flux and terminal current.)
   - What is field current?

3. A 1,000 kW, 60 Hz, eight-pole synchronous motor has a rated terminal voltage of 4,200 V, RMS, line-line. It has a field to phase mutual inductance of $M = 182$ mH, a direct axis synchronous inductance of $L_d = 95.5$ mH and a quadrature axis synchronous inductance of $L_q = 63.7$ mH.
   (a) On no-load (open circuit) test, what value would we expect for field current to achieve rated voltage?
   (b) On short-circuit test, what value would we expect for field current to achieve rated current?

(c) Draw a phasor diagram for this machine operated as a motor at rated output, operated at unity power factor.
(d) Draw a phasor diagram for this machine operated as a motor at rated output (1000 kW), operated at 0.8 power factor, leading (over-excited).
(e) With the field current required for unity power factor, rated output operation, what is the breakdown torque?

4. A 100 MVAR synchronous condenser (a mechanically unloaded synchronous machine) has a rated terminal voltage of 13.8 kV, RMS, line–line, a *d-axis* synchronous reactance of $x_d = 2.0$ per-unit and a q-axis synchronous reactance of $x_q = 1.0$ per-unit. Field current required to achieve rated voltage under open circuit conditions is $I_{fnl} = 100$ A. This is a 60 Hz machine.

(a) What is peak field-phase mutual inductance M?
(b) What are the values of direct and quadrature axis synchronous inductances L_d and L_q?
(c) To reach rated VA at zero power factor (machine is producing VARS), what is required field current?
(d) Because of saliency, this machine can actually carry *some* field current in the negative direction. What value of field current can it carry before it loses synchronism? How much reactive power can it absorb?
(e) Using the results of the previous two parts of this problem, sketch a zero real power Vee curve for this machine, showing the limits to over- and under-excited operation.

5. A large 60 Hz synchronous generator can be considered to be 'round rotor': that is $x_d = x_q = 2.0$ per-unit. Field current to achieve rated voltage while the stator is open circuited is $I_{fnl} = 1,000$ A. Armature resistance is small enough that it can be ignored.

(a) Assume the machine is initially operated as a generator with rated VA at a power factor of 0.8 per-unit, current lagging (over-excited). What field current is required?
(b) Holding power constant, calculate and plot torque-angle δ vs. field current as field current is reduced. At what value of field current does the machine become unstable?
(c) Now assume that this is a 1,000 MW generator with a rated terminal voltage of 24 kV, line-line, RMS. Calculate and plot the RMS magnitude of armature current over the range of field current between the stability limit you calculated in the previous part and the field current for rated operation, over-excited. This is part of the Vee-curve set for this machine.
(d) Calculate and plot a Vee curve for operation with zero real power.
(e) Calculate and plot a Vee curve set for this machine, with each curve consisting of armature current RMS magnitude vs. field current, plotted from the stability limit (lower end of field current range) to the value of field current required for 0.8 power factor operation at rated VA (upper value of field current range). Do this for values of real power of 0, 0.2, 0.4, 0.6 and 0.8 per-unit.

6. A 1,000 kW, 8 pole, 60 Hz synchronous motor has synchronous reactances $x_d = 1.5$ and $x_q = 1.0$ per-unit, respectively. The field current for rated voltage, armature open circuited if $I_{fnl} = 100$ A. The armature phase resistance is $R_a = 0.1\Omega$ and the field resistance is

$R_f = 0.3\Omega$. The motor has core loss of 2 kW and friction and windage losses of 1 kW, both of which you can assume to be constant. Terminal voltage of this motor is 4,200 V, RMS, line-line.

(a) Find required field current for rated operation as a motor at unity power factor.
(b) What is motor efficiency at rated operation?
(c) Find and plot motor efficiency assuming unity power factor operation over a range of loads from 100 kW (10%) to 1,000 kW (100%) output power.

7. Referring to Figure 9.13, if a machine has per-unit reactances $x_d = 2.0$ and $x'_d = 0.4$, and assuming per-unit armature leakage reactance $x_{al} = 0.1$. The open-circuit field time constant is $T'_{do} = 5$ s.

(a) What are x_{ad} and x_{fl}?
(b) What is r_f?
(c) If the machine has base quantities of its rating of 500 MVA and 24 kV, RMS, line-line, and if it is a 60 Hz machine, what are the ordinary values L_{ad} and L_{fl}?
(d) If the field current required for rated voltage with the armature open circuited is $I_{fnl} = 500$ A, find field winding inductance and resistance.

8. A turbogenerator has the following per-unit parameters:

Synchronous reactances	x_d	2.0
	x_q	1.8
Transient reactances	x'_d	0.4
	x'_q	0.4
Armature resistance	r_a	0.01
Transient time constants	T'_{do}	5 S
	T'_{qo}	40 mS
Inertia constant	H	3 S

Assume armature leakage reactance is $x_{al} = 0.1$ per-unit.

(a) In preparation for a simulation, initial conditions must be estimated. Assume the machine is to be operated as a generator at rated VA, 0.95 power factor, current lagging. Find the initial conditions ψ_d, ψ_q, ψ_f, ψ_{kd}, ψ_{kq} and δ.
(b) Show that these initial conditions are right by simulating operation of the machine for about 100 ms.
(c) Using the equal-area criterion, find the 'critical clearing time' for the machine, assuming (somewhat improbably) that its connection to the power system has zero external reactance.
(d) Simulate the transient resulting from disconnecting the machine from the system for 10 ms.
(e) Find, through simulation, the 'critical clearing time'. How does this compare with your equal area criterion results?

10

System Analysis and Protection

Sometimes things go wrong. Designers of electric power systems must ensure that those systems don't damage themselves during either normal operation nor when something happens. Contingencies range from lightning strikes to animal contact and automobile accidents. This chapter deals with methods for analyzing current flows in power systems in abnormal situations, and with methods used for protecting the system for consequences of short circuit contingencies.

Up to this point, the book, has dealt *primarily* with networks that are *balanced*, in which the three voltages (and three currents) are identical but for exact 120° phase shifts. Unbalanced conditions may arise from unequal voltage sources or loads. It *is* possible to analyze some simple types of unbalanced networks using straightforward solution techniques and *wye–delta* transformations. However, power networks can become quite complex and many situations would be very difficult to handle using ordinary network analysis. For this reason, a technique which has come to be called *symmetrical components* has been developed.

Symmetrical components, in addition to being a powerful analytical tool, is also conceptually useful. The symmetrical components themselves, which are obtained from a transformation of the ordinary line voltages and currents, are useful in their own right. Symmetrical components have become accepted as one way of describing the properties of many types of network elements such as transmission lines, motors and generators.

10.1 The Symmetrical Component Transformation

The basis for this analytical technique is a transformation of the three voltages and three currents into a second set of voltages and currents. This second set is known as the *symmetrical components*.

Working in complex amplitudes:

$$\begin{aligned}
v_{\mathrm{a}} &= \mathrm{Re}\{\mathbf{V}_{\mathrm{a}}\mathrm{e}^{j\omega t}\} \\
v_{\mathrm{b}} &= \mathrm{Re}\{\mathbf{V}_{\mathrm{b}}\mathrm{e}^{j(\omega t - \frac{2\pi}{3})}\} \\
v_{\mathrm{c}} &= \mathrm{Re}\{\mathbf{V}_{\mathrm{c}}\mathrm{e}^{j(\omega t + \frac{2\pi}{3})}\}
\end{aligned}$$

Electric Power Principles: Sources, Conversion, Distribution and Use James L. Kirtley

The transformation is defined as:

$$\begin{bmatrix} \mathbf{V}_1 \\ \mathbf{V}_2 \\ \mathbf{V}_0 \end{bmatrix} = \frac{1}{3} \begin{bmatrix} 1 & \mathbf{a} & \mathbf{a}^2 \\ 1 & \mathbf{a}^2 & \mathbf{a} \\ 1 & 1 & 1 \end{bmatrix} \begin{bmatrix} \mathbf{V}_a \\ \mathbf{V}_b \\ \mathbf{V}_c \end{bmatrix} \tag{10.1}$$

where the complex number **a** is a rotation by 120°:

$$\begin{aligned} \mathbf{a} &= e^{j\frac{2\pi}{3}} = -\frac{1}{2} + j\frac{\sqrt{3}}{2} \\ \mathbf{a}^2 &= e^{j\frac{4\pi}{3}} = e^{-j\frac{2\pi}{3}} = -\frac{1}{2} - j\frac{\sqrt{3}}{2} \\ \mathbf{a}^3 &= 1 \end{aligned}$$

This transformation may be used for both voltage and current, and works for variables in *ordinary* form as well as variables that have been normalized and are in *per-unit* form. The inverse of this transformation is:

$$\begin{bmatrix} \mathbf{V}_a \\ \mathbf{V}_b \\ \mathbf{V}_c \end{bmatrix} = \begin{bmatrix} 1 & 1 & 1 \\ \mathbf{a}^2 & \mathbf{a} & 1 \\ \mathbf{a} & \mathbf{a}^2 & 1 \end{bmatrix} \begin{bmatrix} \mathbf{V}_1 \\ \mathbf{V}_2 \\ \mathbf{V}_0 \end{bmatrix}$$

The three component variables $\mathbf{V}_1$, $\mathbf{V}_2$, $\mathbf{V}_0$ are called, respectively, *positive sequence*, *negative sequence* and *zero sequence*. They are called *symmetrical components* because, taken *separately*, they transform into symmetrical sets of voltages. The properties of these components can be demonstrated by transforming each one back into phase variables.

Consider first the *positive sequence* component taken by itself:

$$\begin{aligned} \mathbf{V}_1 &= V \\ \mathbf{V}_2 &= 0 \\ \mathbf{V}_0 &= 0 \end{aligned}$$

yields:

$$\begin{aligned} \mathbf{V}_a &= V & \text{or} \quad v_a &= V\cos\omega t \\ \mathbf{V}_b &= \mathbf{a}^2 V & \text{or} \quad v_b &= V\cos\left(\omega t - \frac{2\pi}{3}\right) \\ \mathbf{V}_c &= \mathbf{a}V & \text{or} \quad v_c &= V\cos\left(\omega t + \frac{2\pi}{3}\right) \end{aligned}$$

This is the familiar *balanced* set of voltages: Phase b lags phase a by 120°, phase c lags phase b and phase a lags phase c.

The same transformation carried out on a *negative sequence* voltage:

$$\mathbf{V}_1 = 0$$
$$\mathbf{V}_2 = V$$
$$\mathbf{V}_0 = 0$$

yields:

$$\mathbf{V}_\mathrm{a} = V \quad \text{or} \quad v_\mathrm{a} = V\cos\omega t$$
$$\mathbf{V}_\mathrm{b} = \mathbf{a}V \quad \text{or} \quad v_\mathrm{b} = V\cos\left(\omega t + \frac{2\pi}{3}\right)$$
$$\mathbf{V}_\mathrm{c} = \mathbf{a}^2V \quad \text{or} \quad v_\mathrm{c} = V\cos\left(\omega t - \frac{2\pi}{3}\right)$$

This is called *negative sequence* because the sequence of voltages is reversed: phase *b* now *leads* phase a rather than lagging. Note that the negative sequence set is still balanced in the sense that the phase components still have the same magnitude and are separated by 120°. The only difference between positive and negative sequence is the phase rotation. This is shown in Figure 10.1.

The third symmetrical component is *zero sequence*. If:

$$\mathbf{V}_1 = 0$$
$$\mathbf{V}_2 = 0$$
$$\mathbf{V}_0 = V$$

Then:

$$\mathbf{V}_\mathrm{a} = V \quad \text{or} \quad v_\mathrm{a} = V\cos\omega t$$
$$\mathbf{V}_\mathrm{b} = V \quad \text{or} \quad v_\mathrm{b} = V\cos\omega t$$
$$\mathbf{V}_\mathrm{c} = V \quad \text{or} \quad v_\mathrm{c} = V\cos\omega t$$

That is, all three phases are varying *together*.

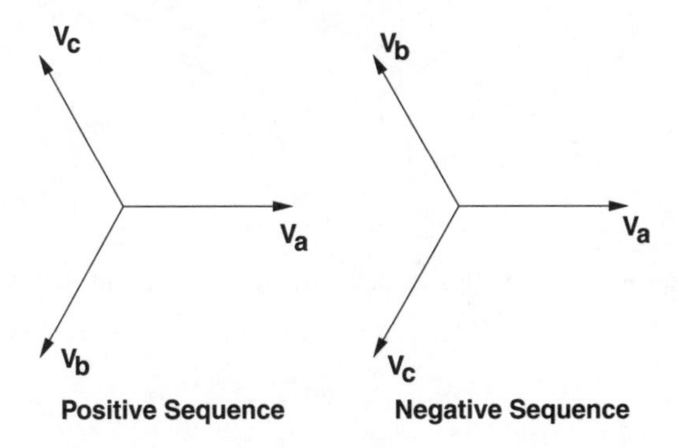

Figure 10.1 Phasor diagram: three phase voltages

Positive and negative sequence sets contain those parts of the three-phase excitation that represent balanced normal and reverse phase sequence. Zero sequence is required to make up the difference between the total phase variables and the two *rotating* components.

The great utility of symmetrical components is that, for most types of network elements, the symmetrical components are independent of each other. In particular, balanced impedances and rotating machines will draw only positive sequence currents in response to positive sequence voltages. It is thus possible to describe a network in terms of sub-networks, one for each of the symmetrical components. These are called *sequence networks*. A completely balanced network will have three entirely separate sequence networks. If a network is unbalanced at a particular spot, the sequence networks will be interconnected at that spot. The key to use of symmetrical components in handling unbalanced situations is in learning how to formulate those interconnections.

10.2 Sequence Impedances

Many different types of network elements exhibit different behavior to the different symmetrical components. For example, as will be seen shortly, transmission lines have one impedance for positive and negative sequence, but an entirely different impedance to zero sequence. Rotating machines have different impedances to all three sequences.

10.2.1 Balanced Transmission Lines

To illustrate the independence of symmetrical components in balanced networks, consider the transmission line illustrated in Section 7.1. The expressions for voltage drop in the lines may be written as a single vector expression:

$$\mathbf{V}_{\mathrm{ph1}} - \mathbf{V}_{\mathrm{ph2}} = j\omega \mathbf{L}_{\mathrm{ph}} \mathbf{I}_{\mathrm{ph}} \tag{10.2}$$

where

$$\mathbf{V}_{\mathrm{ph}} = \begin{bmatrix} \mathbf{V}_{\mathrm{a}} \\ \mathbf{V}_{\mathrm{b}} \\ \mathbf{V}_{\mathrm{c}} \end{bmatrix} \qquad \mathbf{I}_{\mathrm{ph}} = \begin{bmatrix} \mathbf{I}_{\mathrm{a}} \\ \mathbf{I}_{\mathrm{b}} \\ \mathbf{I}_{\mathrm{c}} \end{bmatrix}$$

If the transmission line is balanced:

$$\mathbf{L}_{\mathrm{ph}} = \begin{bmatrix} L & M & M \\ M & L & M \\ M & M & L \end{bmatrix}$$

Note that the symmetrical component transformation (Equation 10.1) may be written in compact form:

$$\mathbf{V}_{\mathrm{s}} = \mathbf{T}\mathbf{V}_{\mathrm{p}} \tag{10.3}$$

where $\mathbf{T}$ is defined by Equation 10.1.

and $\mathbf{V}_s$ is the vector of sequence voltages:

$$\mathbf{V}_s = \begin{bmatrix} \mathbf{V}_1 \\ \mathbf{V}_2 \\ \mathbf{V}_0 \end{bmatrix}$$

Rewriting Equation 10.2 using the inverse of Equation 10.3:

$$\mathbf{T}^{-1}\mathbf{V}_{s1} - \mathbf{T}^{-1}\mathbf{V}_{s2} = j\omega\mathbf{L}_{ph}\mathbf{T}^{-1}\mathbf{I}_s$$

Then transforming to get sequence voltages:

$$\mathbf{V}_{s1} - \mathbf{V}_{s2} = j\omega\mathbf{T}\mathbf{L}_{ph}\mathbf{T}^{-1}\mathbf{I}_s$$

The sequence inductance matrix is defined by carrying out the operation indicated:

$$\mathbf{L}_s = \mathbf{T}\mathbf{L}_{ph}\mathbf{T}^{-1}$$

which is:

$$\mathbf{L}_s = \begin{bmatrix} L - M & 0 & 0 \\ 0 & L - M & 0 \\ 0 & 0 & L + 2M \end{bmatrix}$$

Thus the *coupled* set of expressions which described the transmission line in *phase* variables becomes an *uncoupled* set of expressions in the symmetrical components:

$$\begin{aligned} \mathbf{V}_{11} - \mathbf{V}_{12} &= j\omega(L - M)\mathbf{I}_1 \\ \mathbf{V}_{21} - \mathbf{V}_{22} &= j\omega(L - M)\mathbf{I}_2 \\ \mathbf{V}_{01} - \mathbf{V}_{02} &= j\omega(L + 2M)\mathbf{I}_0 \end{aligned}$$

The *positive*, *negative* and *zero* sequence impedances of the balanced transmission line are then:

$$\begin{aligned} \mathbf{Z}_1 = \mathbf{Z}_2 &= j\omega(L - M) \\ \mathbf{Z}_0 &= j\omega(L + 2M) \end{aligned}$$

So, in analysis of networks with transmission lines, it is now possible to replace the lines with three *independent*, single-phase networks.

10.2.2 Balanced Load

Consider next a three-phase load with its neutral connected to ground through an impedance as shown in Figure 10.2. If the load is 'balanced', $Z_a = Z_b = Z_c = Z$.

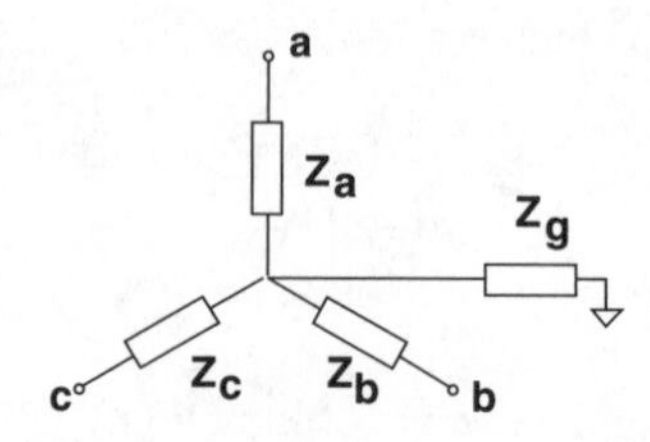

Figure 10.2 Three-phase load with neutral impedance

The symmetrical component voltage–current relationship for this network is found simply, by assuming positive, negative and zero sequence currents and finding the corresponding voltages. If this is done, it is found that the symmetrical components *are* independent, and that the voltage–current relationships are:

$$\begin{aligned}\mathbf{V}_1 &= \mathbf{Z}\mathbf{I}_1 \\ \mathbf{V}_2 &= \mathbf{Z}\mathbf{I}_2 \\ \mathbf{V}_0 &= (\mathbf{Z} + 3\mathbf{Z}_g)\mathbf{I}_0\end{aligned} \tag{10.4}$$

10.2.3 Possibly Unbalanced Loads

The theory of symmetrical components can be used to get a more general picture, even of systems with mutual impedances and of loads that are not balanced. If a four-terminal circuit element has three phase terminals and ground, the voltage–current relationship is:

$$\mathbf{V}_{ph} = \mathbf{Z}_{ph}\mathbf{I}_{ph}$$

Then a similar relationship can be found for the symmetrical components:

$$\mathbf{V}_s = \mathbf{T}\mathbf{V}_{ph} = \mathbf{T}\mathbf{Z}_{ph}\mathbf{I}_{ph} = \mathbf{T}\mathbf{Z}_{ph}\mathbf{T}^{-1}\mathbf{I}_s$$

Thus the symmetrical component impedance is:

$$\mathbf{Z}_s = \mathbf{T}\mathbf{Z}_{ph}\mathbf{T}^{-1}$$

Figure 10.2 shows a somewhat restricted version of a generalized phase impedance picture, in which mutual impedance is the same for all phases. The phase impedance is:

$$\mathbf{Z}_{ph} = \begin{bmatrix} Z_a + Z_g & Z_g & Z_g \\ Z_g & Z_b + Z_g & Z_g \\ Z_g & Z_g & Z_c + Z_g \end{bmatrix}$$

Note that this can be written as:

$$\mathbf{Z}_{\text{ph}} = \begin{bmatrix} Z_a & 0 & 0 \\ 0 & Z_b & 0 \\ 0 & 0 & Z_c \end{bmatrix} + Z_g \begin{bmatrix} 1 & 1 & 1 \\ 1 & 1 & 1 \\ 1 & 1 & 1 \end{bmatrix}$$

Carrying out the transforms on the two parts separately, the two parts of the symmetrical component impedance matrix are:

$$\mathbf{T} \begin{bmatrix} Z_a & 0 & 0 \\ 0 & Z_b & 0 \\ 0 & 0 & Z_c \end{bmatrix} \mathbf{T}^{-1} = \frac{1}{3} \begin{bmatrix} A & B & C \\ C & A & B \\ B & C & A \end{bmatrix}$$

where:

$$\begin{aligned} A &= Z_a + Z_b + Z_c \\ B &= Z_a + \mathbf{a}^2 Z_b + \mathbf{a} Z_c \\ C &= Z_a + \mathbf{a} Z_b + \mathbf{a}^2 Z_c \end{aligned}$$

$$\mathbf{T} Z_g \begin{bmatrix} 1 & 1 & 1 \\ 1 & 1 & 1 \\ 1 & 1 & 1 \end{bmatrix} \mathbf{T}^{-1} = \begin{bmatrix} 0 & 0 & 0 \\ 0 & 0 & 0 \\ 0 & 0 & 3Z_g \end{bmatrix}$$

If the three phase impedances are the same ($Z_a = Z_b = Z_c = Z$), the symmetrical component impedance matrix is:

$$\mathbf{T}_s = \begin{bmatrix} Z_1 & 0 & 0 \\ 0 & Z_2 & 0 \\ 0 & 0 & Z_0 \end{bmatrix}$$

where the impedances are as given by:

$$\begin{aligned} Z_1 &= Z \\ Z_2 &= Z \\ Z_0 &= Z + 3Z_g \end{aligned}$$

Note that the same techniques used for unbalanced loads can also be used for unbalanced transmission lines, for example for lines in which positions of conductors are not transposed so that phase to phase mutual inductances are not all the same.

10.2.4 Unbalanced Sources

Consider the network shown in Figure 10.3. A balanced three-phase resistor is fed by a balanced line (with mutual coupling between phases). Assume that only one phase of the voltage source is working, so that:

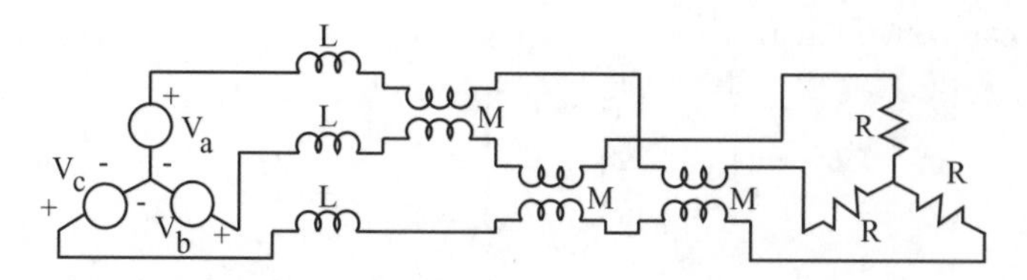

Figure 10.3 Balanced load, balanced line, unbalanced source

$$\mathbf{V}_a = V$$
$$\mathbf{V}_b = 0$$
$$\mathbf{V}_c = 0$$

The objective of this example is to find currents in the three phases.

To start, note that the unbalanced voltage source has the following set of symmetrical components:

$$\mathbf{V}_1 = \frac{V}{3}$$
$$\mathbf{V}_2 = \frac{V}{3}$$
$$\mathbf{V}_0 = \frac{V}{3}$$

Next, the network facing the source consists of the line, with impedances:

$$\mathbf{Z}_1 = j\omega(L - M)$$
$$\mathbf{Z}_2 = j\omega(L - M)$$
$$\mathbf{Z}_0 = j\omega(L + 2M)$$

and the three-phase resistor has impedances:

$$\mathbf{Z}_1 = R$$
$$\mathbf{Z}_2 = R$$
$$\mathbf{Z}_0 = \infty$$

Note that the impedance to zero sequence is infinite because the neutral is not connected back to the neutral of the voltage source. Thus the sum of line currents must always be zero and this in turn precludes any zero sequence current. The problem is thus described by the networks which appear in Figure 10.4.

Currents are:

$$\mathbf{I}_1 = \frac{V}{3(j\omega(L - M) + R)}$$
$$\mathbf{I}_2 = \frac{V}{3(j\omega(L - M) + R)}$$
$$\mathbf{I}_0 = 0$$

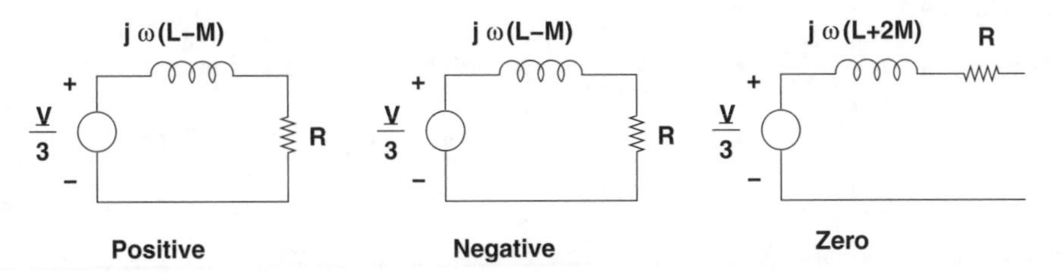

Figure 10.4 Sequence networks for the problem

Phase currents may now be re-assembled:

$$\mathbf{I}_a = \mathbf{I}_1 + \mathbf{I}_2 + \mathbf{I}_0$$
$$\mathbf{I}_b = \mathbf{a}^2\mathbf{I}_1 + \mathbf{a}\mathbf{I}_2 + \mathbf{I}_0$$
$$\mathbf{I}_c = \mathbf{a}\mathbf{I}_1 + \mathbf{a}^2\mathbf{I}_2 + \mathbf{I}_0$$

or:

$$\mathbf{I}_a = \frac{2V}{3(j\omega(L-M)+R)}$$
$$\mathbf{I}_b = \frac{(\mathbf{a}^2+\mathbf{a})V}{3(j\omega(L-M)+R)} = \frac{-V}{3(j\omega(L-M)+R)}$$
$$\mathbf{I}_c = \frac{(\mathbf{a}+\mathbf{a}^2)V}{3(j\omega(L-M)+R)} = \frac{-V}{3(j\omega(L-M)+R)}$$

(Note that $\mathbf{a}^2 + \mathbf{a} = -1$).

10.2.5 Rotating Machines

Some network elements are more readily represented by sequence networks than by ordinary phase networks. This is the case, for example, for synchronous machines. synchronous motors and generators produce a positive sequence *internal* voltage and have terminal impedance. Because of rotation of the physical parts of the machine, the impedance to positive sequence currents is not the same as the impedance to negative or to zero sequence currents. A phase-by-phase representation will not, in many situations, be adequate, but a sequence network representation will. Such a representation is three Thevenin equivalent circuits, as shown in Figure 10.5

10.2.6 Transformers

Transformers provide some interesting features in setting up sequence networks. The first of these arises from the fact that *wye-delta* or *delta-wye* transformer connections produce phase shifts from primary to secondary. Depending on connection, this phase shift may be either plus or minus 30° from primary to secondary for positive sequence voltages and currents. It is

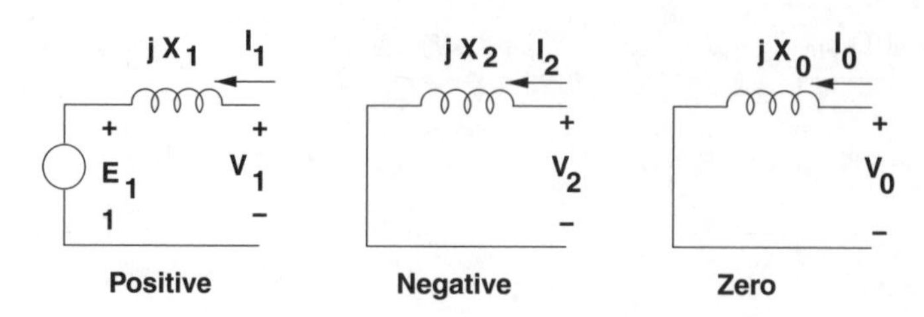

Figure 10.5 Sequence networks for a synchronous machine

straightforward to show that *negative* sequence shifts in the *opposite* direction from *positive*. Thus if the connection *advances* positive sequence across the transformer, it *retards* negative sequence. This does not turn out to affect the setting up of sequence networks, but does affect the re-construction of phase currents and voltages. That is, if positive sequence voltage rotates forward in one direction through the transformer, positive sequence current rotates backwards in the reverse direction. Negative sequence does just the opposite.

A second important feature of transformers arises because *delta* and ungrounded *wye* connections are open circuits to zero sequence at their terminals. A *delta*-connected winding, on the other hand, will provide a short circuit to zero sequence currents induced from a *wye*-connected winding. Thus the zero sequence network of a transformer may take one of several forms. Figures 10.6 through 10.8 show the zero sequence networks for various transformer connections.

10.2.6.1 Example: Rotation of Symmetrical Component Currents

Consider the circuit shown in Figure 10.9. Assume the delta side is connected to a balanced 480 V, line–line, RMS, voltage source. Assume as well that the current source connected line-neutral on the wye side is drawing 30 A at unity power factor. Assume further that the three-phase transformer has a voltage ratio of 1, meaning that $\frac{N_2}{N_1} = \frac{480}{277} = \sqrt{3}$.

This problem can be worked in a fairly straightforward way: there is current in only one of the three transformers: the one that is connected between terminals a and c on the *delta* side and terminals A to neutral on the *wye* side. Because of the turns ratio, current in the *delta* side will have an RMS magnitude equal to $\frac{30}{\sqrt{3}}$. The phase angle of that current will be the same as the phase angle of the voltage on the *wye* side of the transformer, and that can be seen to be $-30°$, because that phase is connected between phases a and c. The resulting current on the delta side will be:

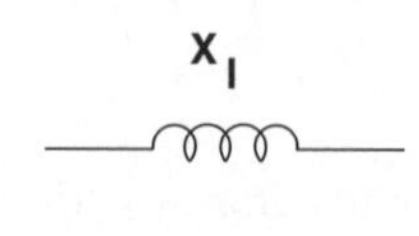

Figure 10.6 Zero sequence network: *wye–wye* connection, both sides grounded

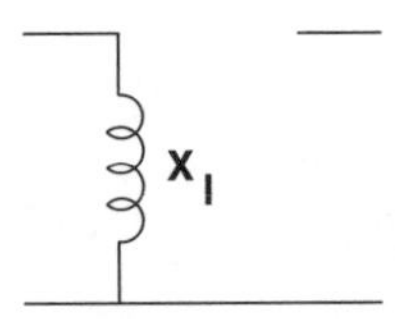

Figure 10.7 Zero sequence network: *wye–delta* connection, *wye* side (left) grounded

$$I_a = -I_c = \frac{30}{\sqrt{3}} e^{-j\frac{\pi}{6}}$$

These are drawn in Figure 10.10.

This problem can also be done using symmetrical components. On the *wye* side, since the voltage phase angle is -30°, $I_a = 30e^{-j\frac{\pi}{6}}$. Then, using the symmetrical component transformation:

$$I_{1Y} = I_{2Y} = I_{0Y} = \frac{1}{3} I_a = 10e^{-j\frac{\pi}{6}}$$

Since positive sequence voltage shifts -30° from the *delta* side to the *wye* side, positive sequence current shifts $+30^\circ$ from the *wye* side to the *delta* side, but negative sequence current shifts in the opposite direction, and zero sequence current does not make it out of the *delta* winding, so that on the *delta* side the currents are:

$$I_{1\Delta} = 10 \qquad I_{2\Delta} = 10e^{-j\frac{\pi}{3}} \qquad I_{0\Delta} = 0$$

Now it is straightforward to re-assemble the phase currents (*a* is a complex constant):

$$I_a = I_{1\Delta} + I_{2\Delta} = 10\left(1 + e^{-j\frac{\pi}{3}}\right) = 10\sqrt{3}e^{-j\frac{\pi}{6}}$$
$$I_b = \mathbf{a}^2 I_{1\Delta} + \mathbf{a} I_{2\Delta} = 10\left(e^{-j\frac{2\pi}{3}} + e^{j\frac{2\pi}{3}} e^{-j\frac{\pi}{3}}\right) = 0$$
$$I_c = \mathbf{a} I_{1\Delta} + \mathbf{a}^2 I_{2\Delta} = 10\left(e^{j\frac{2\pi}{3}} + e^{-j\frac{2\pi}{3}} e^{-j\frac{\pi}{3}}\right) = 10\sqrt{3}e^{j\frac{5\pi}{6}}$$

Noting that $\frac{30}{\sqrt{3}} = 10\sqrt{3}$, this answer agrees with the simple solution that started this example. Of course in many situations the use of symmetrical components is the more straightforward (and certainly systematic) way of solving the problem.

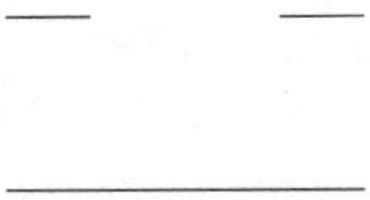

Figure 10.8 Zero sequence network: *wye–delta* connection, *wye* side ungrounded or *delta–delta*

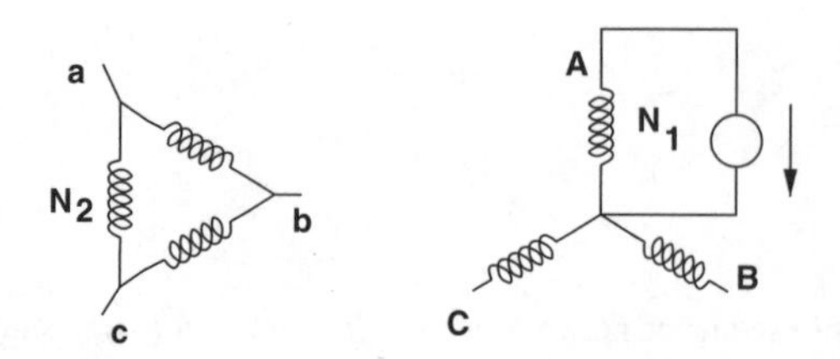

Figure 10.9 *Delta-wye* transformer loaded by a single phase on the *wye* side

10.2.6.2 Example: Reconstruction of Currents

As a further illustration of some of the analytical manipulations involving symmetrical component currents, look at the example of Figure 10.11. This is a *wye-delta* transformer loaded on the *delta* side by a single resistor. Assume that the *wye* side is driven by a symmetrical (positive sequence) voltage source with phase-neutral amplitude V_a. Use notation that V_a is line–neutral voltage on the wye side and V_A is line–neutral voltage on the *delta* side. Because $V_{AC} = \frac{N_2}{N_1} V_a$, the voltage on the *delta* side is:

$$V_A = \frac{N_2}{\sqrt{3}N_1} e^{j\frac{\pi}{6}} V_a$$

Since the resistor is connected across phases A and B, current through it will be:

$$I_A = \frac{V_A - V_B}{R} = \frac{N_2}{\sqrt{3}N_1} e^{j\frac{\pi}{6}} \frac{V_a}{R} \left(1 - e^{-j\frac{2\pi}{3}}\right) = \frac{N_2}{N_1} \frac{V_a}{R} e^{j\frac{\pi}{3}}$$

and $I_B = -I_A$

It is straightforward to compute the symmetrical components of the secondary (delta) side currents:

$$I_{S1} = \frac{1}{3}(I_A + aI_B) = \frac{1}{\sqrt{3}} \frac{N_2}{N_1} \frac{V_a}{R} e^{j\frac{\pi}{2}}$$

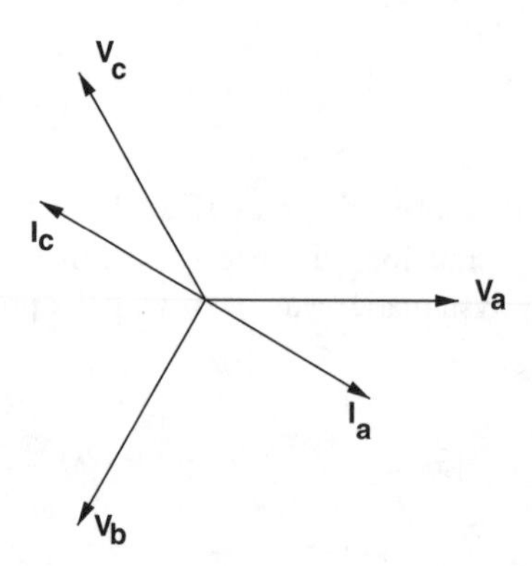

Figure 10.10 Currents from unbalanced load

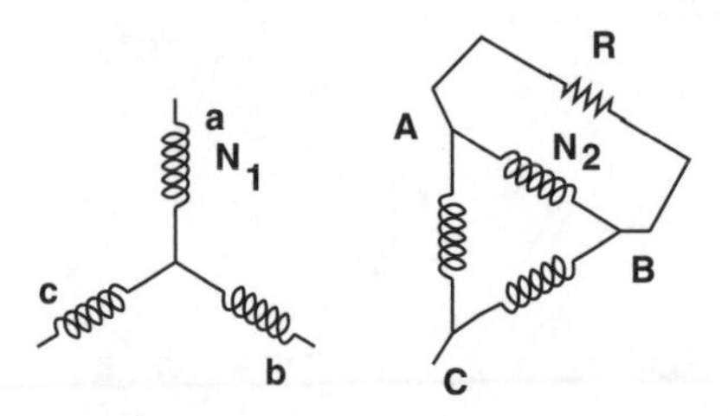

Figure 10.11 Example problem: *wye-delta* connection, single phase load on delta side

$$I_{S2} = \tfrac{1}{3}\left(I_A + a^2 I_B\right) = \frac{j}{\sqrt{3}} \frac{N_2}{N_1} \frac{V_a}{R}$$

The symmetrical component currents are transformed by the inverse of the voltage ratio of the transformer and shifted by 30°, positive sequence in the opposite direction of the voltage shift from primary to secondary, and the negative sequence in the opposite direction from the positive sequence, or 30° *forward*. Thus:

$$I_{P1} = \frac{1}{3}\left(\frac{N_2}{N_1}\right)^2 \frac{V_a}{R}$$

$$I_{P2} = \frac{1}{3}\left(\frac{N_2}{N_1}\right)^2 \frac{V_a}{R} e^{j\frac{2\pi}{3}}$$

Finally, the phase currents can be extracted from the symmetrical components:

$$I_a = I_{P1} + I_{P2} = \frac{1}{3}\left(\frac{N_2}{N_1}\right)^2 \frac{V_a}{R} e^{j\frac{\pi}{3}}$$

$$I_b = a^2 I_{P1} + a I_{P2} = \frac{2}{3}\left(\frac{N_2}{N_1}\right)^2 \frac{V_a}{R} e^{-j\frac{2\pi}{3}}$$

$$I_c = a I_{P1} + a^2 I_{P2} = \frac{1}{3}\left(\frac{N_2}{N_1}\right)^2 \frac{V_a}{R} e^{j\frac{\pi}{3}}$$

These are shown in Figure 10.12.

10.3 Fault Analysis

A very common application of symmetrical components is in calculating currents arising from unbalanced short circuits. For three-phase systems, the possible unbalanced faults are:

1. Single line–neutral,
2. Double line–neutral,
3. Line–line.

These are considered separately.

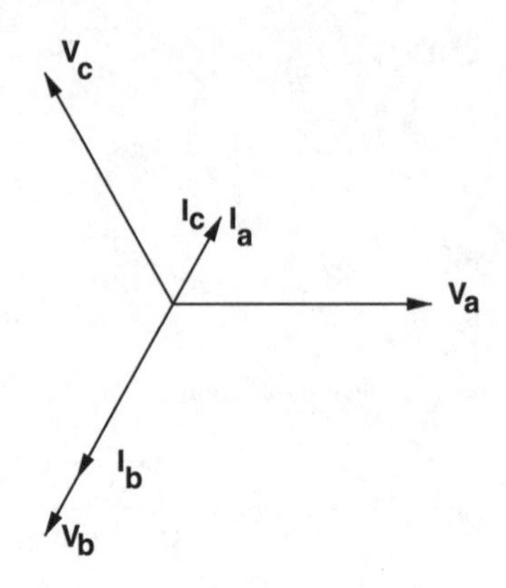

Figure 10.12 Reconstructed currents

10.3.1 Single Line–neutral Fault

The situation is as shown in Figure 10.13.

The *system* in this case consists of networks connected to the line on which the fault occurs. The point of fault itself consists of a set of terminals (which we might call 'a,b,c'). The fault sets, at this point on the system:

$$\mathbf{V}_a = 0$$
$$\mathbf{I}_b = 0$$
$$\mathbf{I}_c = 0$$

Now: using the inverse of the symmetrical component transformation, we see that:

$$\mathbf{V}_1 + \mathbf{V}_2 + \mathbf{V}_0 = 0 \tag{10.5}$$

And using the transformation itself:

$$\mathbf{I}_1 = \mathbf{I}_2 = \mathbf{I}_0 = \frac{1}{3}\mathbf{I}_a \tag{10.6}$$

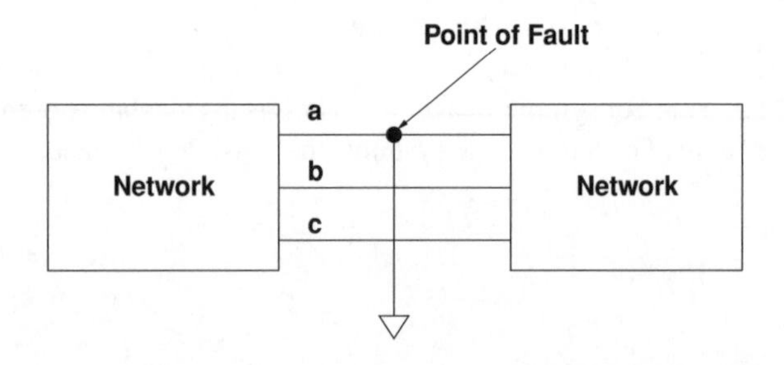

Figure 10.13 Schematic picture of a single line-to-ground fault

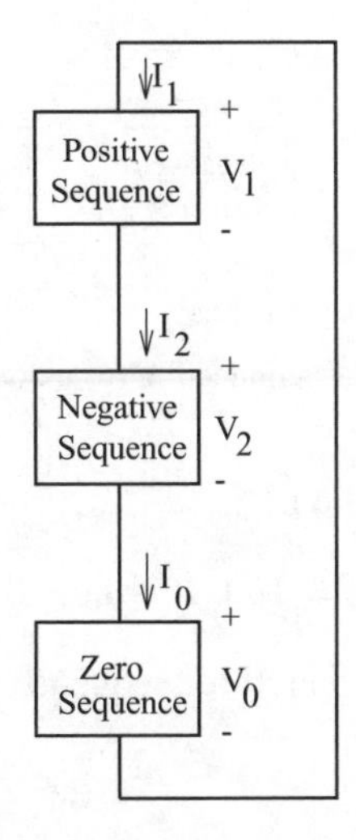

Figure 10.14 Sequence connection for a single line-neutral fault

Together, these two expressions describe the *sequence network* connection shown in Figure 10.14. This connection has all three sequence networks connected in *series*.

10.3.2 Double Line–neutral Fault

The situation is shown in Figure 10.15. The fault is between phases b and c and ground.

The 'terminal' relationship at the point of the fault is:

$$\mathbf{V}_b = 0$$
$$\mathbf{V}_c = 0$$
$$\mathbf{I}_a = 0$$

Then, using the sequence transformation:

$$\mathbf{V}_1 = \mathbf{V}_2 = \mathbf{V}_0 = \frac{1}{3}\mathbf{V}_a$$

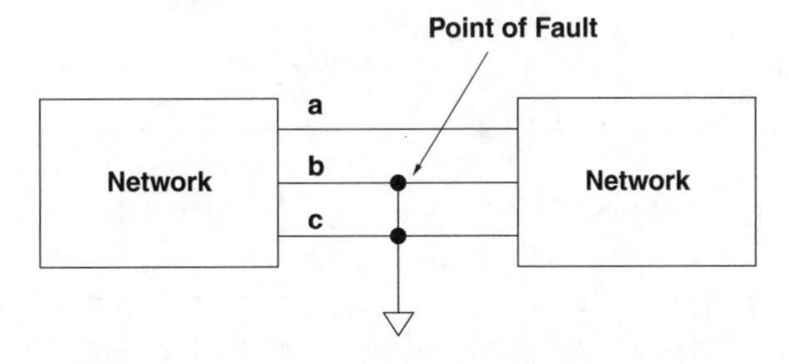

Figure 10.15 Schematic picture of a double line-neutral fault

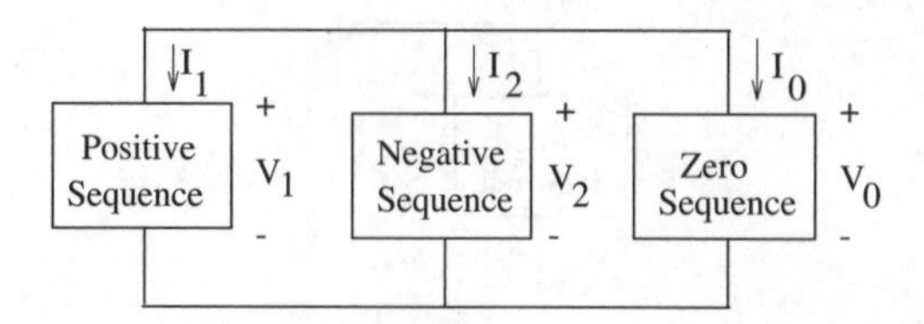

Figure 10.16 Sequence connection for a double line-neutral fault

Combining the inverse transformation:

$$\mathbf{I}_a = \mathbf{I}_1 + \mathbf{I}_2 + \mathbf{I}_0 = 0$$

These describe a situation in which all three sequence networks are connected in parallel, as shown in Figure 10.16.

10.3.3 Line–Line Fault

The situation is shown in Figure 10.17. For convenience it is assumed that the fault is between phases b and c. Ground is not involved.

If phases b and c are shorted together but not grounded,

$$\mathbf{V}_b = \mathbf{V}_c$$
$$\mathbf{I}_b = -\mathbf{I}_c$$
$$\mathbf{I}_a = 0$$

Expressing these in terms of the symmetrical components:

$$\mathbf{V}_1 = \mathbf{V}_2 = \frac{1}{3}\left(a + a^2\right)\mathbf{V}_b$$
$$\mathbf{I}_0 = \mathbf{I}_a + \mathbf{I}_b + \mathbf{I}_c = 0$$
$$\mathbf{I}_a = \mathbf{I}_1 + \mathbf{I}_2 = 0$$

These expressions describe a parallel connection of the positive and negative sequence networks, as shown in Figure 10.18.

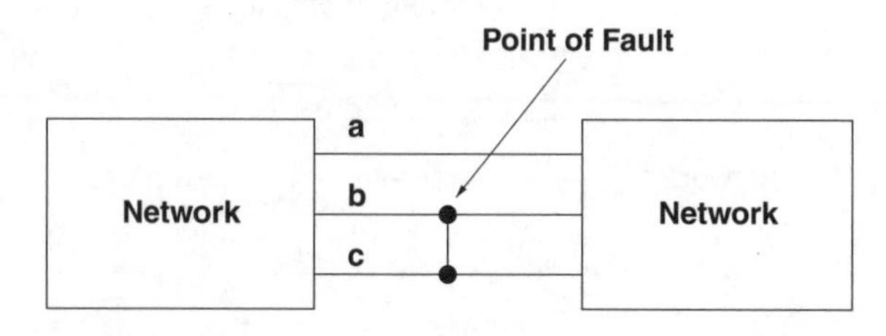

Figure 10.17 Schematic picture of a line-line fault

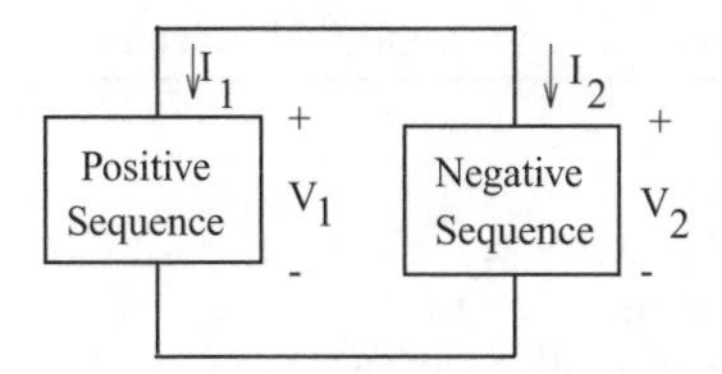

Figure 10.18 Sequence connection for a line-line fault

10.3.4 Example of Fault Calculations

In this example, the objective is to determine maximum current through the breaker **B** due to a fault at the location shown in Figure 10.19. All three types of unbalanced fault, as well as the balanced fault are to be considered. This is the sort of calculation that has to be done whenever a line is installed or modified, so that protective relaying can be set properly. (Note that the transformer connections might be different in any particular power system.)

Parameters of the system are given in Table 10.1. The fence-like symbols at either end of the figure represent 'infinite buses', or positive sequence voltage sources.

The first step in the fault calculation is to find the sequence networks. These are shown in Figure 10.20. Note that they are exactly like what one would expect to have drawn for equivalent single-phase networks. Only the positive sequence network has sources, because the infinite bus supplies only positive sequence voltage. The zero sequence network is open at the right hand side because of the *delta–wye* transformer connection there.

10.3.4.1 Symmetrical Fault

For a symmetrical (three-phase) fault, only the positive sequence network is involved. The fault shorts the network at its position, so that the current is:

$$\mathbf{i}_1 = \frac{1}{j0.15} = -j6.67\text{per-unit}$$

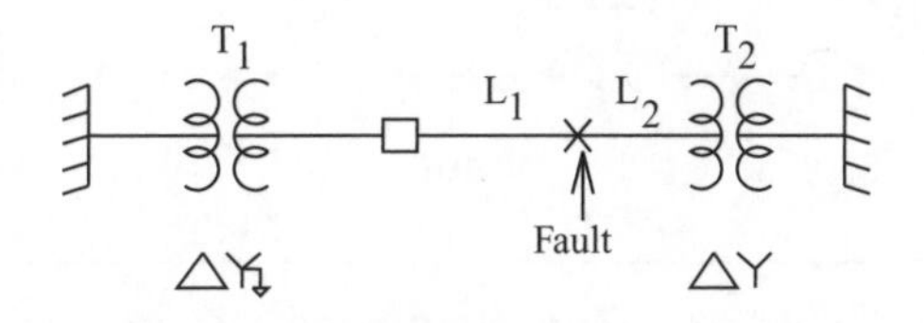

Figure 10.19 One-line diagram for example fault

Table 10.1 Parameters for example fault calculation

System base voltage	138 kV
System base power	100 MVA
Transformer T_1 leakage reactance	0.1 per-unit
Transformer T_2 leakage reactance	0.1 per-unit
Line L_1 positive and negative sequence reactance	$j0.05$ per-unit
Line L_1 zero sequence impedance	$j0.1$ per-unit
Line L_2 positive and negative sequence reactance	$j0.02$ per-unit
Line L_2 zero sequence impedance	$j0.1$ per-unit

10.3.4.2 Single Line–Neutral Fault

For this situation, the three networks are in series and the situation is as shown in Figure 10.21.

The current **i** shown in Figure 10.21 is a *sequence* current, and is given by:

$$\mathbf{i} = \frac{1}{2 \times (j0.15 || j0.12) + j0.2} = -j3.0$$

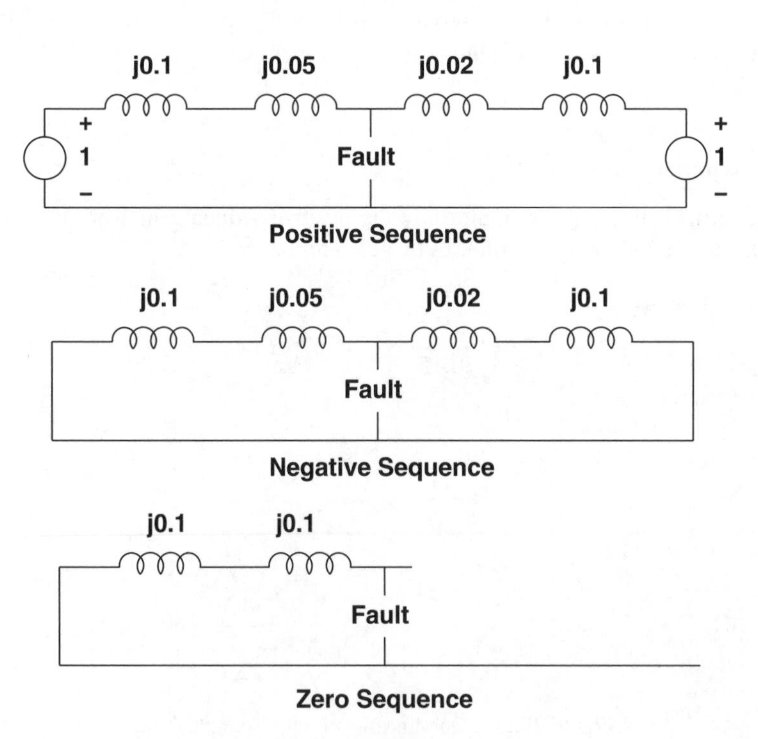

Figure 10.20 Sequence networks

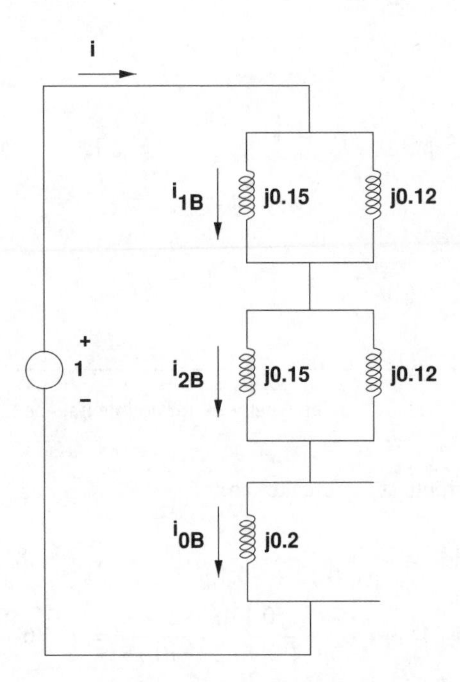

Figure 10.21 Completed network for single line–neutral fault

Then the *sequence currents* at the breaker are:

$$\mathbf{i}_{1\mathrm{B}} = \mathbf{i}_{2B} = \mathbf{i} \times \frac{j0.12}{j0.12 + j0.15} = -j1.33$$

$$\mathbf{i}_{0\mathrm{B}} = \mathbf{i} \quad = -j3.0$$

The phase currents are re-constructed using:

$$\begin{aligned}
\mathbf{i}_{\mathrm{a}} &= \mathbf{i}_{1\mathrm{B}} + \mathbf{i}_{2\mathrm{B}} + \mathbf{i}_{0\mathrm{B}} &= -j5.66 \quad \text{per-unit} \\
\mathbf{i}_{\mathrm{b}} &= \mathbf{a}^2\mathbf{i}_{1\mathrm{B}} + \mathbf{a}\mathbf{i}_{2\mathrm{B}} + \mathbf{i}_{0\mathrm{B}} &= -j1.67 \quad \text{per-unit} \\
\mathbf{i}_{\mathrm{c}} &= \mathbf{a}\mathbf{i}_{1\mathrm{B}} + \mathbf{a}^2\mathbf{i}_{2\mathrm{B}} + \mathbf{i}_{0\mathrm{B}} &= -j1.67 \quad \text{per-unit}
\end{aligned}$$

10.3.4.3 Double Line–Neutral Fault

For the double line–neutral fault, the networks are in parallel, as shown in Figure 10.22.

To start, find the source current **i**:

$$\mathbf{i} = \frac{1}{j(0.15||0.12) + j(0.15||0.12||0.2)} = -j8.57$$

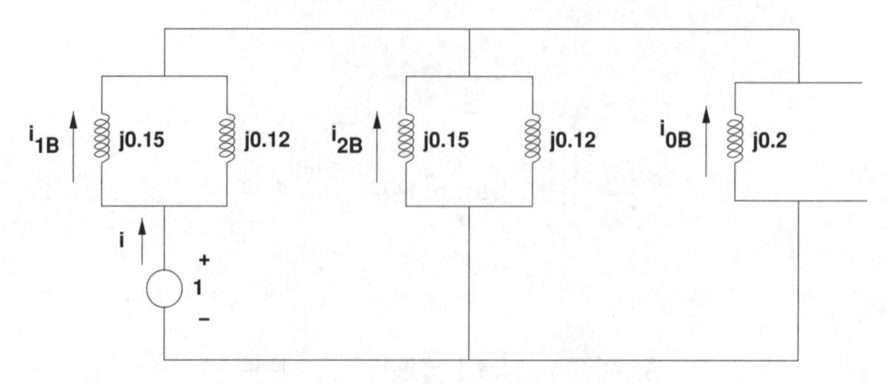

Figure 10.22 Completed network for double line–neutral fault

Then the *sequence* currents at the breaker are:

$$\mathbf{i}_{1B} = \mathbf{i} \times \frac{j0.12}{j0.12 + j0.15} = -j3.81$$

$$\mathbf{i}_{2B} = -\mathbf{i} \times \frac{j0.12||j0.2}{j0.12||j0.2 + j0.15} = j2.86$$

$$\mathbf{i}_{0B} = \mathbf{i} \times \frac{j.12||j0.15}{j0.2 + j0.12||j0.15} = j2.14$$

Reconstructed phase currents are:

$$\mathbf{i}_a = j1.19$$

$$\mathbf{i}_b = \mathbf{i}_{0B} - \frac{1}{2}(\mathbf{i}_{1B} + \mathbf{i}_{2B}) - \frac{\sqrt{3}}{2} j(\mathbf{i}_{1B} - \mathbf{i}_{2B}) = j2.67 - 5.87$$

$$\mathbf{i}_c = \mathbf{i}_{0B} - \frac{1}{2}(\mathbf{i}_{1B} + \mathbf{i}_{2B}) + \frac{\sqrt{3}}{2} j(\mathbf{i}_{1B} - \mathbf{i}_{2B}) = j2.67 + 5.87$$

$$|\mathbf{i}_a| = 1.19 \quad \text{per-unit}$$

$$|\mathbf{i}_b| = 6.43 \quad \text{per-unit}$$

$$|\mathbf{i}_c| = 6.43 \quad \text{per-unit}$$

10.3.4.4 Line-Line Fault

The situation is shown in Figure 10.23

The source current **i** is:

$$\mathbf{i} = \frac{1}{2 \times j(0.15||0.12)} = -j7.50$$

and then:

$$\mathbf{i}_{1B} = -\mathbf{i}_{2B} = i\frac{j0.12}{j0.12 + j0.15} = -j3.33$$

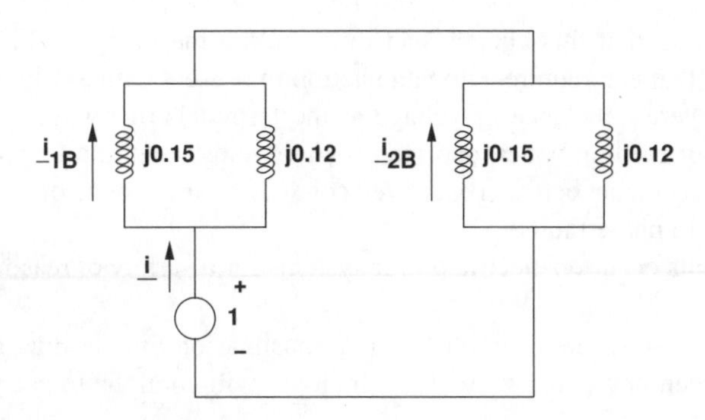

Figure 10.23 Completed network for line–line fault

Phase currents are:

$$\mathbf{i}_a = 0$$

$$\mathbf{i}_b = -\frac{1}{2}(\mathbf{i}_{1B} + \mathbf{i}_{2B}) - j\frac{\sqrt{3}}{2}(\mathbf{i}_{1B} - \mathbf{i}_{2B})$$

$$|\mathbf{i}_b| = 5.77 \quad \text{per-unit}$$

$$|\mathbf{i}_c| = 5.77 \quad \text{per-unit}$$

10.3.4.5 Conversion To Amperes

Base current is:

$$I_B = \frac{P_B}{\sqrt{3}V_{Bl-l}} = 418.4\text{A}$$

Then current amplitudes at the breaker are, in amperes, RMS:

	Phase A	Phase B	Phase C
Three-phase fault	2791	2791	2791
Single line–neutral, ϕ_a	2368	699	699
Double line–neutral, ϕ_b,ϕ_c	498	2690	2690
Line–line, ϕ_b,ϕ_c	0	2414	2414

10.4 System Protection

Electric power systems can deliver quite a lot of power. In certain circumstances such as short circuit faults, a power system could deliver enough current to cause severe damage, not only at the fault site but also to the system itself. Thus the need for system protection, which is the detection of abnormal conditions and automatic action to reduce damage caused by those abnormal conditions. Whole books are written about the various techniques for system protection, so that this chapter can only outline the basics. While there are many system conditions that can require automatic action to prevent damage, including under- and over-frequency, severe unbalance in voltage at the terminals of rotating electric machines, high temperature of cooling water, and many others, most attention is paid to short circuit faults: unintentional contact between energized conductors and ground or with other energized conductors (phase to phase faults).

Short circuit faults occur on electric power systems for a variety of reasons.

1. Lightning can cause severe overvoltage on a transmission line, leading to an arc between phases or between phase and ground. Such an arc will continue to carry current ('burn') until the line is isolated.
2. Other types of weather events such as ice storms can generate loads on transmission structures that can lead to energized conductors falling on the ground.
3. Contact between energized conductors and trees produces ground faults.
4. Animal or bird contact with energized conductors can and does cause short circuit faults.
5. Accidents, often involving motor vehicles, can sometimes cause conductors to fall on the ground.
6. Failure of system components can cause short circuits.

In the case of short circuits, the current must be interrupted. For 'temporary' faults such as those caused by lightning or some kinds of animal contact, the system can be re-energized quickly. For other, 'permanent' faults, the source of the fault must be removed before the part of the system can be re-energized.

10.4.1 Fuses

The simplest system protection device is a fuse. This is a (generally short) piece of metal with a low melting point that is sized so that a current larger than some threshold will cause it to melt and open the circuit. Fuses generally have characteristics similar to the diagram of Figure 10.24. The fuse will melt if current is above some threshold (the current rating of the fuse). The fuse will melt (or 'blow') faster for higher currents, as shown in the figure.

Fuses have historically been used in consumer products to protect against overloads and short circuits. At one time they were widely used as protection in residences, but are not now used for two reasons:

1. Fuses do not give any protection for personal contact with energized conductors (shock), and that is now provided with ground fault interrupting circuit breakers.
2. A blown fuse is a nuisance, and sometimes, particularly when a replacement fuse is not available, a careless individual would defeat the purpose of the fuse by wiring around it or some other measure. (In the United States, a 1 cent piece was sometimes used to bypass the commonly used branch circuit fuse. This was called 'a penny in the socket'. This was a dangerous practice that caused a number of fires.)

For these reasons, circuit breakers are now used to protect branch circuits. However, they are still used within appliances and in some distribution circuits.

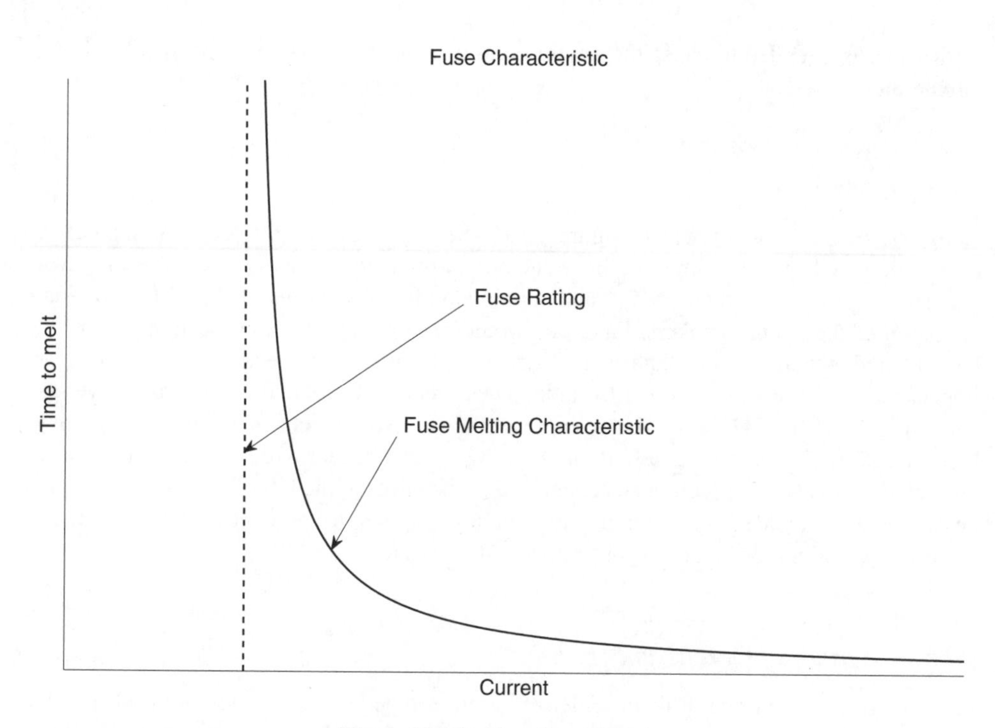

Figure 10.24 Fuse characteristic

10.5 Switches

In the context of electric power systems, a switch is a mechanical device that can control the flow of current in a circuit. A switch is 'closed' when it allows current to flow and 'open' when current cannot flow through it. Switches come in a variety of capabilities. Some switches, called disconnect switches, are used to isolate sections of a system but cannot interrupt load currents. Some switches, called load break switches, can interrupt load currents but not fault currents, and circuit breakers are generally rated to handle fault currents (that is, currents due to short circuits). All switches, including circuit breakers, have limits on how much current they can safely interrupt, and applications engineers must be careful to ensure that circuit breakers are capable of interrupting the currents resulting from the most severe faults that can occur in a given situation. Failure of any switch to interrupt current is generally spectacular and sometimes has severe consequences.

A circuit breaker is a switch that has the capacity to interrupt load and fault currents. Circuit breakers are usually controlled by systems that include 'relays', or sensing devices that can sense overcurrents or other dangerous conditions.

In low-voltage branch circuit application, circuit breakers have built in current sensors that operate much as do fuses: they have a characteristic for time to 'trip' that is inversely related to the overcurrent. This type of circuit breaker, with built in sensors, is used in many types of low voltage (up to about 1 kV) systems with rated currents into the range of a few thousand amperes. Sometimes the time inverse characteristic is modified with an 'instantaneous' trip setting that is intended to cause the circuit breaker to open quickly on sensing a hard short circuit fault.

10.6 Coordination

Consider the situation shown in Figure 10.25. Several load buses (loads not shown) are connected together by transmission lines and circuit breakers. If this system is fed only from the left, the system is called 'radial' and certain simple measures can be used to coordinate operation of the circuit breakers. First, one would expect that a fault at location 'F3' would have a smaller fault current than would a fault at location 'F1' because there is a higher impedance from the source at that location. Thus breaker 'B5' should be set to a lower trip value than breakers 'B3' and 'B1'. Because of the inverse time characteristic of the various breakers, if 'B5' fails to clear a fault at 'F3', the other breakers will back it up, but if 'B5' does manage to clear the fault, loads connected to the buses to the left of that breaker will not be disconnected (aside from the momentary voltage dip due to the fault itself). This kind of protection coordination can be accomplished with fuses too.

10.6.1 Ground Overcurrent

Faults to ground are often difficult to detect because impedance of the ground itself is often fairly high, resulting in fault current that may not be very high (but still damaging). By adding together all of the phase currents in the relay element, one can find the ground current. Generally in a polyphase system in normal operation, ground current will be substantially smaller than any one phase current because polyphase loads will be nearly balanced and normal system design measures will attempt to balance single phase loads. Thus a ground overcurrent relay, operating with the sum of all phase currents, can be set to be quite a bit more sensitive than the individual phase relays (Figure 10.26).

10.7 Impedance Relays

A more sophisticated type of relay is built to use both voltage and current to measure apparent impedance. The simplest characteristic of one of these is shown in Figure 10.27. This type of relay senses the apparent impedance (ratio of voltage to current) and, if that impedance has a magnitude that is smaller than a given value (that is, the impedance $Z = R + jX$ is within a circle), the relay indicates a 'trip' condition. If the impedance is outside that circle it 'blocks' or does not trip. This type of relay can be used with a timer, and so accomplish relay coordination. For example, in the radial distribution system of

Figure 10.25 Radial arrangement

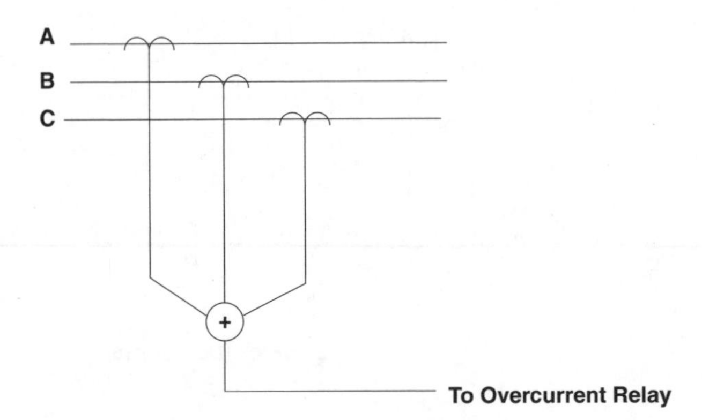

Figure 10.26 Ground overcurrent relay

Figure 10.25, timer settings would be made progressively shorter for breakers further from the source. Thus breakers to the left can serve as backup for breakers to the right but won't trip un-necessarily.

10.7.1 Directional Elements

Transmission lines are inductive, so an impedance relay can sense if the fault is 'in front of' or 'behind' by sensing the sign of the reactive part of the apparent reactance, as shown in Figure 10.28. This can be matched up with an impedance magnitude relay to produce a characteristic as shown in Figure 10.29. If a fault occurs on a transmission line, the impedance to the fault will be the transmission line impedance per unit length times the distance to the fault. The further the fault, the larger the impedance. With the directional element, a relay can be made to ignore faults that occur 'behind' it (although other relays will be set to see those faults).

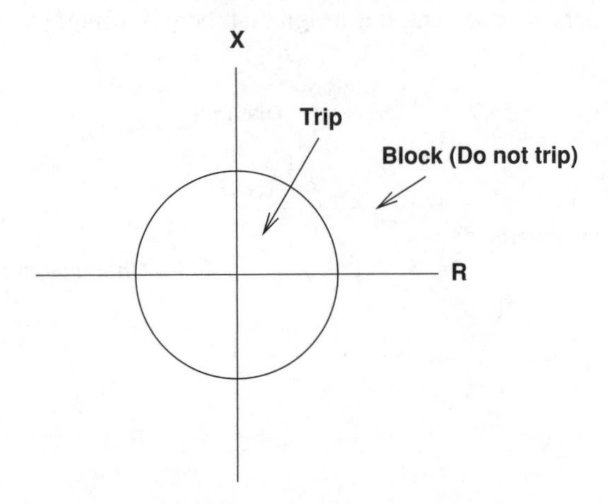

Figure 10.27 Impedance relay characteristic

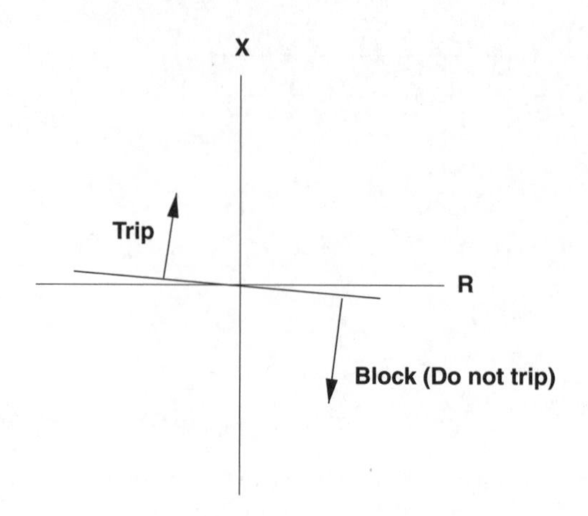

Figure 10.28 Directional element

Multiple relays, set to different distances, can be set with multiple timers to achieve breaker coordination. A relay set to fairly high impedance would be set up with a large time delay to serve as backup to a relay down the line that would be set with a lower impedance and a lower delay. A system of relays like this could be built in both directions, with the directional elements achieving coordination in both directions.

10.8 Differential Relays

Distance relays are difficult, if not impossible, to coordinate with short distances and very low impedances such as one would have in a substation, and for this purpose, differential relays are used. The idea of a differential relay is straightforward: the sum of currents into a node should be zero. If the sum of currents is not zero, this means that current is going somewhere it should not. Figure 10.30 shows the simplest version of this. If the currents measured by the two current transformers are different, it is a sign that there is a short on the bus in between.

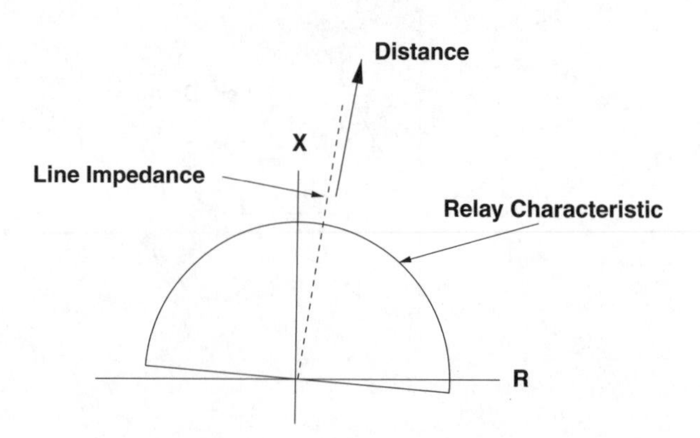

Figure 10.29 Impedance relay with directional element

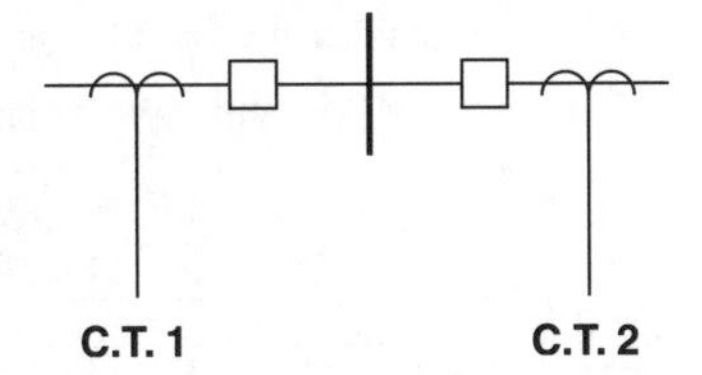

Figure 10.30 Differential relay arrangement

To make a differential relay system sensitive over a wide range of load current, a restraint is often imposed, based on the magnitude of the current carried. Thus a larger difference in measured currents is required for a 'trip' with high currents than with low currents. The characteristic for a simple two-current differential is shown in Figure 10.31. Note that this principle can be used with an arbitrary number of current measurements.

Differential relay protection can be used for transformers by using current transformers with appropriate current ratios to match primary and secondary currents. They are also used for protection of generators, matching currents at the ends of the armature windings and thus protecting against ground faults.

10.8.1 Ground Fault Protection for Personnel

A kind of differential relay that is widely used in the United States for personnel protection in low voltage branch circuits is shown in Figure 10.32. Both wires in a single-phase circuit are passed through a magnetic core (generally a toroid), with a multiple turn sensing coil. Generally, if the currents 'going to' and 'coming from' a load are equal, there will be zero net current through the core. However, if any current leaves the circuit (as in if a person contacts the 'hot' wire and is also in contact with ground), the current through the core will become unbalanced and the sensing coil will develop a non-zero voltage. These devices can be made quite sensitive: on the order of the threshold for feeling a current (on the order of 0.5 mA). This kind of integrated sensor and circuit breaker is now required to be

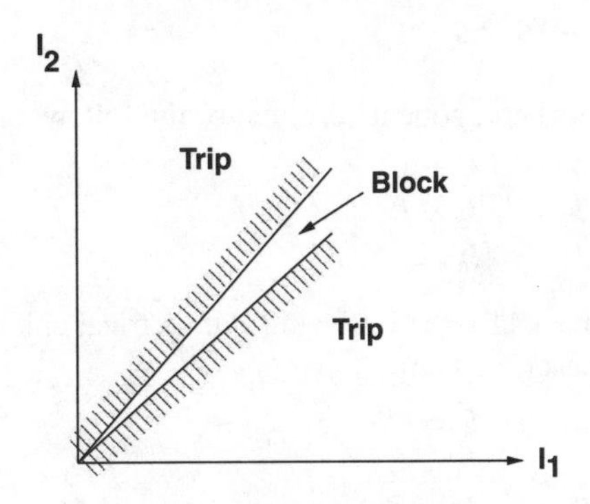

Figure 10.31 Characteristic of a differential relay

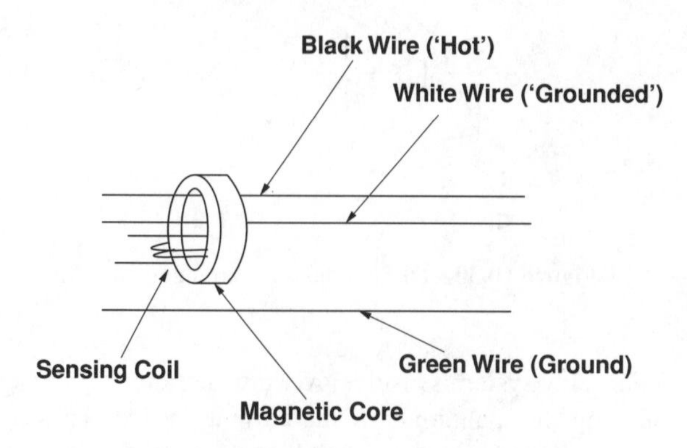

Figure 10.32 Basic ground fault interruptor

used in bathrooms and outdoor branch circuit locations where the shock hazard to people is particularly acute.

10.9 Zones of System Protection

System protection typically involves several different types of protective relays and a number of 'zones', geographic parts of the system. Transformers and busbars as well as generators are protected by differential relays. Transmission lines are protected by impedance ('distance') relays. Distribution lines, which typically are radial in nature, tend to be protected by overcurrent relays or, sometimes, fuses. In industrial settings, equipment is often protected by circuit breakers integrated with overcurrent and ground overcurrent relays. Generally the zones of protection overlap, with circuit breakers in the overlap zone, so that a failure that takes place within the circuit breaker will be interrupted by both zones. Backup is usually provided for breaker failure through coordination between zones so that the backup breaker (in another zone) takes a bit longer to sense and interrupt a fault than would the primary breaker for the zone in which the fault is sensed.

10.10 Problems

1. What are the symmetrical component currents for the following phase current sets:

   (a) $I_a = I, \quad I_b = 0, \quad I_c = 0$
   (b) $I_a = 0, \quad I_b = I, \quad I_c = 0$
   (c) $I_a = 0, \quad I_b = I, \quad I_c = -I$

2. What are the phase currents for the following symmetrical current sets? You should draw phasor diagrams that describe them:

   (a) $I_1 = I, \quad I_2 = 0, \quad I_0 = 0$
   (b) $I_1 = 0, \quad I_2 = I, \quad I_0 = 0$
   (c) $I_1 = 0, \quad I_2 = 0, \quad I_0 = I$

3. In the circuit of Figure 10.33, the resistor is $R = 10\ \Omega$ and the source is balanced, 480 V, RMS, line–line. What are the symmetrical component currents drawn from the source?

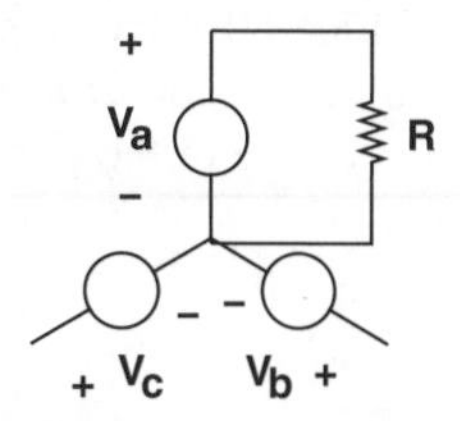

Figure 10.33 Phase a is loaded

4. In the circuit of Figure 10.34, the resistor is $R = 10\ \Omega$ and the source is balanced, 480 V, RMS, line–line. What are the symmetrical component currents drawn from the source?

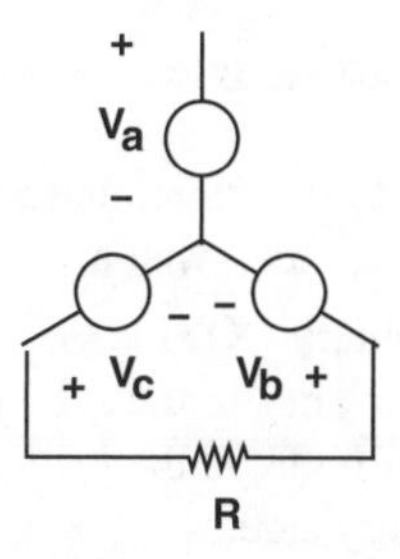

Figure 10.34 Circuit loaded phases b to c

5. In the circuit of Figure 10.35, a 2400:480 V transformer connected in *delta–wye* is loaded on the low-voltage side by a 10 Ω resistor from the phase A terminal to the neutral. In this connection, the 'X' side of the transformer voltages *lead* the voltages on the primary side.

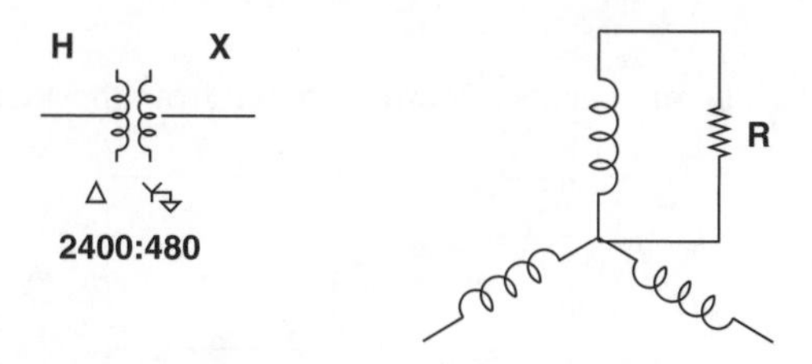

Figure 10.35 Phase a is loaded on the secondary of a *delta–wye* transformer

Ignoring any impedance in the transformer, and assuming the high voltage side is fed by a balanced three-phase voltage source, what are these currents:

(a) Phase currents on the secondary (X) side.

(b) Symmetrical component currents on the secondary side.
(c) Symmetrical component currents on the primary side.
(d) Phase currents on the primary side.

6. In the circuit of Figure 10.36, a 2400:480 V transformer connected in *delta–wye* is loaded on the low voltage side by a 10 Ω resistor from the phase B terminal to the phase C terminal. In this connection, the 'X' side of the transformer voltages *lead* the voltages on the primary side.

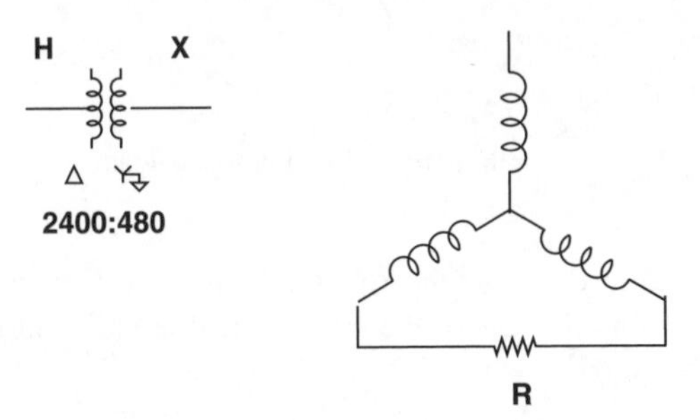

Figure 10.36 Resistor loads phases B to C on the secondary of a *delta–wye* transformer

Ignoring any impedance in the transformer, and assuming the high voltage side is fed by a balanced three-phase voltage source, what are these currents:

(a) Phase currents on the secondary (X) side.
(b) Symmetrical component currents on the secondary side.
(c) Symmetrical component currents on the primary side.
(d) Phase currents on the primary side.

7. In the circuit of Figure 10.37, the source is balanced, 480 V, line–line, RMS. The resistors are:

$$R_a = 10\ \Omega$$
$$R_b = 12\ \Omega$$
$$R_c = 12\ \Omega$$

Find the symmetrical components of current drawn from the source.

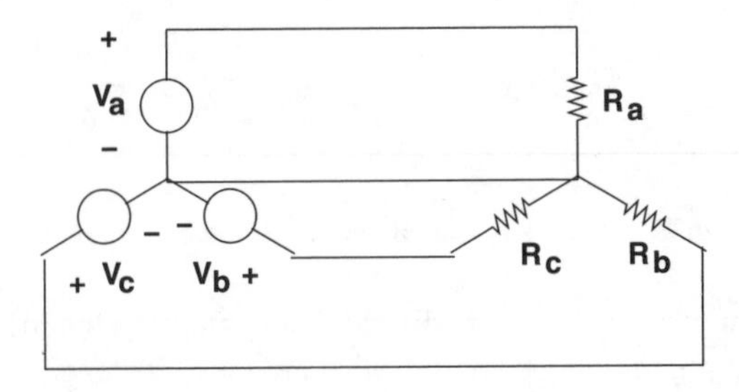

Figure 10.37 Three-phase source connected to three-phase resistor

8. In the circuit of Figure 10.38, the source is balanced, 480 V, line–line, RMS. The resistors are:

$$R_a = 10\ \Omega$$
$$R_b = 12\ \Omega$$
$$R_c = 12\ \Omega$$

Note the star point of the resistors is not grounded.

(a) Find the complex amplitude of the voltage between the star point of the resistors and the center point of the voltage source.
(b) Now find currents in the resistors.
(c) What are the symmetrical component currents drawn from the source?

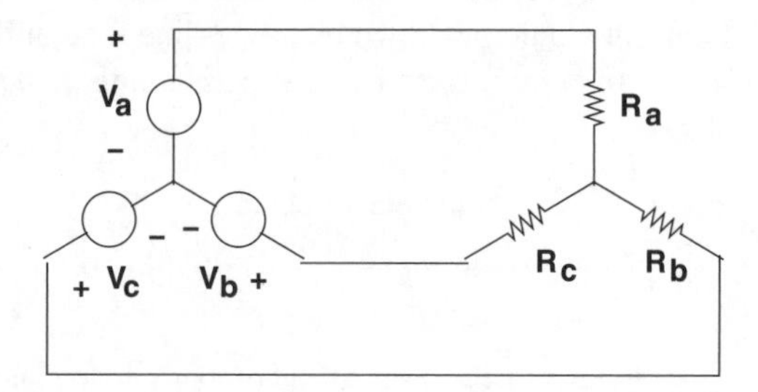

Figure 10.38 Three-phase source connected to ungrounded three-phase resistor

9. Referring back to the circuit of Figures 10.30 and 10.31, the resistors are all the same, making a balanced load ($R_a = R_b = R_c = 10\ \Omega$), but the source is unbalanced. Working in RMS amplitudes:

$$\mathbf{V}_a = 280$$
$$\mathbf{V}_b = 277e^{-j\frac{2\pi}{3}}$$
$$\mathbf{V}_c = 277e^{j\frac{2\pi}{3}}$$

what are the symmetrical component currents if the neutral point of the resistors is:

(a) Grounded (as in Figure 10.30)
(b) Ungrounded (as in Figure 10.31)

10. This problem is shown in Figure 10.39 and concerns an untransposed transmission line. This line has different mutual reactances between phases. Self reactance of each phase is $X_a = 20\ \Omega$. Reactance between the center phase (b) and the two outer phases is: $X_{ab} = X_{bc} = 8\ \Omega$. Reactance between the two outer phases is $X_{ac} = 5\ \Omega$. This is a 115 kV, line-line, RMS line. The objective of this problem is to find the negative sequence currents provoked by the line unbalance.

(a) First, find the symmetrical component impedance matrix for the line.
(b) Assuming the sources at each end of the line are both balanced voltage sources, find the phase angle required for the line to be carrying 100 MW. For the purposes of this problem, ignore line resistance.

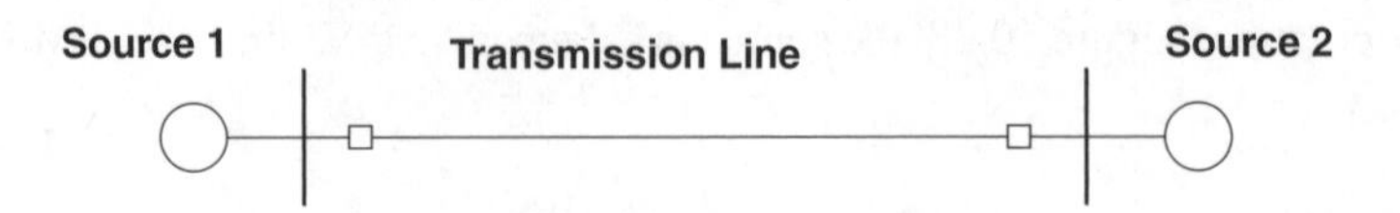

Figure 10.39 Transmission line between two sources

(c) Now using the voltage difference from the previous part, which is purely positive sequence, find negative sequence current in the line.

11. A fault situation is shown in Figure 10.40. Assume the transmission line has reactances $x_1 = x_2 = 0.25$ per-unit and $x_0 = 0.4$ per-unit. Assume the source is a voltage source with low impedance and that the source is grounded at its center point.

(a) What per-unit current would be drawn by a line–neutral fault at the position indicated?
(b) What per-unit current would be drawn by a line–line fault at the position indicated?
(c) Translate these per-unit currents into amperes assuming a 100 MVA, 138 kV base.

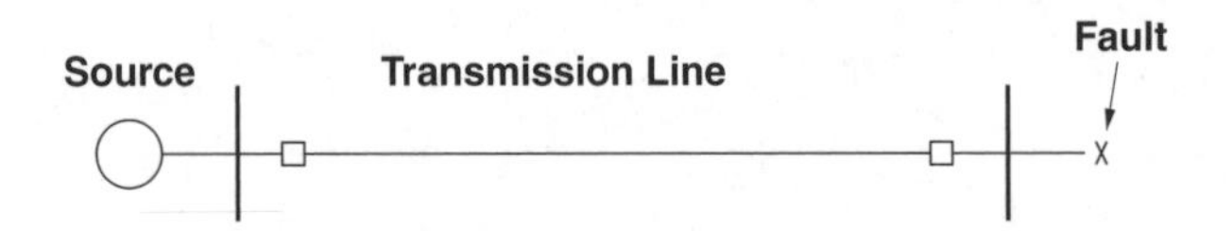

Figure 10.40 Fault at end of transmission line

12. A fault situation is shown in Figure 10.41. Assume the transmission line has reactances $x_1 = x_2 = 0.25$ per-unit and $x_0 = 0.4$ per-unit. The source is a generator with reactances to the fault that are $x_1 = x_2 = 0.25$ per-unit. The transformer can be assumed to present a reactance of $x_t = 0.05$ per-unit. Note the transformer is delta-wye connected, with the wye side being grounded. For both line-ground and line-line faults:

(a) Find current at the fault, in per-unit.
(b) Find current in the generator leads, also in per-unit.
(c) Now, assuming the generator and transmission line were both normalized to a base of 100 MVA, with the generator base voltage being 13.8 kV, line-line, RMS and the transmission line base voltage being 138 kV, what are the currents derived in the previous questions, in amperes?

13. Using the same fault situation as shown in Figure 10.41, here is some data on the system: The generator is rated at 600 MVA, 24 kV, line-line, RMS. Its reactances are, on its base, $x_1 = x_2 = 0.25$ per-unit. The transformer is rated at 600 MVA, 24 kV:345 kV,

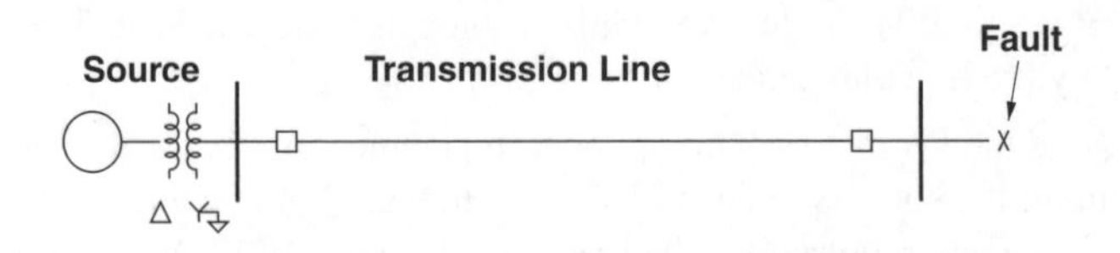

Figure 10.41 Fault at end of transmission line

and on that base it's reactance is 0.07 per-unit. The transmission line has reactances of $x_1 = x_2 = 0.25,\ x_0 = 0.4$ per-unit on a base of 100 MVA, 345 kV. For line–neutral and line–line faults, find:

(a) Per-unit currents at the fault,
(b) Per-unit currents in the machine leads,
(c) Currents in amperes at the fault,
(d) Currents in amperes in the machine leads.

14. A fault situation is shown in Figure 10.42. Assume the transmission line has reactances $x_1 = x_2 = 0.25$ per-unit and $x_0 = 0.4$ per-unit. The source is a generator with reactances to the fault that are $x_1 = x_2 = 0.25$ per-unit. The unit transformer can be assumed to present a reactance of $x_t = 0.05$ per-unit. Note the transformer is *delta–wye* connected, with the wye side being u*n*grounded. The transformer on the right, also delta–wye, but with the wye side grounded, is rated at 50 MVA, 345 kV:34.5 kV, and it has a reactance of 0.03 per-unit on its base.
For both line–neutral and line–line faults:

(a) Find current at the fault, in per-unit.
(b) Find current in the transmission line, in per-unit.
(c) Find current in the generator leads, also in per-unit.
(d) Now, assuming the generator and transmission line were both normalized to a base of 100 MVA, with the generator base voltage being 13.8 kV, line-line, RMS and the transmission line base voltage being 138 kV, what are the currents derived in the previous questions, in amperes?

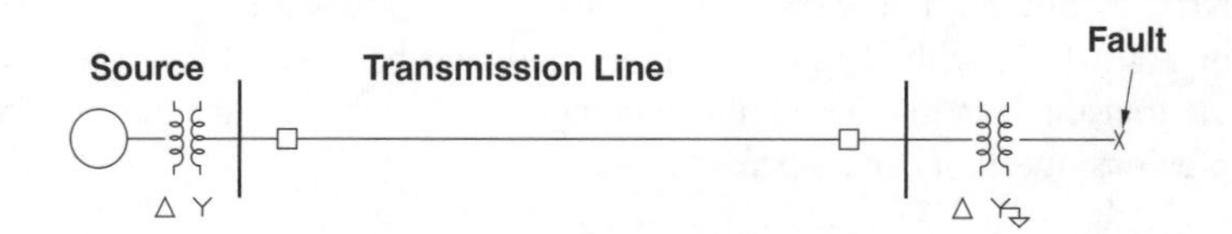

Figure 10.42 Fault at end of transmission line

15. The fault situation is shown in Figure 10.42. The transmission line impedance is, on a base of 100 MVA, 345 kV:

$$z_1 = z_2 = j0.25 + 0.02 \text{ per-unit}$$
$$z_0 = j0.4 + 0.2 \text{ per-unit}$$

The source is a generator with reactances to the fault that are $x_1 = x_2 = 0.25$ per-unit. The unit transformer can be assumed to present a reactance of $x_t = 0.05$ per-unit. Note the transformer is *delta–wye* connected, with the wye side being u*n*grounded. The transformer on the right, also *delta–wye*, but with the wye side grounded, is rated at 50 MVA, 345 kV:34.5 kV, and it has a reactance of 0.03 per-unit on its base.
For line–neutral and line–line faults, find:

(a) Per-unit currents at the fault,
(b) Per-unit currents in the transmission line,

(c) Per-unit currents in the machine leads,
(d) Currents in amperes at the fault,
(e) Currents in amperes in the transmission line,
(f) Currents in amperes in the machine leads.

16. A fault situation is shown in Figure 10.43. Assume the transmission line has reactances $x_1 = x_2 = 0.25$ per-unit and $x_0 = 0.4$ per-unit. The source is a generator with reactances to the fault that are $x_1 = x_2 = 0.25$ per-unit. The unit transformer can be assumed to present a reactance of $x_t = 0.05$ per-unit. Note the transformer is *delta–wye* connected, with the wye side being grounded. The transformer on the right, also *delta–wye*, but with the wye side grounded, is rated at 50 MVA, 345 kV:34.5 kV, and it has a reactance of 0.03 per-unit on its base. The fault occurs exactly at the midpoint of the transmission line.

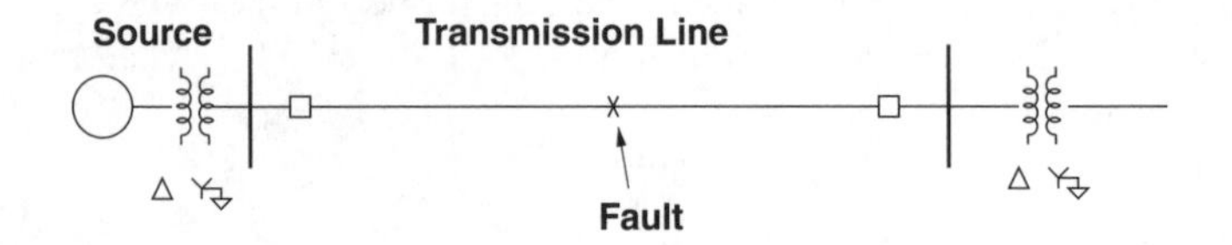

Figure 10.43 Fault in the middle of a transmission line

For both line–neutral and line–line faults:

(a) Find current at the fault, in per-unit.
(b) Find current in the transmission line, in per-unit.
(c) Find current in the generator leads, also in per-unit.
(d) Now, assuming the generator and transmission line were both normalized to a base of 100 MVA, with the generator base voltage being 13.8 kV, line–line, RMS and the transmission line base voltage being 138 kV, what are the currents derived in the previous questions, in amperes?

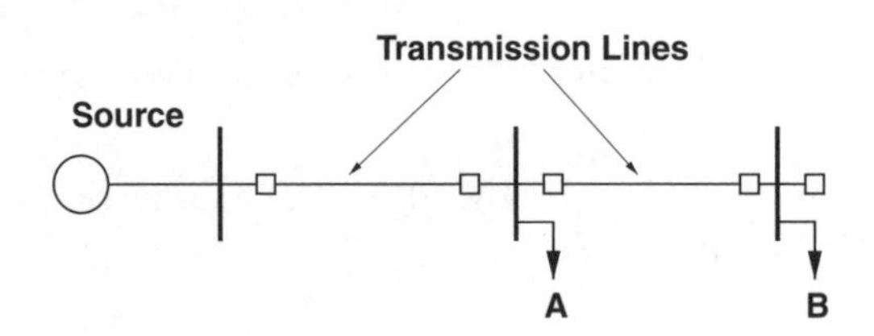

Figure 10.44 Two fault locations

Faults might occur at two possible locations, and the objective is to calculate faults so that relays can be properly coordinated. The source can be assumed to be a voltage source of unity with a reactive impedance of 0.3 per-unit. Assume the source is solidly grounded. There are two transmission line segments, each with the same impedances: $x_1 = x_2 = 0.25$ per-unit and $x_0 = 0.4$ per-unit. The two possible fault locations are marked as 'A' and 'B'. For line–line and line–neutral faults at these two locations, calculate the per-unit currents at the source location.

11

Load Flow

The purpose of an electric power system is to deliver, as the name implies, electric power to a consumer. That electric power is made by a producer and carried through power system components, generally over geographic distance from the producer to the consumer. Electric power systems very often have numerous paths over which power can flow, and the term *load flow* refers to techniques to understand how load flows and over what pathways.

Load flow techniques have been developed so that currents flowing in conductors may be evaluated and so that voltage distributions over the network can be determined. Load flow evaluations are required in advance of any system expansion to ensure that the expanded system will work adequately. And load flow calculations are required as part of generation dispatch to ensure that lines are not overloaded. In fact, optimal dispatch (use of generation resources) must take into account transmission and distribution system losses, which are also determined by load flow calculations. For these reasons, it is necessary to understand how electric power flows through the wires, transformers and other components that make up the electric power system.

Even though electric power networks are composed of components which are (or can be approximated to be) *linear*, electric power flow, real and reactive, is a *non-linear* quantity. The calculation of load flow in a network is the solution to a set of nonlinear equations.

There is still quite a bit of activity in the professional literature concerning load flow algorithms. The reason for this is that electric utility networks are often quite large, having thousands of buses, so that the amount of computational effort required for a solution is substantial. A lot of effort goes into doing the calculation efficiently. This discussion, and the short computer program at the end of this chapter, uses possibly the crudest possible algorithm for this purpose. It does, however, serve to illustrate the problem.

11.1 Two Ports and Lines

A transmission line, as has been shown, may be modeled as a two-port network, with voltages and currents defined at each end of the line. The series impedance of the line is, typically, dominated by inductance with some small series resistance. There is also some shunt admittance

Electric Power Principles: Sources, Conversion, Distribution and Use James L. Kirtley

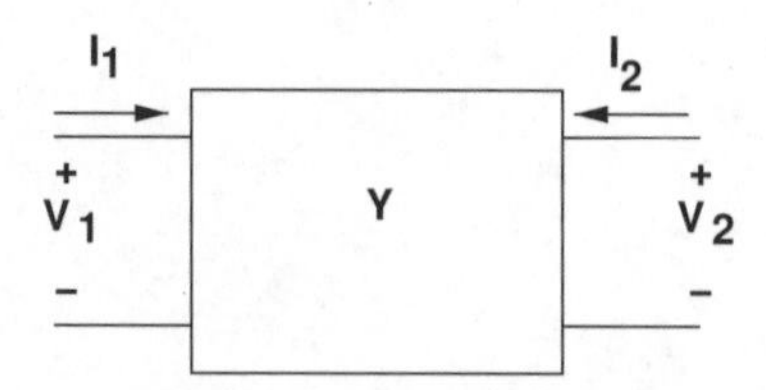

Figure 11.1 Two-port network

that is capacitive. It is convenient to characterize the transmission line as a two-port network that is described by its admittance parameters.

Consider the two-port network shown in Figure 11.1. The admittance parameters determine currents:

$$\mathbf{I}_1 = \mathbf{Y}_{11}\mathbf{V}_1 + \mathbf{Y}_{12}\mathbf{V}_2$$
$$\mathbf{I}_2 = \mathbf{Y}_{21}\mathbf{V}_1 + \mathbf{Y}_{22}\mathbf{V}_2$$

If this two-port network is constrained by voltage sources at both sets of terminals, *or* if the voltages are known, power flow at each set of terminals is:

$$P_1 + jQ_1 = \mathbf{V}_1\mathbf{I}_1^* = \mathbf{Y}_{11}^*|\mathbf{V}_1|^2 + \mathbf{Y}_{12}^*\mathbf{V}_1\mathbf{V}_2^*$$
$$P_2 + jQ_2 = \mathbf{V}_2\mathbf{I}_2^* = \mathbf{Y}_{21}^*\mathbf{V}_2\mathbf{V}_1^* + \mathbf{Y}_{22}^*|\mathbf{V}_2|^2$$

Figure 11.2 is a simple transmission line model that includes line reactance $X = \omega(L - M)$ and series resistance R as well as shunt capacitive reactance $X_c = -\frac{1}{\omega C}$. For this line model, the admittance parameters are:

$$\mathbf{Y}_{11} = \frac{1}{-jX_c} + \frac{1}{jX + R} \tag{11.1}$$

$$\mathbf{Y}_{12} = -\frac{1}{jX + R} \tag{11.2}$$

$$\mathbf{Y}_{21} = -\frac{1}{jX + R} \tag{11.3}$$

$$\mathbf{Y}_{22} = \frac{1}{-jX_c} + \frac{1}{jX + R} \tag{11.4}$$

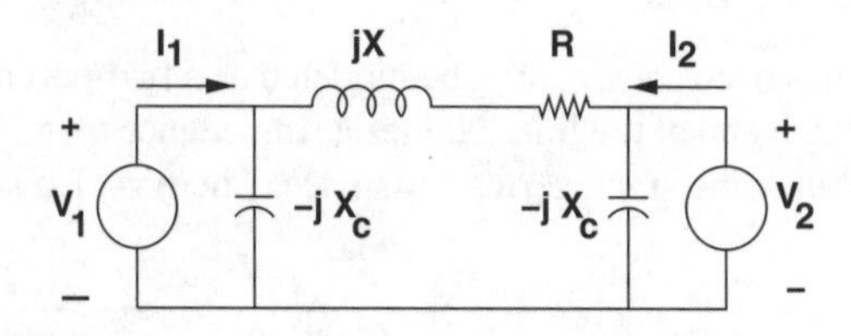

Figure 11.2 Transmission line modelled as a two-port

11.1.1 Power Circles

If, as is common with transmission lines, the voltages at the two ends of a line are characterized by magnitude and relative phase angle:

$$\mathbf{V}_1 = V_1 e^{j\delta}$$
$$\mathbf{V}_2 = V_2$$

The power input to the line at each end is described by a *circle* in the P-Q plane with the relative phase angle δ as a parameter. The centers of those circles are, for the line model described by Equations 11.1 through 11.4, the center of the sending-end power circle is:

$$(P + jQ)_{\text{center}} = |V_1|^2 \mathbf{Y}_{11}^*$$

and the radius of that circle is:

$$|V_1||V_2||\mathbf{Y}_{12}|$$

The receiving end is similar, but often the sign convention used for the receiving end complex power is to consider that positive power at the receiving end is coming *out* of the terminals. In this case, the center of the receiving end circle is in the third quadrant (negative real and reactive power), and so the center of the receiving end circle is:

$$(PY + jQ)_{\text{center}} = -|V_2|^2 \mathbf{Y}_{22}^*$$

and the radius of that circle is:

$$|V_1||V_2||\mathbf{Y}_{21}|$$

Note that the vectors that describe the circle would indicate zero real power flow through the line if the angle δ is zero. The sending end vector rotates in the positive angular direction (counter-clockwise) and the receiving end vector rotates in the other direction. Figure 11.3 shows how one would construct sending and receiving end power circles in the situation of non-zero line series resistance but zero shunt admittance and with the magnitudes of sending and receiving voltage being the same. Note with zero phase angle, no current flows in the line and both real and reactive power are zero at both ends. Figure 11.4 shows the same construction, but in this figure there is non-zero shunt capacitance. With zero phase angle there is zero real power flow into both ends of the line (assuming voltage magnitudes are the same), but the shunt capacitances supply reactive power, offsetting the power circles.

One lesson to be gleaned from exercising power circles is that, through a line with dominantly reactive impedance (the norm), real power flow responds primarily to phase angle while reactive power flow responds primarily to voltage differences between sending and receiving ends. This fact is important in line operation and power systems engineers often take advantage of it in making large load flow problems more efficient.

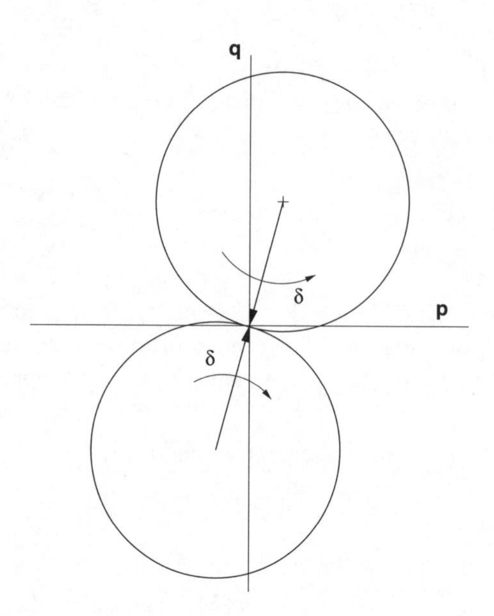

Figure 11.3 Construction of power circle with zero shunt elements

11.2 Load Flow In A Network

If all voltage magnitudes and phase angles are known, calculation of real and reactive power flow in a network is simple, since each line can be treated as a two-port with known admittance parameters. Usually, however, what is known about the network is power flows at the terminals

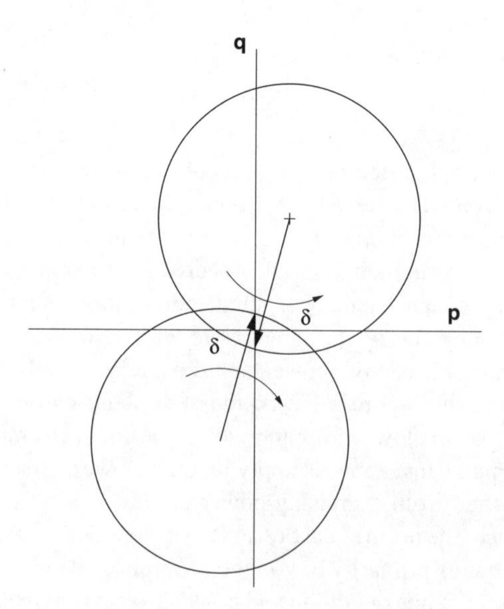

Figure 11.4 Construction of power circle with capacitive shunt elements

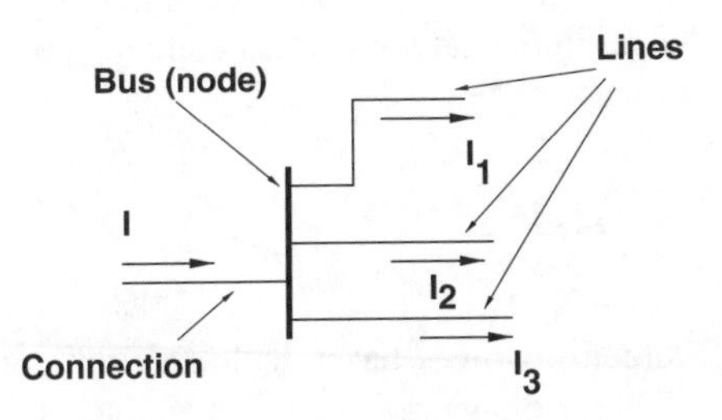

Figure 11.5 Network node is a connection of lines

of the network and, sometimes, voltage magnitudes at those terminals. Voltage magnitudes and angles at internal nodes of the network are often not controlled or known a priori.

The electric power network consists of lines (wires) with admittance parameters (inductive and resistive for transmission lines, capacitive for shunt elements), connected together at buses. Those buses are also the terminals of the network, so that currents can flow from or to the network at those buses. Power plants, loads, reactive power sources are all connected to the network at buses, often referred to as nodes. A node is illustrated in Figure 11.5. Current into the network at this node is $\mathbf{I} = \mathbf{I}_1 + \mathbf{I}_2 + \mathbf{I}_3$. Of course it is possible that there is no external connection to a node, in which case the sum of line currents into that node must be zero.

Power flow in a network is determined by the voltage at each bus of the network and the impedances of the lines between buses. Power flow into and out of each of the buses that are network terminals is the sum of power flows of all of the lines connected to that bus. The load flow problem consists of finding the set of voltages: magnitude and angle, that, together with the network admittances, produces the load flows that are known to be correct at the system terminals. To start, the power system is represented as a collection of *buses*, connected together by *lines*. At each of the buses, referred to as *nodes*, is connected equipment that will supply or remove real and reactive power.

Complex power flow *into* the network at node k is:

$$\mathbf{S}_k = P_k + jQ_k = \mathbf{V}_k \mathbf{I}_k^*$$

Network currents are given by:

$$\mathbf{I} = \mathbf{Y}\,\mathbf{V}$$

Where $\mathbf{I}$ is the vector of bus currents (that is, those currents *entering* the network at its buses. $\mathbf{V}$ represents the bus voltages and $\mathbf{Y}$ is the *bus admittance matrix*.

A formal method for estimating the bus admittance matrix appears in Section 11.4. For the moment, note that an individual bus current is given by:

$$\mathbf{I}_k = \sum_{j=1}^{N} \mathbf{Y}_{jk} \mathbf{V}_j$$

where N is the number of buses in the network. Then complex power flow at a node is:

$$\mathbf{S}_k = \mathbf{V}_k \sum_{j=1}^{N} \mathbf{Y}_{jk}^* \mathbf{V}_j^* \tag{11.5}$$

The typical load flow problem involves buses with different constraints. It is possible to specify six quantities at each bus: voltage magnitude and angle, current magnitude and angle, real and reactive power. These are, of course, inter-related so that any two of these are specified by the other four, and the network itself provides two more constraints. Thus it is necessary to, in setting up a load flow problem, specify two of these six quantities. Typical combinations are:

Generator bus: Real power and terminal voltage magnitude are specified.

Load bus: Real and reactive power are specified.

Fixed impedance: A fixed, linear impedance connected to a bus constrains the relationship between voltage and current. Because it constrains both magnitude and angle, such an impedance constitutes two constraints. A fixed linear impedance can usually be represented in the same fashion as a line, but connected to only the one node, with the other end at ground.

Infinite bus: This is a voltage source, of constant magnitude and phase angle.

The load flow problem consists of solving Equation 11.5 as constrained by the terminal relationships.

One bus in a load flow problem is assigned to be the 'slack bus' or 'swing bus' (the terms are synonymous). This bus, which is taken to be an 'infinite bus', since it does not have real nor reactive power constrained, accommodates real power dissipated and reactive power consumed or produced by network lines. This bus is necessary because these losses are not known *a priori*. Further, one phase angle needs to be specified, to serve as an origin for all of the others.

11.3 Gauss–Seidel Iterative Technique

This is one of many techniques for solving the non-linear load flow problem. It should be pointed out that this solution technique, while straightforward to use and easy to understand, has a tendency to use a lot of computation, particularly in working large problems. It is also quite capable of converging on incorrect solutions (that is a problem with non-linear systems). As with other iterative techniques, it is often difficult to tell when the correct solution has been reached. Despite these shortcomings, Gauss–Seidel can be used to get a good feel for load flow problems without excessive numerical analysis baggage.

Suppose an initial estimate (guess) for network voltages exists. Equation 11.5 may be partitioned as:

$$\mathbf{S}_k = \mathbf{V}_k \sum_{j \neq k} \mathbf{Y}_{jk}^* \mathbf{V}_j^* + \mathbf{V}_k \mathbf{Y}_{kk}^* \mathbf{V}_k^* \tag{11.6}$$

Noting that $\mathbf{S}_k = P_k + jQ_k$, is the complex power flow into bus k, voltage at that bus $\mathbf{V}_k^*$ can be solved for and, taking the complex conjugate of that, an expression for $\mathbf{V}_k$ is derived in terms of all of the voltages of the network, and the real and reactive power at that bus (k), P_k and Q_k:

$$\mathbf{V}_k = \frac{1}{\mathbf{Y}_{kk}} \left(\frac{P_k - jQ_k}{\mathbf{V}_k^*} - \sum_{j \neq k} \mathbf{Y}_{jk} \mathbf{V}_j \right) \tag{11.7}$$

Equation 11.7 is a better estimate of $\mathbf{V}_k$ than the original estimate. The solution to the set of non-linear equations consists of carrying out this expression, repeatedly, for all of the buses of the network, until the calculated load flows at the buses are 'close enough' to the actual load flows. Here, 'close enough' is determined by a criterion that, typically calculates the mean squared error, or difference between calculated and actual load flows and compares that with a 'tolerance'. When the error is less than the *a priori* tolerance the solution is said to have converged and the iteration terminates.

An iterative procedure in which a correction for each of the voltages of the network is computed in one step, and the corrections applied all at once is called *Gaussian iteration*. If, on the other hand, the improved variables are used immediately, the procedure is called *Gauss–Seidel iteration*. This sort of iterative procedure is flexible and relatively easy to implement. It is also typically slow and is quite inefficient for large systems. More typical of load flow procedures is to use a variation on Newton's method. In such procedures a Jacobian matrix is calculated: entries in this matrix are, at each bus, variations of P and Q with voltage magnitude and angle at all of the buses. This is a square matrix that may be inverted to give variations of voltage and current with errors in real and reactive power. Newton's method is, of course, still iterative but usually requires far fewer iterations before it terminates. It can be made even faster by noting that real power depends primarily on voltage angle while reactive power depends primarily on voltage magnitude, making a system in which there are two Jacobian matrices that are only one quarter as large. Bergen and Vittal (2000) gives a good summary of how this is done.

Note that Equation 11.7 uses as its constraints P and Q for the bus in question. Thus it is useable directly for *load* type buses. For other types of bus constraints, modifications are required.

For *generator* buses, usually the *real* power and terminal voltage *magnitude* are specified. At each time step it is necessary to come out with a terminal voltage of specified magnitude: voltage phase angle and reactive power Q are the unknowns. One way of handling this situation is to:

1. Generate an *estimate* for reactive power Q, then
2. Use Equation 11.7 to generate an estimate for terminal voltage, and finally,
3. Holding voltage phase angle constant, adjust magnitude to the constraint.

At any point in the iteration, reactive power is:

$$Q_k = \text{Im} \left\{ \mathbf{V}_k \sum_{j=1}^{N} \mathbf{Y}_{jk}^* \mathbf{V}_j^* \right\} \tag{11.8}$$

For buses loaded by fixed impedance, it is sufficient to lump the load impedance into the network. That is, the *load admittance* goes directly in parallel with the *driving point admittance* at the node in question.

These three bus constraint types, generator, load and constant impedance are sufficient for handling most problems of practical importance.

11.4 Bus Admittance

This section discusses ways of estimating the 'bus admittance matrix' **Y**, the description of the line interconnections between buses in a network.

The network consists of a number N_b of buses and another number N_l of lines. Each of the lines will have some (generally complex) impedance Z. The *line admittance matrix* is formed by placing the admittance (reciprocal of impedance) of each line on the appropriate spot on the main diagonal of an $N_l \times N_l$ matrix:

$$\mathbf{Y}_l = \begin{bmatrix} \frac{1}{Z_1} & 0 & 0 & \cdots \\ 0 & \frac{1}{Z_2} & 0 & \cdots \\ 0 & 0 & \frac{1}{Z_3} & \cdots \\ \vdots & \vdots & & \ddots \end{bmatrix} \tag{11.9}$$

11.4.1 Bus Incidence

Interconnections between buses is described by the *bus incidence matrix*. This matrix, which has N_l columns and N_b rows, has two entries for each line, corresponding to the buses at each end. A 'direction' should be established for each line, and the entry for that line, at location (n_b, n_l) in the *node incidence matrix*, is a 1 for the 'sending' end and a -1 at the 'receiving' end. Actually, it is not important which end is which. An admittance element that describes a fixed impedance connected to a node is represented in the node incidence matrix by a bus incidence matrix for the network described by a single entry (a '1') at the bus to which it is connected.

The bus incidence matrix for the example network shown in Figure 11.6 is:

$$\mathbf{N} = \begin{bmatrix} 1 & 0 & 1 & 0 & 0 & 0 & 0 & 0 \\ -1 & 1 & 0 & 0 & 0 & 0 & 0 & -1 \\ 0 & 0 & 0 & 0 & 1 & 1 & 0 & 0 \\ 0 & 0 & 0 & 0 & 0 & 0 & -1 & 0 \\ 0 & -1 & -1 & 1 & -1 & 0 & 0 & 0 \\ 0 & 0 & 0 & -1 & 0 & -1 & 1 & 1 \end{bmatrix}$$

With this formulation, it is possible to find voltages across each of the lines in the matrix by:

$$\mathbf{V}_l = N'\mathbf{V}_B$$

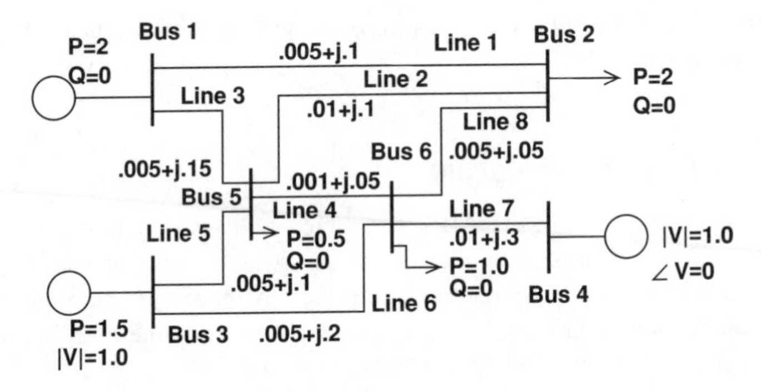

Figure 11.6 Load flow example

Similarly, currents into each bus being the sum of currents in lines leaving that bus:

$$\mathbf{I}_{\mathrm{B}} = N\,\mathbf{I}_{\mathrm{l}}$$

Then the *bus admittance* matrix is given by the easily computed expression:

$$\mathbf{Y} = \mathbf{N}\,\mathbf{Y}_{\mathrm{l}}\,\mathbf{N}'$$

The elements of the bus admittance matrix, the self- and mutual-admittances, are all of the following form:

$$\mathbf{Y}_{jk} = \frac{\mathbf{I}_k}{\mathbf{V}_j}$$

with all other voltages equal to zero.

11.4.2 Alternative Assembly of Bus Admittance

Thus an alternative way to estimate the bus admittance matrix is to:

- Assume that all nodes (buses) are shorted to ground.
- Assume that one node is unshorted and connected to a voltage source.
- Calculate all node currents resulting from that one source.
- Do this for each node as driving point.

One may observe:

- Reciprocity holds:

$$\mathbf{Y}_{jk} = \mathbf{Y}_{kj}$$

- Driving point admittance $\mathbf{Y}_{kk}$ is just the sum of all admittances of lines connected to bus k, including any fixed impedances connected from that bus to ground.

- Mutual admittance $\mathbf{Y}_{jk}$ is *minus* the sum of the admittances of all lines connected *directly* between buses j and k.

11.5 Example: Simple Program

Appended to this chapter is a MATLAB script which will set up and carry out the Gauss–Seidel procedure for networks with the simple constraints described here. The script is self-explanatory and carries out the load flow described by the simple example of Figure 11.6.

Note that, as with many non-linear equation solvers, success sometimes requires having an initial guess for the solution which is reasonably close to the final solution.

11.5.1 Example Network

Consider the system shown in Figure 11.6. This simple system has six buses (numbered 1 through 6) and eight lines. Two of the buses are connected to generators, three to loads and bus 4 is the 'swing bus', represented as a voltage source.

Line impedances are represented, in per-unit on some base in the figure. Real power and voltage magnitude are specified at one generator bus, real and reactive at the other generator bus and at the load buses:

- Bus 1: Generator treated like a load bus: Real power is 2, Reactive power is zero.
- Bus 2: Load bus: Real power is −2, Reactive power is zero.
- Bus 3: Generator bus: Real power is 1.5, voltage is 1.00 per-unit
- Bus 4: Swing bus: Voltage is 1.0 with zero angle.
- Bus 5: Load bus: Real power is −0.5, Reactive power is zero.
- Bus 6: Load bus: Real power is −1, Reactive power is zero.

Note that load power is taken to be negative, for this simple program assumes all power is measured *into* the network.

The script in the appendix does the heavy lifting here. Among items of importance are the bus voltages:

```
Voltage at Bus( 1) = 0.967718
Voltage at Bus( 2) = 0.967027
Voltage at Bus( 3) = 1
Voltage at Bus( 4) = 1
Voltage at Bus( 5) = 0.975345
Voltage at Bus( 6) = 0.975036
```

Also important are line currents, because often load flows are intended to determine if lines are to be overloaded. Here:

```
Current in Line( 1) = 1.40697
Current in Line( 2) = 0.431514
Current in Line( 3) = 0.661446
```

```
Current in Line( 4) = 0.62755
Current in Line( 5) = 0.929723
Current in Line( 6) = 0.619909
Current in Line( 7) = 0.0852025
Current in Line( 8) = 0.275765
```

11.6 MATLAB Script for the Load Flow Example

```
% Simple-Minded Load Flow Example
% This is a self contained file
% Line impedances

Z_L = [.005+j*.1 .01+j*.1 .005+j*.15 .001+j*.05...
       .005+j*.1 .005+j*.2 .01+j*.3 .005+j*.05];

%
fprintf('Simple Minded Load Flow Problem\n')
fprintf('Line Impedances:\n')
for k = 1:length(Z_L)
    fprintf('Z(%2.0f) = %g + j %g\n', k, real(Z_L(k)), imag(Z_L(k)))
end
Y_L = 1 ./ Z_L;

fprintf('Line Admittances:\n')
for k = 1:length(Z_L)
    fprintf('Y(%2.0f) = %g + j %g\n', k, real(Y_L(k)), imag(Y_L(k)))
end

% This is the node-incidence Matrix
NI=[1 0 1 0 0 0 0 0;
    -1 -1 0 0 0 0 0 -1;
    0 0 0 0 1 1 0 0;
    0 0 0 0 0 0 -1 0;
    0 1 -1 1 -1 0 0 0;
    0 0 0 -1 0 -1 1 1];
% This is the vector of voltage magnitudes
VNM = [0 0 1.0 1.0 0 0]';
% And the vector of known voltage angles
VNA = [0 0 0 0 0 0]';
% and this is the "key" to which are actually known
KNM = [0 0 1 1 0 0]';
KNA = [0 0 0 1 0 0]';
% and which are to be manipulated by the system
KUM = 1 - KNM;
KUA = 1 - KNA;
% Here are the known power flows (positive is INTO network
% Use zeros for unknowns
P=[2 -2 1.5 0 -0.5 -1.0]';
```

```
Q=[0 0 0 0 0 0]';
% and here are the corresponding vectors to indicate
% which elements should be checked in error checking
PC = [1 1 1 0 1 1]';
QC = [1 1 0 0 1 1]';
Check = KNM + KNA + PC + QC;
% Unknown P and Q vectors
PU = 1 - PC;
QU = 1 - QC;

Y = zeros(length(Y_L));
fprintf('Here is the line admittance matrix:\n');
for k = 1:length(Y_L)
    Y(k,k) = Y_L(k);
end
Y
% Construct Node-Admittance Matrix
fprintf('And here is the bus admittance matrix\n')
YN=NI*Y*NI'
% Now: here are some starting voltage magnitudes and angles
VM = [1 1 1 1 1 1]';
VA = [.0965 .146 .00713 .0261 0 0]';
% Here starts a loop
Error = 1;
Tol=1e-16;

N = length(VNM);
 % Construct a candidate voltage from what we have so far
VMAG = VNM .* KNM + VM .* KUM;
VANG = VNA .* KNA + VA .* KUA;
V = VMAG .* exp(j .* VANG);

% and calculate power to start
 I = (YN*V);
 PI = real(V .* conj(I));
 QI = imag(V .* conj(I));
%pause
while(Error>Tol);
 for i=1:N, % Run through all of the buses
                        % What we do depends on what bus!
                        % don't know voltage magnitude or angle
   if (KUM(i) == 1) & (KUA(i) == 1),
      pvc= (P(i)-j*Q(i))/conj(V(i));
      for n=1:N,
        if n ~=i, pvc = pvc - (YN(i,n) * V(n)); end
      end
      V(i) = pvc/YN(i,i);
                        % know magnitude but not angle
   elseif (KUM(i) == 0) & (KUA(i) == 1),
      % first must generate an estimate for Q
```

```
      Qn = imag(V(i) * conj(YN(i,:)*V));
      pvc= (P(i)-j*Qn)/conj(V(i));
      for n=1:N,
        if n ~=i, pvc = pvc - (YN(i,n) * V(n));  end
      end
      pv=pvc/YN(i,i);
      V(i) = VNM(i) * exp(j*angle(pv));
   end % this list of cases is not exhaustive
 end % one shot through voltage list: check error

 % Now calculate currents indicated by this voltage expression
 I = (YN*V);
 % For error checking purposes, compute indicated power
 PI = real(V .* conj(I));
 QI = imag(V .* conj(I));
 % Now we find out how close we are to desired conditions
 PERR = (P-PI) .* PC;
 QERR = (Q-QI) .* QC;

Error = sum(abs(PERR) .^2 + abs(QERR) .^2);
end
fprintf('Here are the voltages\n')
for k = 1:length(V)
    fprintf('Voltage at Bus(%2.0f) = %g\n', k, abs(V(k)))
end
fprintf('Real Power at the buses\n')
PI
fprintf('Reactive Power at the buses\n')
QI
fprintf('Line Voltages are\n')
Vline = NI'*V
fprintf('Line Currents are\n')
Iline = abs(Y*Vline)
for k = 1:length(Iline)
    fprintf('Current in Line(%2.0f) = %g\n', k, Iline(k))
end
```

11.7 Problems

1. A single phase line is modeled as an inductive reactance as shown in Figure 11.7. In this case, 10 Ω. The voltage sources at each end have the same voltage magnitude of 1000 Ω, RMS. Assume that the sending end is at the left and the receiving end is at the right.
   (a) What is the maximum amount of real power that can flow from one end to the other?
   (b) What is real power flow when the phase angle between ends is 30°?
   (c) With the phase angle of 30°, what is reactive flow at each end?
   (d) What phase angle is required for 75 kW to flow?

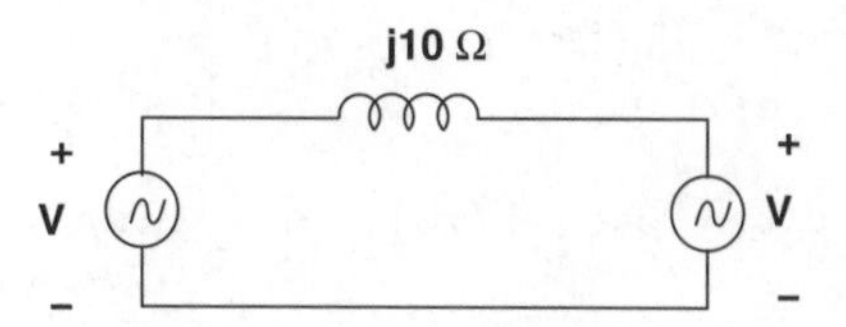

Figure 11.7 Single-phase line

2. A three-phase transmission line is shown in one-line format in Figure 11.8. The buses at each end, represented as voltage sources, have the same voltage magnitude: 138 kV, RMS, line-line. For now, model the line as having an impedance to balanced currents of $Z_l = j40\,\Omega$. Assume the sending end has a phase advance over the receiving end of angle δ.
   (a) What is real power flow through the line for $\delta = 10°$?
   (b) What is real power flow through the line for $\delta = 30°$?
   (c) What angle δ results in a real power flow of 100 MW?
   (d) With 100 MW flowing through the line, what is reactive power at the sending and receiving ends?

Figure 11.8 Three-phase line

3. This is about the same line, shown in one-line format in Figure 11.8. The buses at each end, represented as voltage sources, have the same voltage magnitude: 138 kV, RMS, line-line. Now, however, model the line as having an impedance to balanced currents of $Z_l = j40 + 4\,\Omega$. Assume the sending end has a phase advance over the receiving end of angle δ.
   (a) What is real power flow into the receiving end of the line for $\delta = 10°$?
   (b) What is real power flow out of the sending end of the line for $\delta = 10°$?
   (c) What is real power flow into the receiving end of the line for $\delta = 30°$?
   (d) What is real power flow out of the sending end of the line for $\delta = 30°$?
   (e) What angle δ results in a real power flow into the receiving end of 100 MW?
   (f) When 100 MW is flowing into the receiving end of the line, how much real power is flowing out of the sending end?
   (g) With 100 MW flowing into the receiving end, what is reactive power at the sending and receiving ends?

4. This is about the same line, shown in one-line format in Figure 11.8. The buses at each end, represented as voltage sources, have the same voltage magnitude: 138 kV, RMS, line-line. Model the line as having an impedance to balanced currents of $Z_l = j40\,\Omega$. In this case, shunt capacitances of 6.6 μF have been installed from all three phases to ground at each end of the line. Assume the sending end has a phase advance over the receiving end of angle δ. Frequency is 60 Hz.

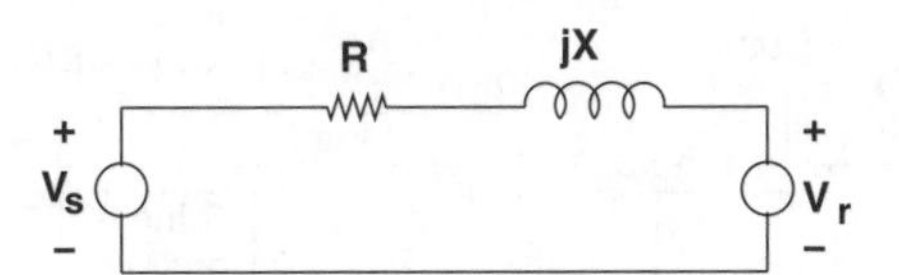

Figure 11.9 Lossy transmission line

(a) With zero phase angle between ends of the line, how much reactive power is produced at each end of the line.
(b) What are real and reactive power flows at both ends of the line for $\delta = 30°$?
(c) With 100 MW flowing through the line, what is reactive power at the sending and receiving ends?

5. This is about the same line, shown in one-line format in Figure 11.8. The buses at each end, represented as voltage sources, have the same voltage magnitude: 138 kV, RMS, line-line. Now, however, model the line as having an impedance to balanced currents of $Z_1 = j40 + 4\,\Omega$. Assume the sending end has a phase advance over the receiving end of angle δ. Sketch and label a power circle diagram for real and reactive power flow out of the sending end source and into the receiving end source of the line.

6. A lossy transmission line problem is shown in Figure 11.9. Assume that the magnitude of voltage at the sending and receiving ends is the same: $|V_s| = |V_r| = 1000$V, RMS, and that the resistance R is $5\,\Omega$ and the reactance X is $50\,\Omega$. The phase shift from sending to receiving end is δ.

(a) Construct and sketch the sending end and receiving end power circle (P vs. Q with $0 < \delta < \pi/2$). Use the convention that complex power flow is from the source to the line at the sending end and from the line to the source at the receiving end.
(b) What is the phase shift δ across the line when 10 kW is the real power flow at the receiving end? What is power at the sending end? What are reactive flows for that case?

7. This is a single-phase transmission line problem. The situation is shown in Figure 11.10. The 8 kV 60 Hz (RMS) source is feeding a $80\,\Omega$ load through a transmission line which may be represented as a simple inductance.

(a) Assuming both capacitances to be zero, draw a phasor diagram showing sending and receiving end voltages and voltage across the transmission line. What is power dissipated in the resistance?
(b) Size the receiving end capacitance C_r so that receiving end voltage is equal in magnitude to sending end voltage.

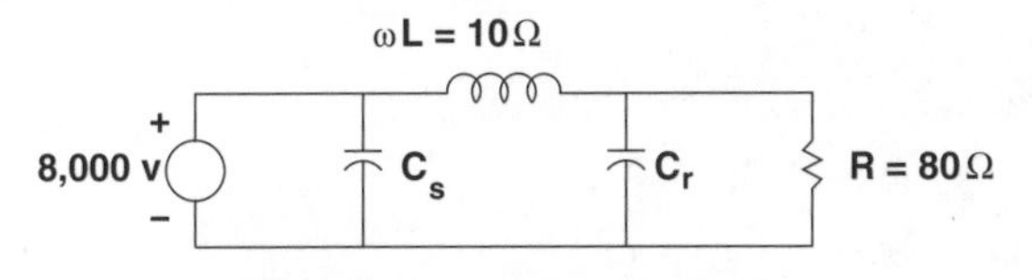

Figure 11.10 Compensated load

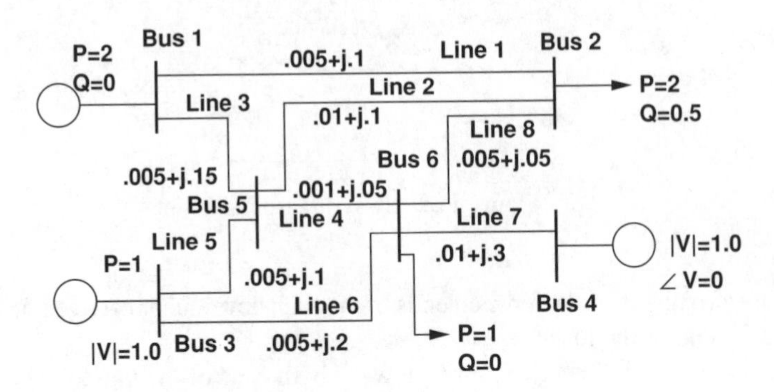

Figure 11.11 Small load flow example

(c) Calculate and plot, over the range of $1\,\mu F < C_r < 6\,\mu F$, receiving end voltage. You will probably want to use MATLAB to do the hard work here.

8. A small load flow problem, similar to the one in the text, is shown in Figure 11.11. Note the voltages, impedances and flows are all noted in per-unit.

(a) Write a Gauss–Seidel based program to solve for the voltage magnitudes and angles that result in the power flows described in the figure. Note that bus 4 here is the 'swing bus', which means its angle is the reference point.

(b) Note the voltage source at bus 3 is producing quite a lot of reactive power. How does this change if you inject $Q = 0.5$ per-unit into bus 5?

(c) What happens to the line flows if line 1 is removed?

Reference

Bergen, A.R. and Vittal, V. (2000) *Power System Analysis*. Englewood Cliffs: Prentice Hall.

12

Power Electronics and Converters in Power Systems

Power electronics is becoming more important for electric power systems as both device improvements and circuit techniques have developed. There are many applications of power electronics in electric power systems and applications, including:

1. Power electronics are essential for converting the power from solar cells or fuel cells (really, any source of electric power that is DC) into AC.
2. Bidirectional converters are used in the doubly fed induction machines that serve as generators in wind turbines.
3. Phase controlled converters are used at the terminations of high-voltage DC transmission lines.
4. Phase controlled devices are used in lighting systems.
5. Uncontrolled rectifiers and boost converters are used in 'unity power factor' front ends for lighting devices.
6. Combined rectifiers and inverters are used in variable speed drives for industrial applications and transportation, including electric and hybrid electric vehicles.
7. Active converters are used to construct voltage regulating and phase shifting circuits in what is called 'flexible AC transmission systems' (FACTS).

This is not a book about power electronics, but this chapter will give an introduction to how some kinds of power electronic devices and circuits are applied to electric power systems and to devices that use electric power. A more thorough discussion of power electronic devices and circuits is given in a textbook on this topic written by Mohan *et al.* (1995).

12.1 Switching Devices

A number of active devices are used for power electronic circuits, but in virtually all applications deemed to be 'power electronics', those devices are used a switches. That is, in application they are always either 'on' or 'off'. Some of the most important devices are:

Electric Power Principles: Sources, Conversion, Distribution and Use James L. Kirtley

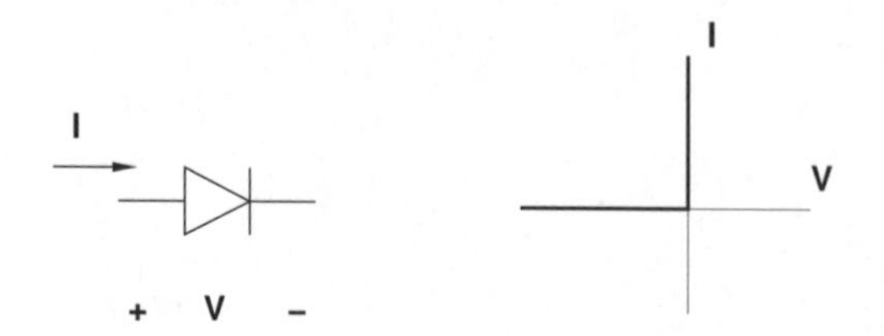

Figure 12.1 Diode symbol and idealized characteristic

12.1.1 *Diode*

A circuit symbol and idealized characteristic for the device called a diode is shown in Figure 12.1. The diode is very much like a check valve: it lets current flow in one direction. When the voltage across the diode is negative there is no current flow. When the current is flowing there is no voltage. This is, of course, an idealization of the real situation. A closer approximation to actual diode behavior is:

$$I = I_0 \left(e^{\frac{V}{\frac{kT}{q}}} - 1 \right)$$

where I_0, the 'leakage current' is very small. k is Boltzman's constant, about 1.38×10^{-23} J/K, q is the charge on an electron, about 1.6×10^{-19} C. Thus the quantity $\frac{kT}{q}$ is, at room temperature (about 300 K), about 25 mV. In normal operation a diode will have a forward voltage drop of perhaps 0.5 to 1 V and negligible reverse leakage current.

12.1.2 *Thyristor*

A thyristor, shown in Figure 12.2, is a little bit like a diode in that it can carry current in only one direction. In order for it to carry forward current it must be turned on, or triggered, by injecting some small current into its gate lead. Once turned 'on' the thyristor will continue to carry current as long as there is positive current to carry. If the current goes to zero and the thyristor has reverse voltage, it turns 'off' and stays 'off' until the voltage is positive and the device is triggered again. This device is useful for some kinds of circuits, such as phase controlled rectifiers and is the basis for high voltage DC circuits. Thyristors are made in ratings

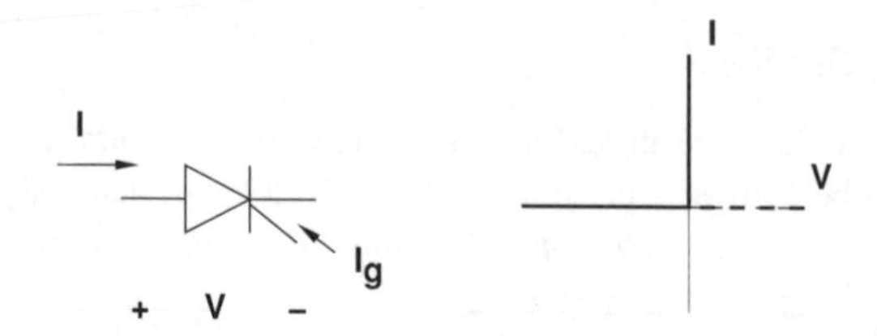

Figure 12.2 Thyristor symbol and idealized characteristic

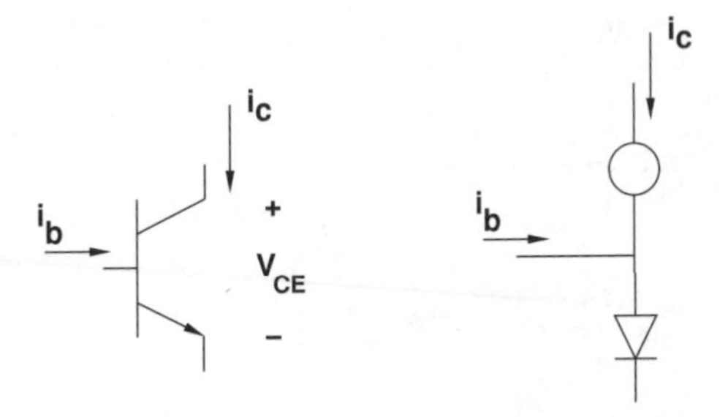

Figure 12.3 Circuit symbol and simple equivalent circuit for a bipolar transistor

from fairly small to very large. Individual devices can be made with voltage ratings of a few thousand volts and current ratings of several thousand amperes.

To meet the need for devices that can turn 'off' as well as 'on', devices called 'gate turn off' thyristors, or simply GTO have been developed. A characteristic of a GTO is the 'turn off gain', the ratio of the current being carried by the device to negative gate current. According to Mohan, this may be as small as about 3.

12.1.3 Bipolar Transistors

The bipolar transistor is a very widely used device in many different types of electronic circuits, not just power electronics. Shown in Figure 12.3 is the circuit symbol for one variant of a transistor, called 'NPN' because of its structure. In this device, if the voltage V_{CE} is positive, the device behaves as shown in the right hand side of Figure 12.3. A positive current in the 'base' lead of the device causes a substantially larger current to flow in the 'collector' lead:

$$i_c = \beta i_b$$

where the current 'gain' β can be quite large (on the order of 100 or more). To illustrate why, in power electronics applications, these devices are used in switching mode, consider the very simple circuit of Figure 12.4. A bipolar transistor is connected to fixed value voltage source through a resistor (which in this case would be considered the 'load'.

In this circuit, if $i_c = \beta i_c$, then voltage across the device is:

$$V_{CE} = V_s - Ri_c = V_s - \beta R i_b$$

and power dissipated in the transistor is:

$$P_d = V_{CE} i_c = \beta i_b V_s - (\beta i_b)^2 R \quad 0 < \beta i_b < \frac{V_s}{R}$$

Dissipation in the transistor is zero when there is no current through it ($i_b = 0$) or when there is no voltage across it (when i_b is sufficient to 'saturate' the transistor). This occurs when:

$$i_b > \frac{V_s}{\beta R}$$

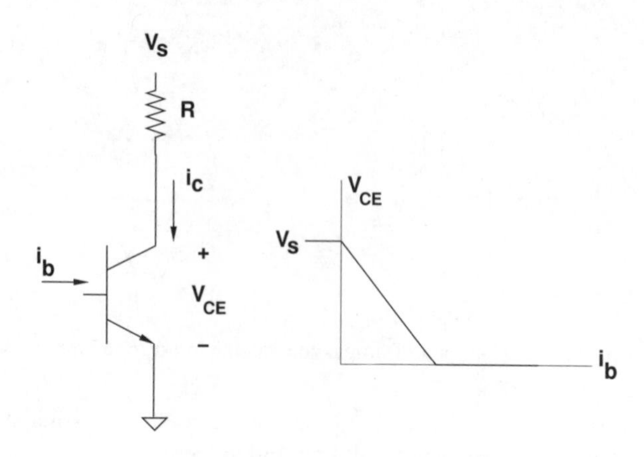

Figure 12.4 Transistor circuit

In between the zero current condition and saturation there is dissipation in the device that reaches a maximum at about half the current required for saturation.

There is, of course, some dissipation in the transistor because v_{CE} does not reach zero. Typical saturation voltage for a power transistor depends on its voltage rating, and can range fro a few tenths of a volt for devices with small voltage ratings to a few volts for high (e.g. 1000 V) voltage devices.

There *is* dissipation in a transistor when it is actively switching because the current does not change instantaneously, and during the switching interval the transistor swings through its 'active' region. This turns out to be a major limitation on some kinds of power electronic circuits.

There are other types of transistor that are used in power electronic circuits. Some of these are shown in Figure 12.5. The Darlington connection is not a new device but rather an interconnection of two bipolar transistors to achieve higher gain. They are sometimes produced on a single die and so are 'monolithic' devices.

Field effect transistors (FETs) are voltage-controlled devices and so have low gate power requirements. They often have very fast switching speed. They tend to have relatively large 'on' state resistance and so are not built in very large power ratings.

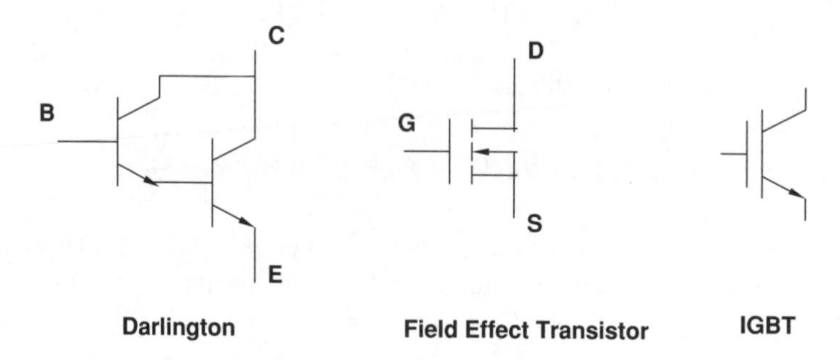

Figure 12.5 Other transistors

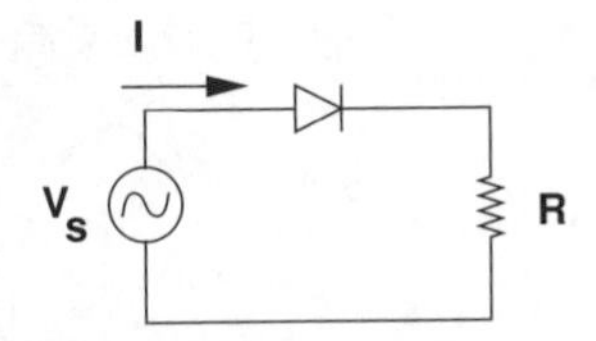

Figure 12.6 Half wave rectifier circuit

Insulated gate bipolar transistors (IGBTs) are also voltage-controlled devices that combine the input characteristics of FETs with the forward voltage of bipolar transistors. They are not as fast at switching as FETs but can be built in larger ratings. IGBTs are used in many medium to relatively high power circuits.

12.2 Rectifier Circuits

These are the simplest of power electronic circuits to understand, so the development starts there. Consider the circuit shown in Figure 12.6. A diode is inserted between a voltage source and a resistor. Operation of this is as follows:

1. If the voltage source is positive, current flows through the diode in the direction shown. The voltage across the diode is small (it is idealized in most analyses to zero).
2. If the voltage source is negative, no current flows at all: the diode is 'back biased' and permits no current to flow. All of the source voltage appears across the diode.

Current drawn by this 'half-wave' rectifier is shown in Figure 12.7. This circuit is not widely used in power circuitry because it imposes a current waveform that is problematic on the AC side of the rectifier. Not only does the current waveform have a DC component, but it also has a spectrum of time harmonics that can cause a variety of mischief.

12.2.1 Full-Wave Rectifier

The full wave rectifier is more often used to convert AC to DC. A single phase full-wave diode rectifier is shown in Figure 12.8. This arrangement of four diodes is a very common circuit, often made into a four terminal 'block'.

12.2.1.1 Full-Wave Bridge with Resistive Load

Driven by AC with a resistive load, the four-diode bridge does an operation similar to taking the absolute value of a waveform. Consider Figure 12.9, in which the AC side of the bridge is driven by a sine wave of voltage and the DC side is loaded by a resistor. If the voltage V_s is positive, diodes D1 and D4 conduct and diodes D2 and D3 are reversed biased and therefore OFF. If the voltage is negative, the opposite is true: current flows through diodes D2 and D3. The current through the resistor is shown in Figure 12.10. The current on the AC side is a sine wave, a replica of the source, scaled by the value of the load resistor. Thus the input of

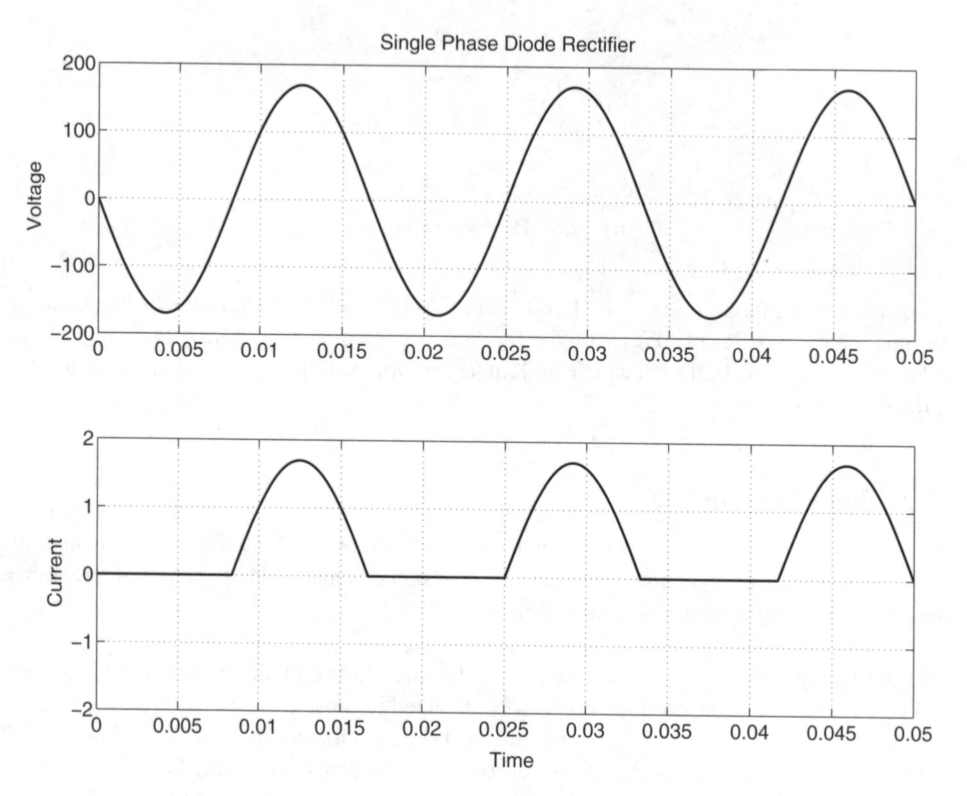

Figure 12.7 Voltage and current input to half wave rectifier

the full-wave bridge 'looks' to the source as if it were simply the resistor. To the resistor the bridge 'looks' like a voltage source with the absolute value of the actual sine wave source.

12.2.1.2 Phase-Control Rectifier

By using two thyristors in the full-wave bridge rectifier, a control of output may be introduced. Consider the circuit of Figure 12.11. This circuit not only rectifies the AC voltage, but it also

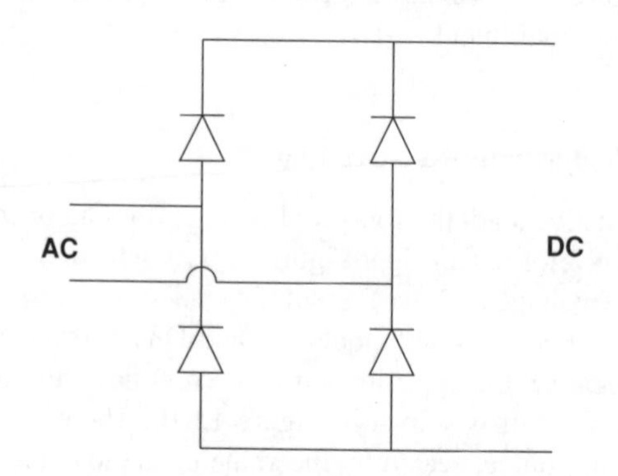

Figure 12.8 Full wave diode bridge

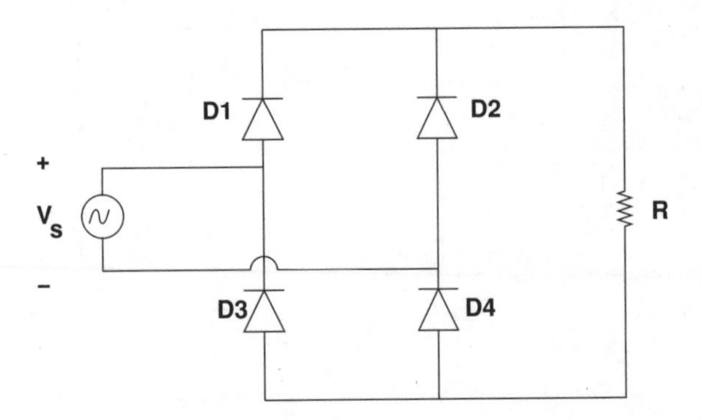

Figure 12.9 Full-wave bridge with resistive DC side load

controls the average value of its output. When the thyristors are conducting, operation of this bridge is exactly the same as operation of the diode bridge shown in Figure 12.9. Control is exercised by turning ON the thyristors some time after a diode would normally turn ON. The phase angle at which devices T4 (positive half cycle) or T3 (negative half cycle) are turned on control the output voltage, as shown in Figure 12.12.

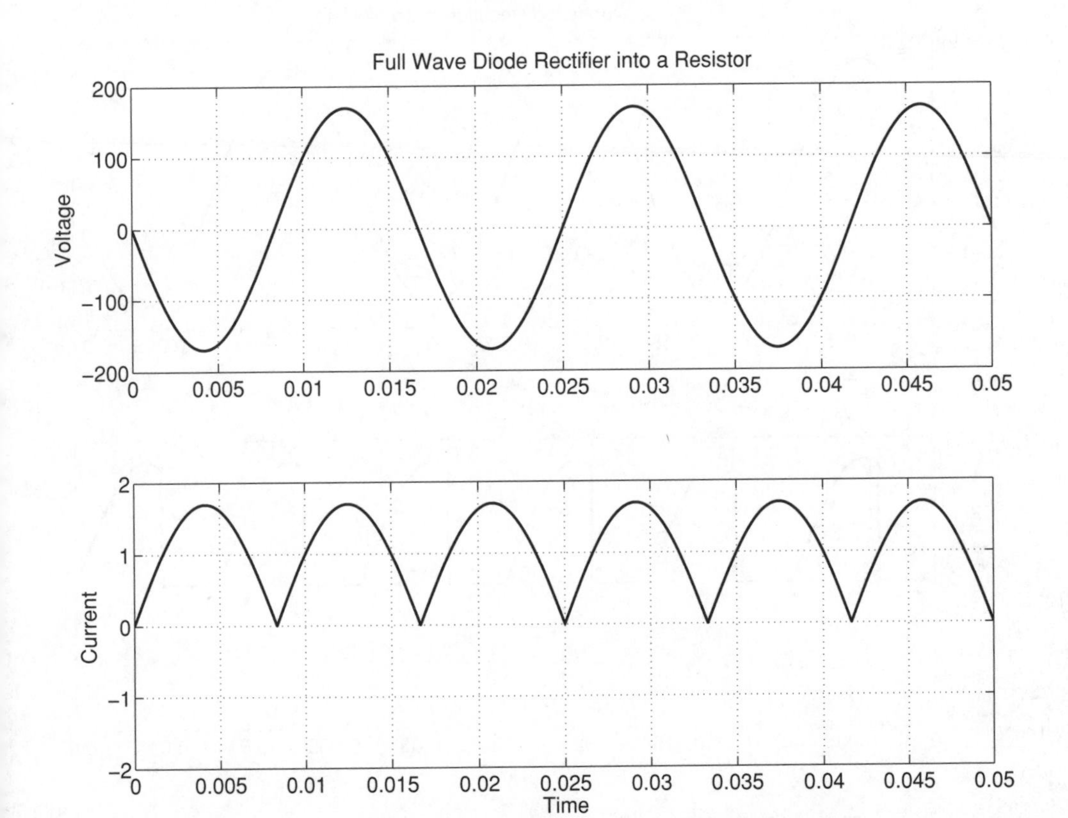

Figure 12.10 Voltage input and current output of a full wave bridge loaded by a resistor

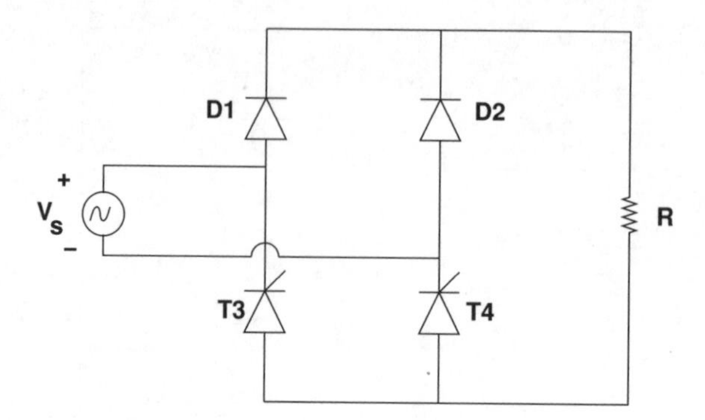

Figure 12.11 Phase-control rectifier circuit

It is clear that increasing the phase angle, corresponding to the time delay between the zero crossing at which the current could start flowing in one of the controlled device and when the device is turned on will reduce the time average output voltage. There are at least two measures of time average voltage: there is the 'average' or mean voltage. If instantaneous input voltage is $V_s \sin \omega t$:

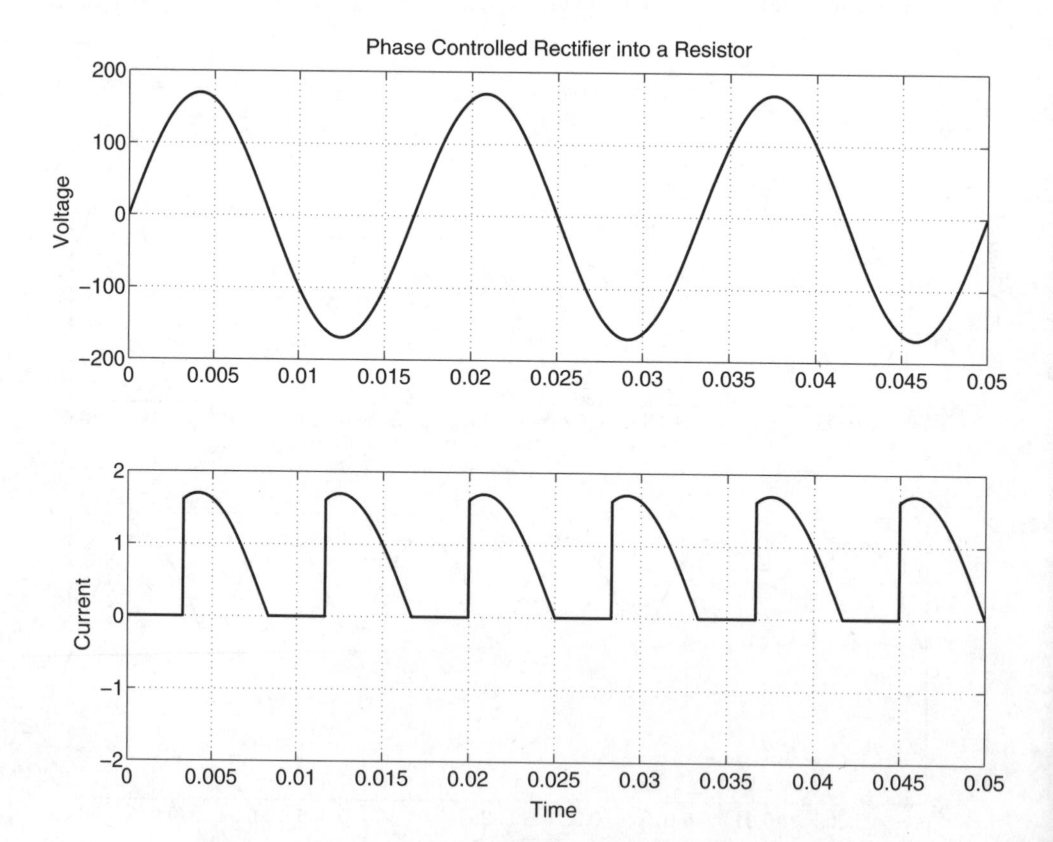

Figure 12.12 Phase control rectifier voltage input and current output

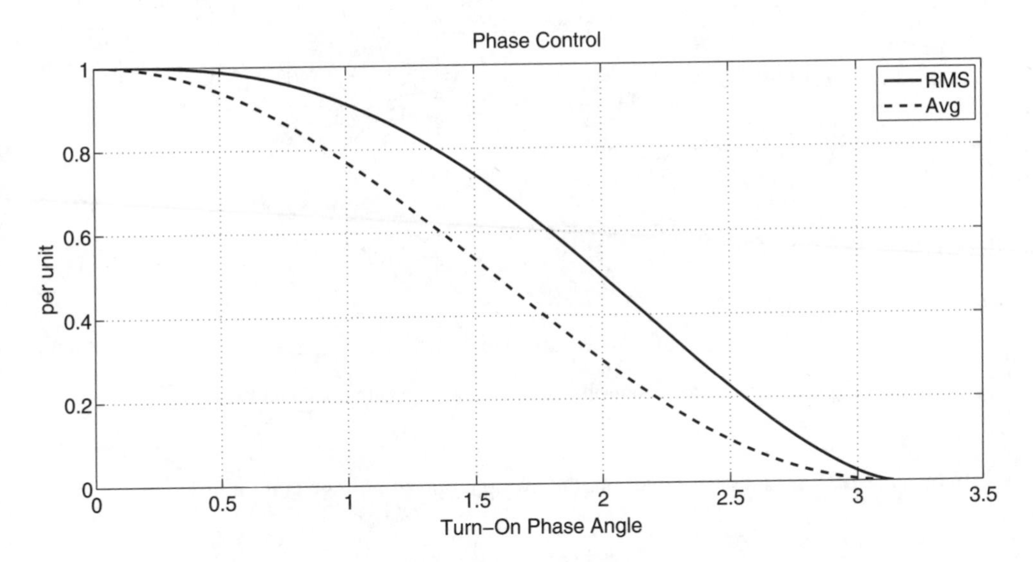

Figure 12.13 Average and RMS reduction with phase control in a rectifier

$$< V_{\text{out}} > = \frac{1}{\pi}\int_{\alpha}^{\pi} \sqrt{2}V_s \sin\theta d\theta = \frac{\sqrt{2}}{\pi}V_s(\cos\alpha + 1)$$

Another measure is the RMS voltage, which is more useful when contemplating, for example, the effect on heating or incandescent lighting equipment:

$$V_{\text{RMS}} = \sqrt{\frac{1}{\pi}\int_{\alpha}^{\pi} 2V_s^2 \sin^2\theta d\theta} = V_s\sqrt{\frac{\pi - \alpha}{\pi} + \frac{\sin 2\alpha}{2\pi}}$$

These two measures, normalized to their base values (when phase angle $\alpha = 0$) are shown in Figure 12.13.

12.2.1.3 Phase Control into an Inductive Load

The circuit of Figure 12.14 illustrates another important principle of phase-controlled power electronic circuitry. In this circuit there are four thyristors arranged in a standard 'H' bridge configuration. In this configuration, the thyristors are fired in pairs: T1 and T4 are ON at the same time and T2 and T3 are ON when the other two are OFF. If the inductor is large enough that current is continuous, the current will transfer between one pair of thyristors and the other only when the second pair of thyristors is fired. The output voltage would look something like what is shown in Figure 12.15. Actually, this figure assumes the inductance is very large so that current in the load resistor is nearly constant. This condition is more than is necessary for continuous current, which is what is necessary for the voltage waveform to be as shown.

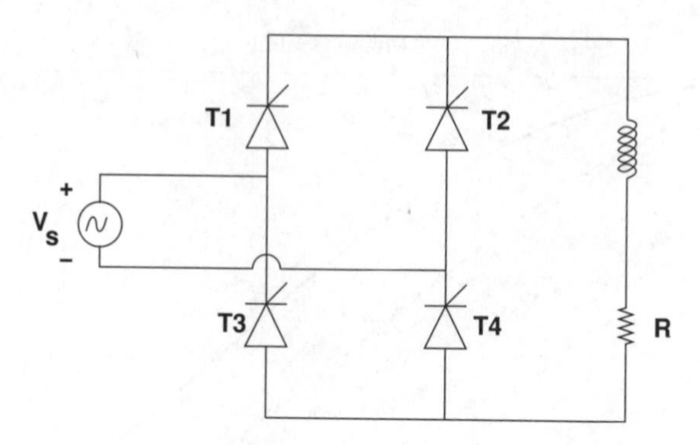

Figure 12.14 Phase controlled full wave bridge with continuous current

In the case of continuous conduction, the average output voltage is found by:

$$< V_{\text{DC}} >= \frac{1}{\pi}\int_{\alpha}^{\alpha+\pi} \sqrt{2}V_s \sin \omega t \mathrm{d}\omega t = \frac{2\sqrt{2}}{\pi}\cos\alpha$$

where α is the firing (delay) angle. This is shown in Figure 12.16.

What this shows is that the output voltage of a phase-controlled rectifier with continuous current can be negative. This would be useful for, say, discharging a large inductance. But much more important is that such a circuit can actually be an inverter. This is actually how high-voltage DC transmission lines work.

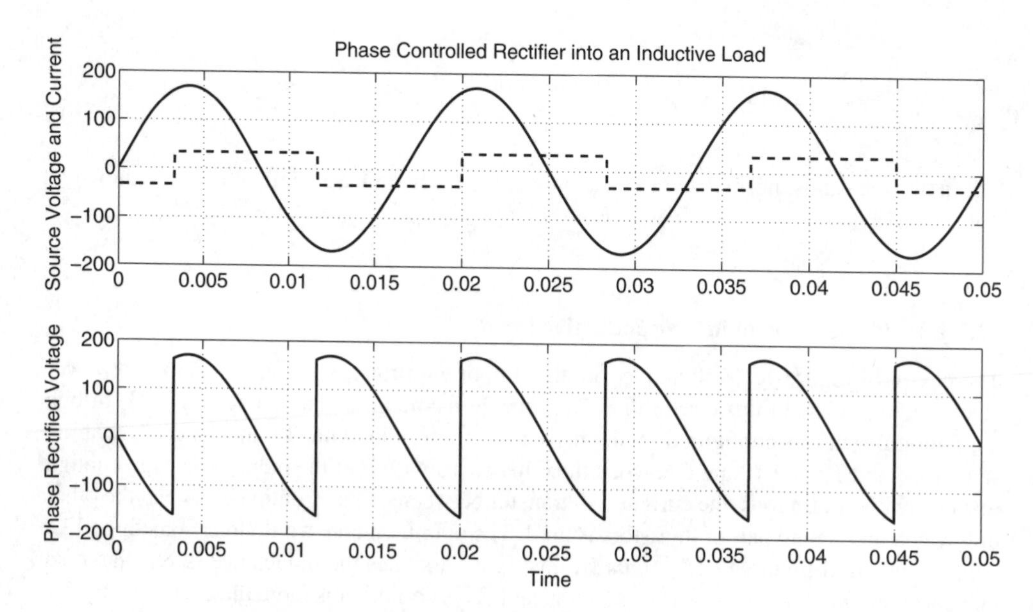

Figure 12.15 Phase controlled rectifier: continuous current

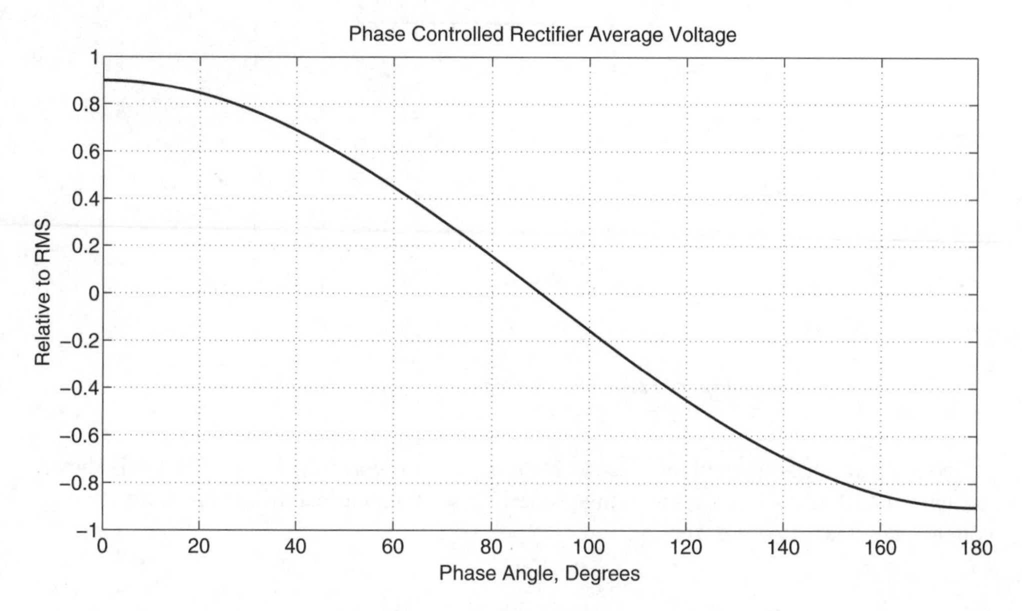

Figure 12.16 DC voltage relative to RMS

Another important point about this circuit is to note that the input current waveform lags the input voltage by the firing angle α. Thus the circuit absorbs reactive power. This will continue to be true even if the circuit is turned into an inverter and the current reverses sign. The inverter will be absorbing reactive power from the AC system even as it supplies real power to that system.

12.2.1.4 AC Phase Control

A similar application of phase control is shown in a circuit that does not rectify in Figure 12.17. This circuit is used in application such as light dimmers, in which all that is required is to control the RMS output voltage. This is such a useful circuit that a special device, called a Triac, has been developed. The Triac behaves just like the back-to-back thyristors in the circuit of Figure 12.17 but with a single gate input.

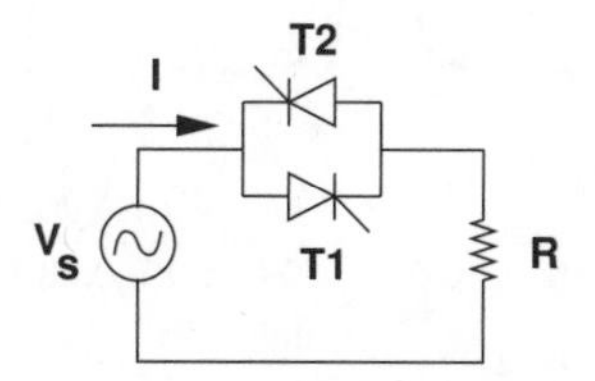

Figure 12.17 AC phase control circuit

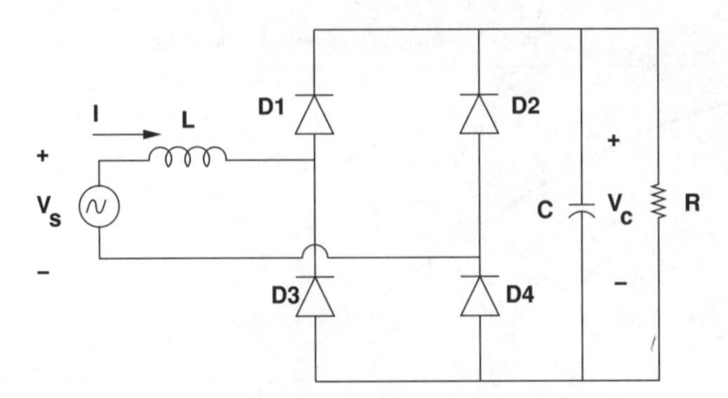

Figure 12.18 Full wave rectifier to capacitor

The RMS voltage output of the phase control circuit of Figure 12.17 is the same as is shown in Figure 12.13. If the circuit is working properly, with symmetrical firing pulses, the average voltage output is, of course, zero.

12.2.1.5 Rectifiers for DC Power Supplies

Generally, the output of an AC/DC converter (rectifier) does not go directly to a resistance, but rather is filtered so that the output is very nearly constant voltage DC. A common circuit is shown in Figure 12.18.

While this type of rectifier draws only alternating current from the alternating voltage source, that is, there is no DC component to the current drawn, it can still be quite far from a sine wave. Figures 12.19 and 12.20 are simulations of this situation for a rectifier drawing

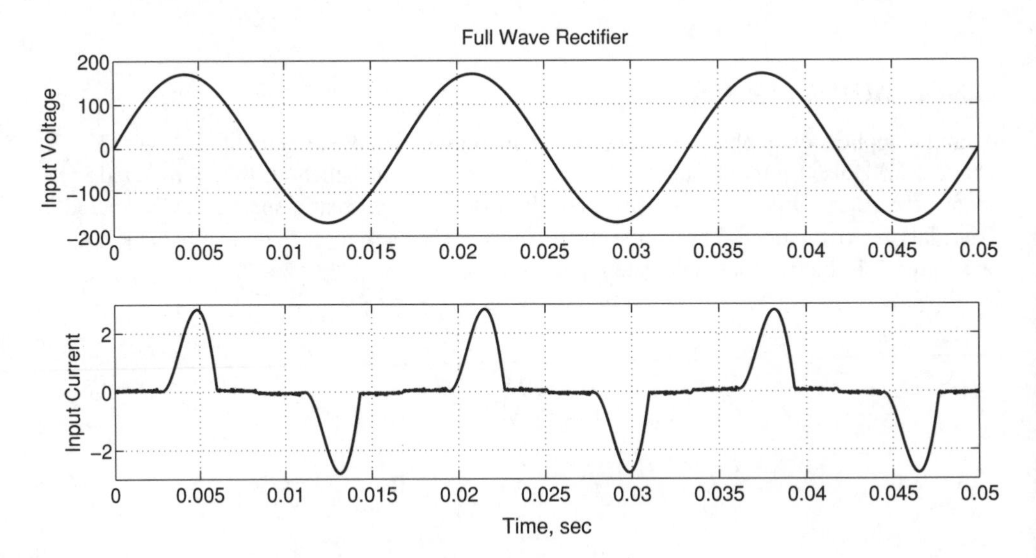

Figure 12.19 Input voltage and current to full wave bridge rectifier

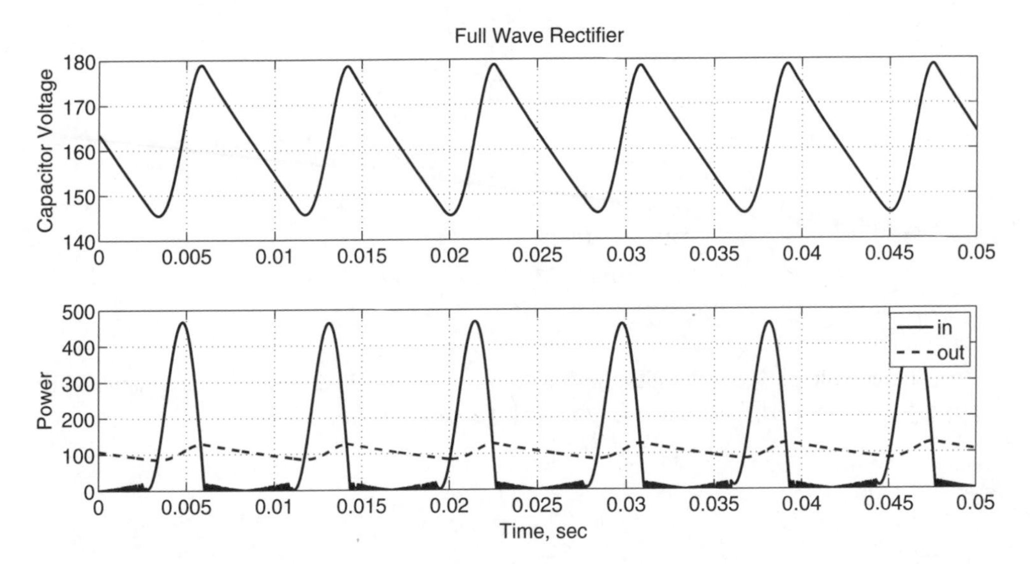

Figure 12.20 Capacitor voltage and input and output power for a full wave bridge rectifier

about 100 W from a 120 V, RMS, 60 Hz. line, with a capacitor of 100 μF and with a source inductance of 10 mH. As can be seen from Figure 12.19, the rectifier tends to draw relatively short pulses of current. This is because the capacitor stores charge, so that for the first part of each half cycle, all of the diodes in the rectifier bridge are back-biased. No current is drawn until the source voltage is larger than the voltage on the capacitance. When the source voltage exceeds the capacitor voltage the current starts to rise, limited by the source inductance. The capacitor voltage will increase until after the source voltage peaks, and then when the source voltage becomes less than the capacitor voltage, current through he source inductance falls to zero. This happens every half cycle.

Figure 12.20 shows the capacitor voltage, which has some ripple because it is feeding the load continuously but is charged through the rectifier bridge only in short pulses. Input and output power are also shown: the capacitance makes the output power much steadier than the input power, but they have the same average.

There are better ways of building full-wave rectified sine wave input power supplies, and a discussion of that will appear after the section on DC–DC converters.

12.3 DC–DC Converters

Consider the 'buck converter' circuit shown in Figure 12.21. The switch could be implemented with any of the transistors described above. The strategy is to open and close the switch periodically with some period T. The switch is closed, or ON, for a time $t_{ON} = DT$ and then open, or OFF, for a time $t_{OFF} = (1 - D)T$, as shown in Figure 12.22.

Examine the circuit and its switching pattern. Assuming that there is current in the inductor, if the switch is OFF, current must be flowing in the diode. In that case, the voltage across the diode v_d must be zero. Similarly, if the switch is ON, voltage across the diode must be equal

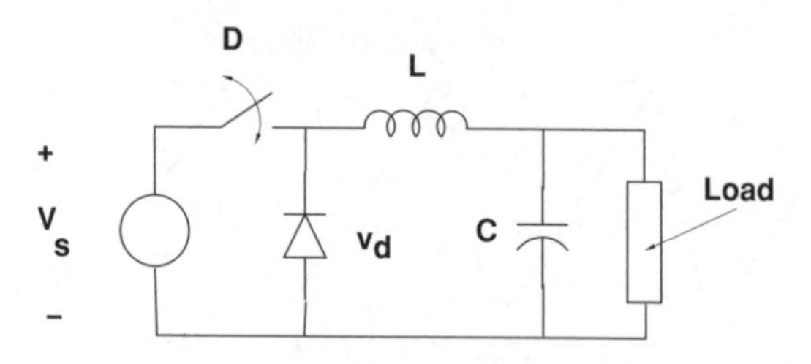

Figure 12.21 Buck converter

to the source voltage V_S, and under those circumstances the diode is back biased and can be carrying no current. The average voltage across the diode is:

$$< v_d >= \frac{V_S DT}{T} = DV_S$$

Assuming the capacitor is large enough that the voltage across it does not vary substantially, once the system is in steady state, the average voltage across the capacitance must be the same as the average voltage across the diode, because the inductance can support no DC voltage.

Figure 12.23 shows a simulation of the operation of a buck converter with a 48 volt supply, an inductance of $L = 10$ mH and a capacitance of $C = 10$ μF supplying a 10 Ω load. The switching frequency is 10 kHz and the duty cycle assumed is 0.5. Thus one would expect an output voltage of about 24 V. Ripple current in the inductor (peak to peak) is found quite simply (if capacitor voltage is nearly constant) by:

$$I_r = \frac{(V_s - V_c)DT}{L}$$

This works out to be, in this case, about 0.12 A, and this is consistent with the picture shown in Figure 12.24, which shows a few cycles of operation once the converter is in steady state.

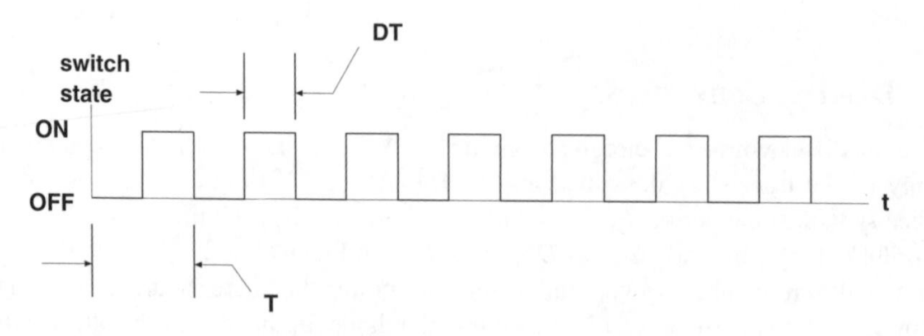

Figure 12.22 PWM switching pattern

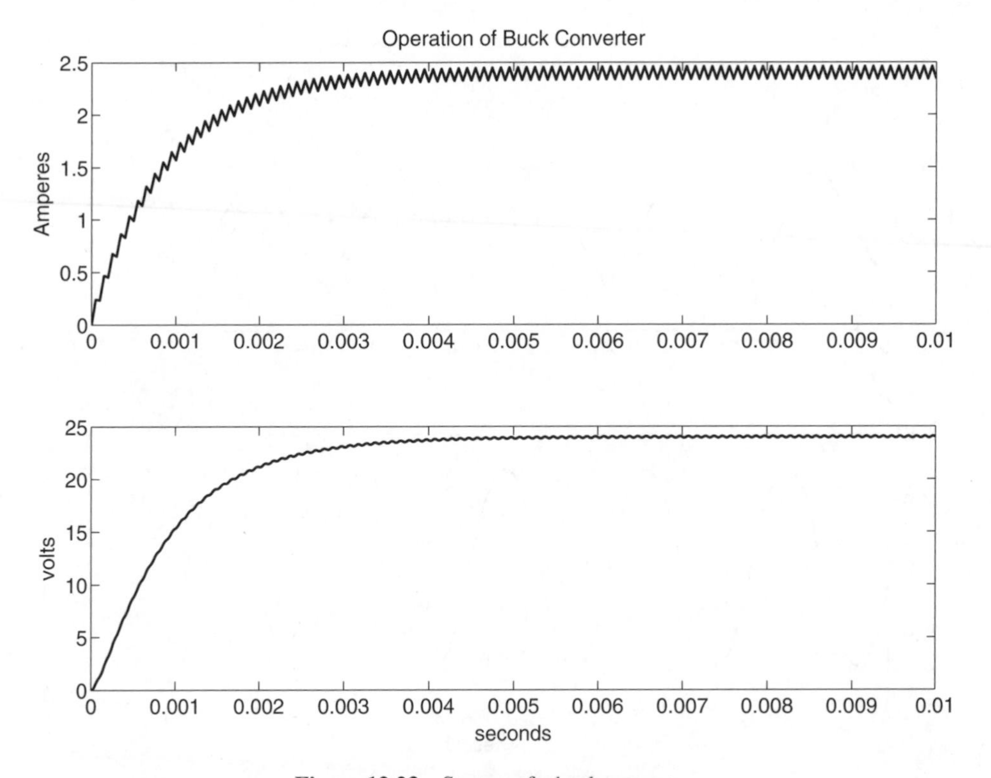

Figure 12.23 Startup of a buck converter

12.3.1 Pulse Width Modulation

The buck converter gives a good opportunity to show how pulse width modulation works. In the example given here, the output voltage can be varied by changing the duty cycle of the converter; that is, by varying how long the switch is ON during each cycle. One can make a high power amplifier this way. A straightforward way of generating the correct ON signal for the switch is to start with a triangle wave or sawtooth with a frequency that is the same as the intended switching frequency and then to compare that with the input signal, as shown in block diagram form in Figure 12.25. The output, or ON signal for the converter switch is then all of the intervals when the input signal is larger than the triangle.

An illustration of the kinds of signals one would expect is in Figure 12.26. This is a PWM signal generated by a comparator as shown in Figure 12.25 with an offset sine wave input signal. The triangle wave frequency shown in this illustration is substantially smaller than one would expect in practice, to make the picture legible. However, one can see that the PWM duty cycle is high when the input signal is high and low when the input is low.

12.3.2 Boost Converter

A basic boost converter is shown in Figure 12.27. This circuit differs from the buck converter in that it is primarily a current controlling circuit. In looking at this circuit, it is necessary to

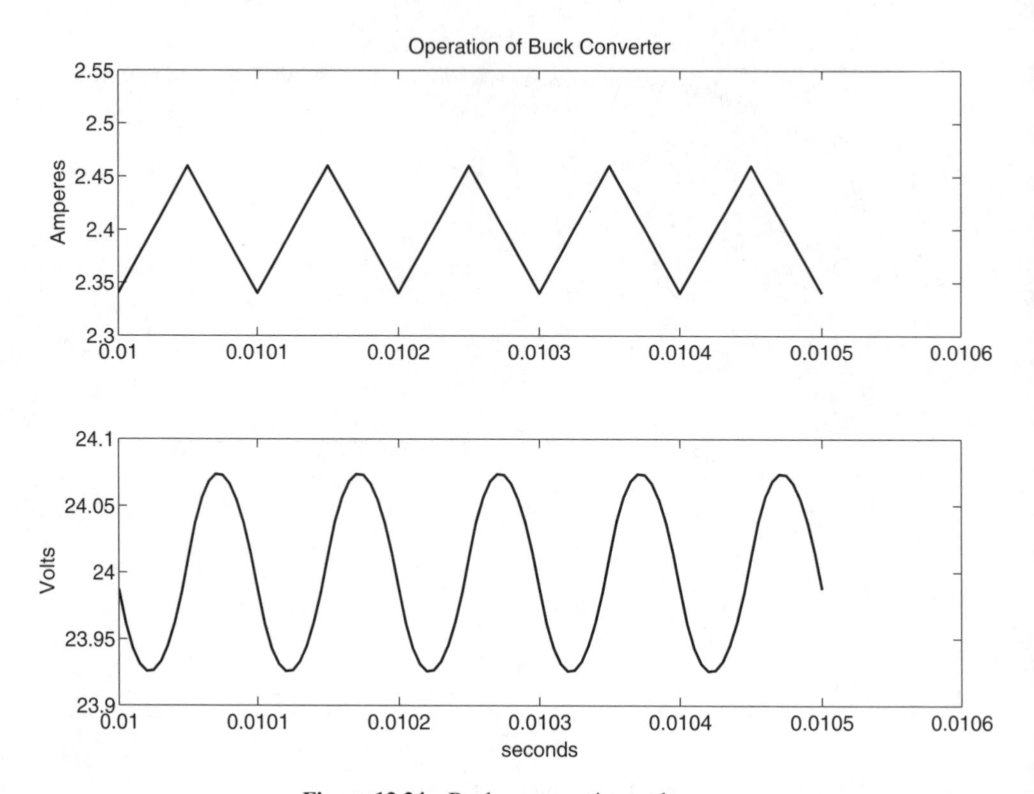

Figure 12.24 Buck converter in steady state

consider two different modes of operation. In the first of these modes the current through the inductor is *continuous*, in that it does not fall to zero. When the switch is ON, the input voltage is impressed directly across the inductance and the current I increases. If there is positive voltage on the capacitance (there will be), the diode is back biased and no current flows in it. When the switch is turned OFF, the current in the inductance flows through the diode and charges the capacitance.

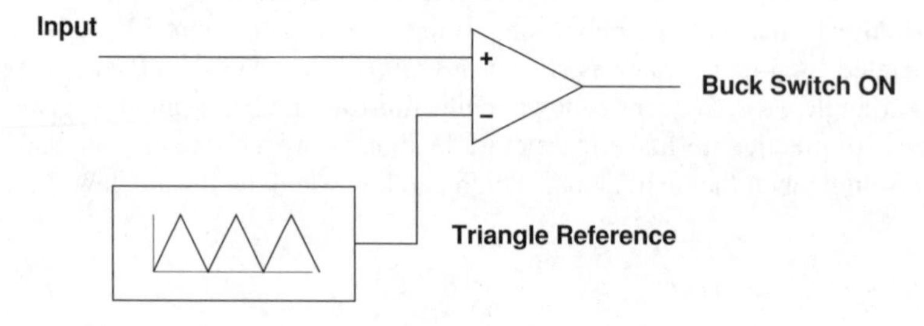

Figure 12.25 Generation of PWM signal

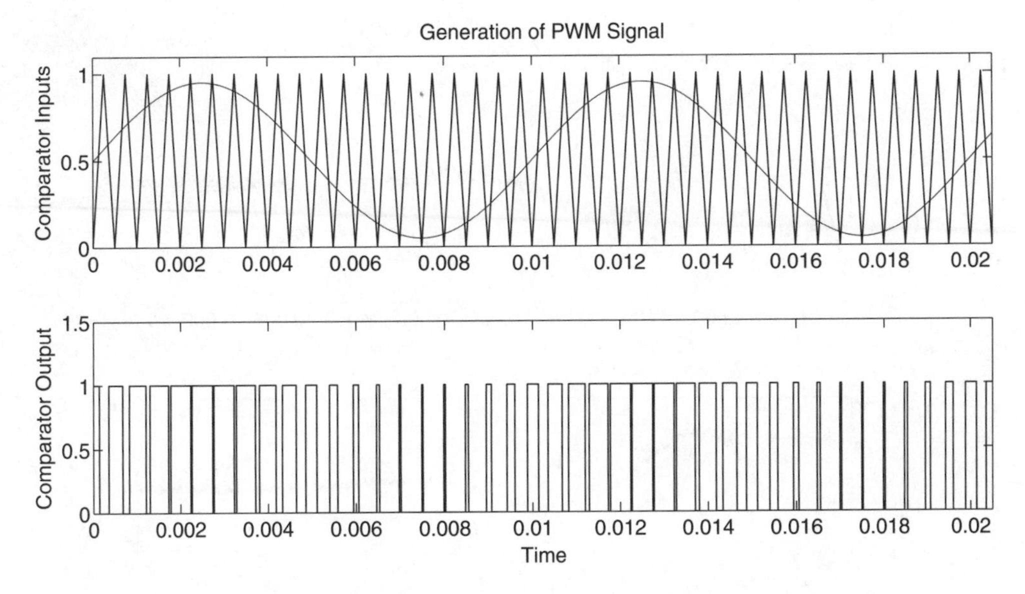

Figure 12.26 PWM signal generated by comparator with triangle wave

12.3.2.1 Continuous Conduction

Continuous conduction operation of the boost converter means that current in the inductor is never zero. This makes the converter straightforward to analyze. During the ON period, voltage at the switch is

$$V_{sw} = 0$$

During the OFF period,

$$V_{sw} = V_c$$

Assuming for the moment that voltage across the capacitor does not vary substantially, so that it may be assumed constant, the average voltage

$$< V_{sw} >= \frac{V_c(1 - D)T}{T} = V_c(1 - D)$$

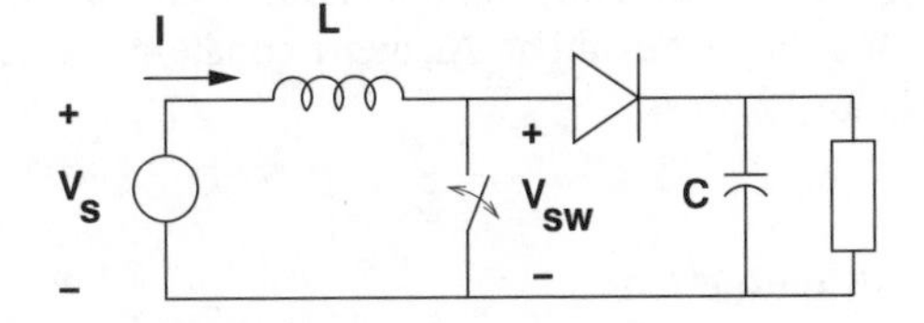

Figure 12.27 Boost converter

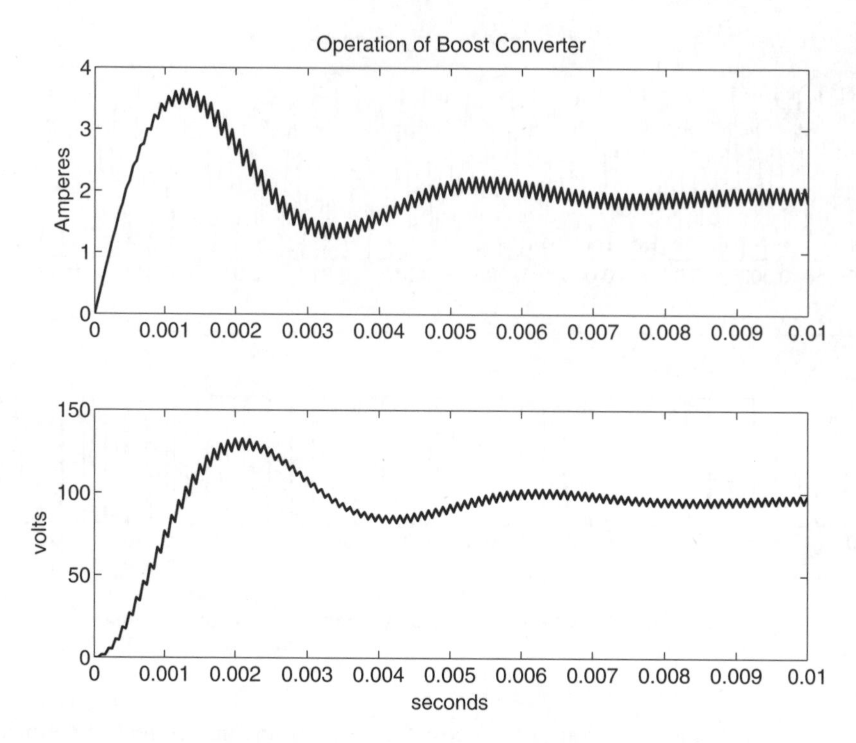

Figure 12.28 Startup transient of a boost converter

or, since the average voltage across the inductance must be zero, at least in steady state operation, the output voltage will be:

$$< V_c >= \frac{V_s}{1 - D}$$

Simulation of voltage buildup of a boost converter with a source voltage of 48 V, and inductance of 10 mH and a capacitance of 10 μF with a load resistance of 100 Ω is shown in Figure 12.29. The switching frequency is 10 kHz with a duty cycle of 1/2. Steady state operation of the same circuit is shown in Figure 12.29.

Operation of the boost converter as illustrated in Figures 12.28 and 12.29 assumes continuous conduction in the inductance. If the boost ratio becomes too high, conduction in the inductance becomes discontinuous: current in the inductance is driven down to zero on every cycle of the switch. In this discontinuous current mode the operation of the circuit takes discrete bits of charge and pumps them through to the output. An approximate shape of the current waveform would be as shown in Figure 12.30.

12.3.2.2 Discontinuous Conduction

In discontinuous conduction, the current in the inductor does go to zero. This would happen, for example, if the boost converter is operating at high boost ratio so that each pulse of current

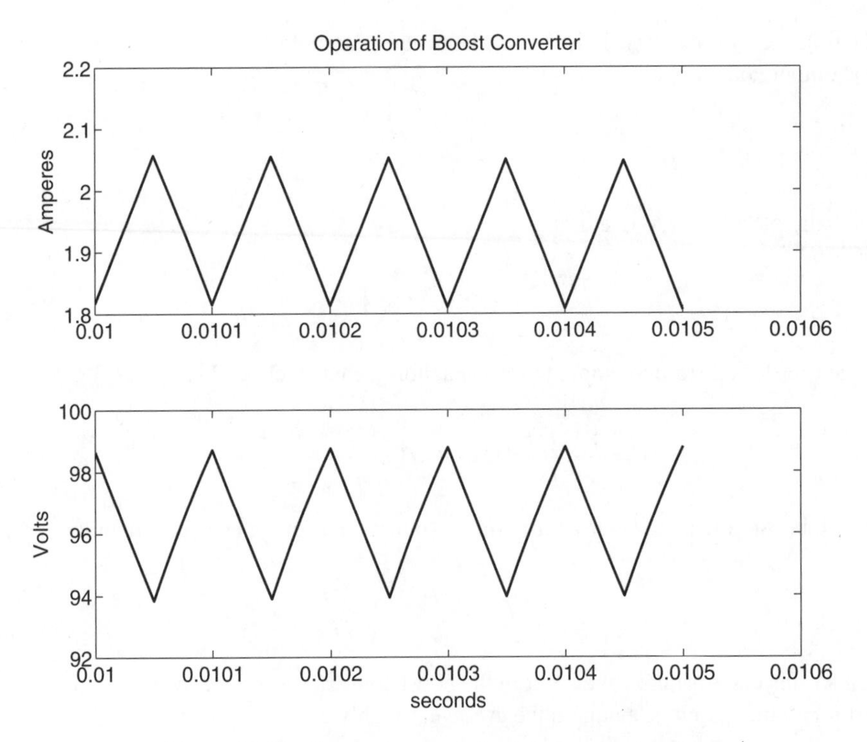

Figure 12.29 Steady state operation of a boost converter

drawn through the inductor by turning on the switch is rapidly driven to zero by the high capacitor voltage. Under discontinuous conduction, the peak of current in the inductor is:

$$I_p = \frac{V_s DT}{L}$$

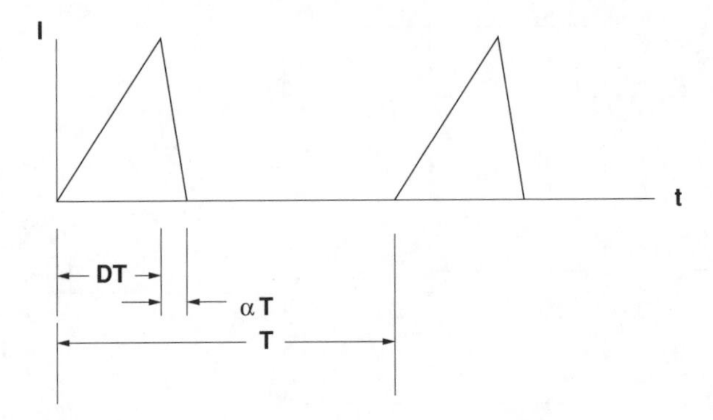

Figure 12.30 Current waveform in discontinuous conduction

If the time for current to rise is DT, the time for the current to fall back to zero must be related to the output voltage:

$$I_{\text{p}} = \frac{V_{\text{c}}\alpha T}{L}$$

or the ratio of fall time to rise time is:

$$\alpha = D\frac{V_{\text{s}}}{V_{\text{c}}}$$

Charge transferred from the input to the capacitor in each cycle will be:

$$q = \frac{1}{2}\frac{V_{\text{s}}DT}{L}DT(1+\alpha) = \frac{V_{\text{s}}D^2T^2(1+\alpha)}{2L}$$

For high boost ratios, the value of α is small, so that the average current is about:

$$< I >= \frac{q}{T} \approx V_{\text{s}}\frac{D^2T}{2L}$$

Under this circumstance, the input to the boost converter is, on the average, equivalent to a resistor and the output is, again on the average, roughly a current source.

12.3.2.3 Unity Power Factor Supplies

It is possible to take advantage of the fact that a boost converter operating in discontinuous conduction 'looks like' a resistance to make a power supply 'front end' that draws current that is more nearly a sinusoid than would otherwise be drawn by a full-wave rectifier. The circuit shown in Figure 12.31 shows a boost converter connected to the output of a four-diode, full-wave bridge. If the components are selected so that the boost converter is operating in discontinuous conduction mode, this thing will draw a waveform that is similar to what is shown in Figure 12.32, which was produced by simulation of a circuit fed by 120 V, RMS.

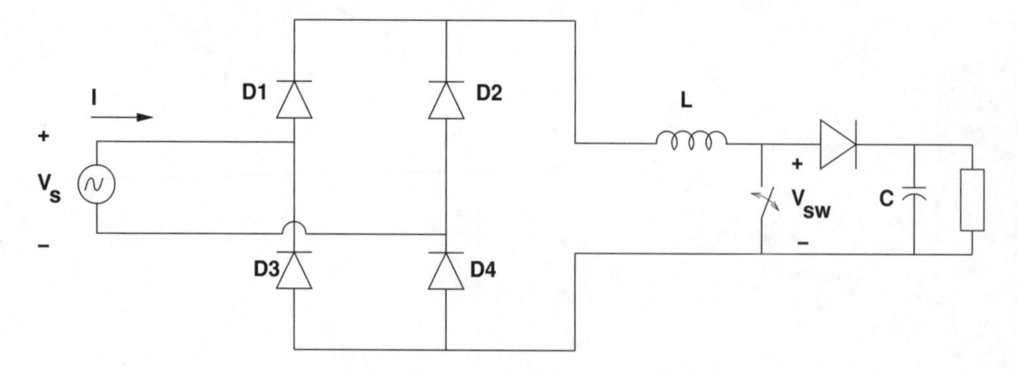

Figure 12.31 Unity power factor front end

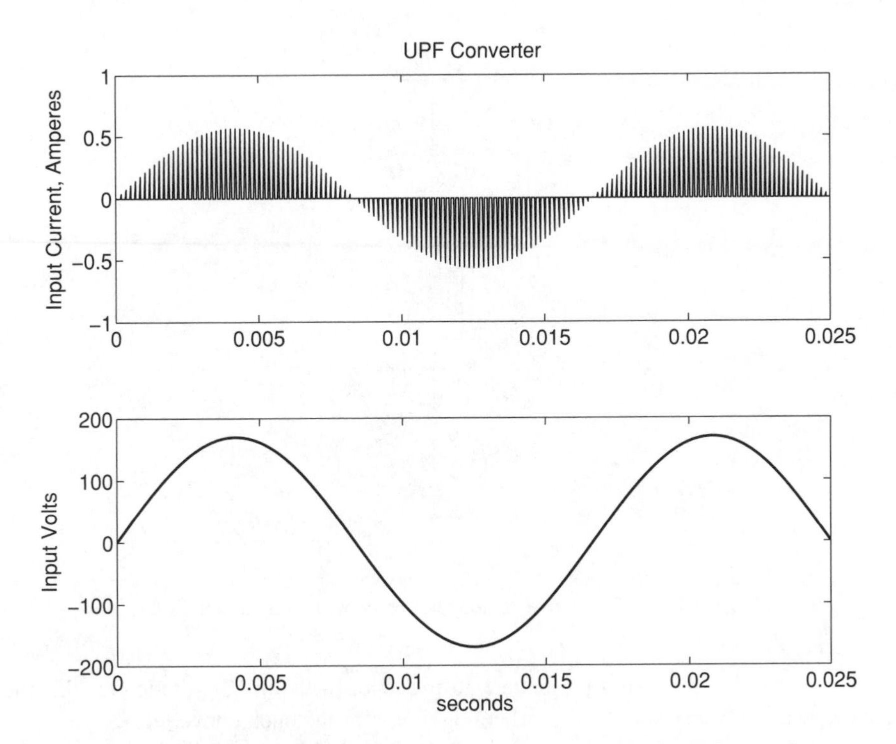

Figure 12.32 Current and voltage input to a unity power factor front end

The inductance is 10 mH, the capacitance is 10 μF, and the circuit is drawing about 9 W with a capacitor voltage of about 300 V, DC. The current drawn is in short 'spikes' with high amplitude. By itself that is not close to a sine wave. However, filtering this waveform with a relatively small capacitance is effective at making it very much like a sine wave. This circuit is commonly used in consumer products such as compact fluorescent lights.

12.4 Canonical Cell

The circuit shown in Figure 12.33 is a very commonly used 'totem pole' circuit, with two transistors and two diodes connected in reverse parallel. The figure shows the circuit built using IGBTs, but it could be constructed with just about any type of controllable switch. This connection can be used for a variety of DC–DC converters and for inverters as well.

12.4.1 Bidirectional Converter

Looking at Figure 12.34, which shows the canonical totem pole circuit embedded between a voltage source (shown as a battery) and some kind of load (shown as an inductance in series with a voltage source). Note that if the device Q1 is switched in PWM mode and Q2 is left off, this circuit serves as a buck converter. Assuming the load voltage V_l is not too large, this

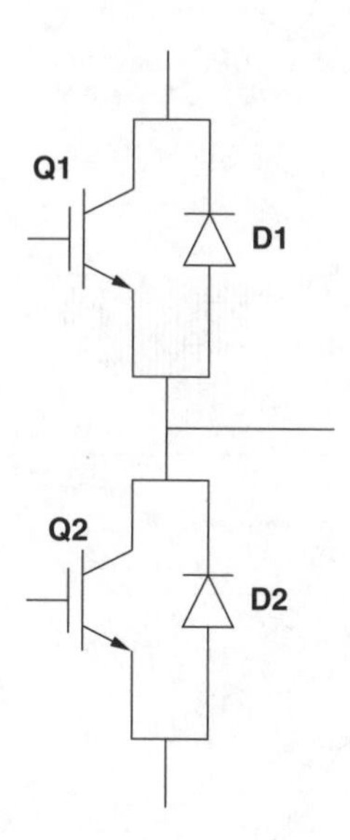

Figure 12.33 Two active device totem pole with antiparallel diodes

circuit can control current from the battery to the load. In this mode, diode D2 (the one in antiparallel with Q2) serves as the freewheeling diode for the buck converter.

In the other direction, if Q1 is turned off and Q2 is switched in PWM mode, the circuit serves as a boost converter, with the load inductance L storing energy when Q2 is ON and freewheeling through diode D1 to the battery. This mode of operation could be used, for example, to recover braking energy from a traction motor, if V_l is the back voltage of that motor.

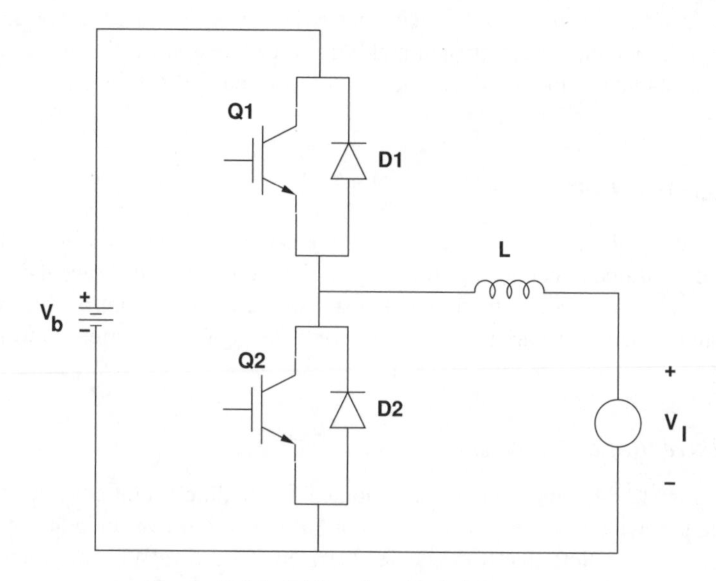

Figure 12.34 Bidirectional DC–DC converter

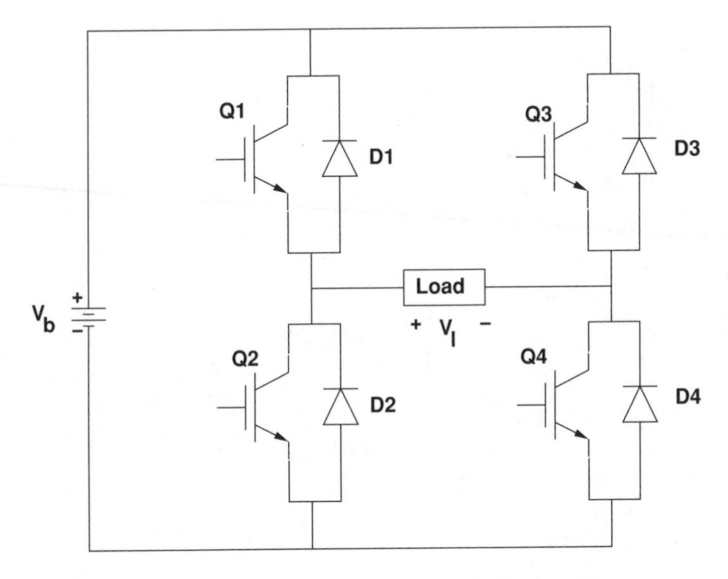

Figure 12.35 H-bridge voltage source single phase inverter

12.4.2 H-Bridge

A single-phase inverter circuit is shown in Figure 12.35. Here, two totem poles are connected, with the top and bottom of the totem poles connected to the source (shown as a battery), and with the two midpoints connected to either end terminal of a single phase load. The 'Load' is shown as a generic two terminal box. It could be almost any load, but it might be, for example, one phase of a motor, or it could be a DC motor that needs to be driven in either direction.

Consider how this might work. To establish a positive voltage V_l on the load, switches Q1 and Q4 would be turned on. For negative voltage out, switches Q2 and Q3 would be turned on. If all that is required is a square wave, this switching strategy is all that is needed.

If a square wave of current is required, putting a fairly large inductance in series with the inverter as shown in Figure 12.36 can generate a good approximation to a current source.

For an inverter that generates an approximation to a sine wave, pulse width modulation (PWM) would be used, as has been described in Section 12.3.1. There are two strategies that might be employed, with different levels of complexity of the control circuit. First, one could simply switch the two devices in each totem pole in opposition: in one state, Q1 and Q4 are ON and Q2 and Q3 are OFF. In the other state the opposite is true. In this way the voltage across the load is either $+V_b$ or $-V_b$. The result is as shown in Figure 12.37. Note that designers of this kind of circuit must take care that at no time are both top and bottom switches of either totem pole ON at the same time, as that would short the power supply. Generally a small delay is inserted between any OFF signal and the complementary ON signal to ensure that no direct shorting path is established between the two supply rails.

Alternatively, the devices could be switched in approximately the same scheme, but only the two top switches or the two bottom switches would be used for PWM. This does require more than the one comparator to generate signals, but results in fewer switching operations.

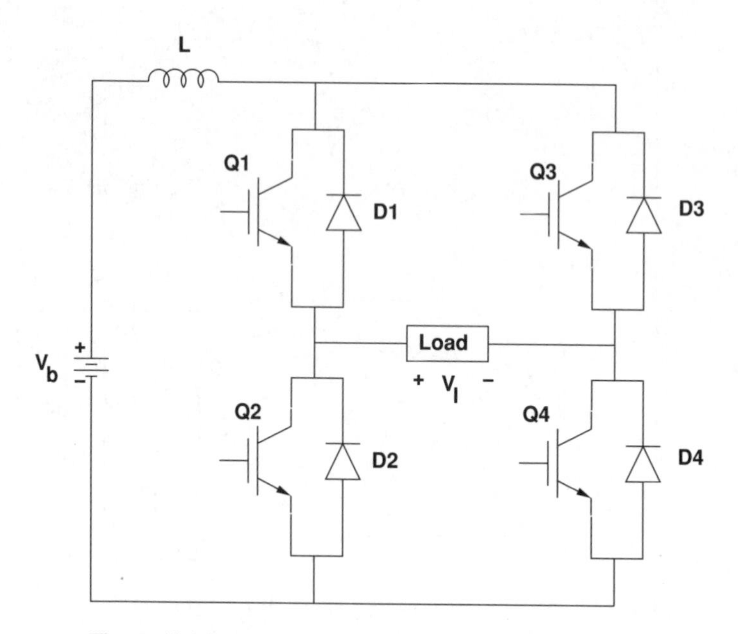

Figure 12.36 H-bridge current source single phase inverter

For positive load voltage, devices Q1 and Q4 would be turned on, but only one of the two devices would be used for generation of the PWM signal. That might be the top switch, Q1. Note that when Q1 is OFF, if there is still positive current in the load, it will freewheel through D2. Then, for negative load voltage, Q2 would be turned ON and Q3 would generate the pulse width modulation. The resulting voltage waveform is as shown in Figure 12.38.

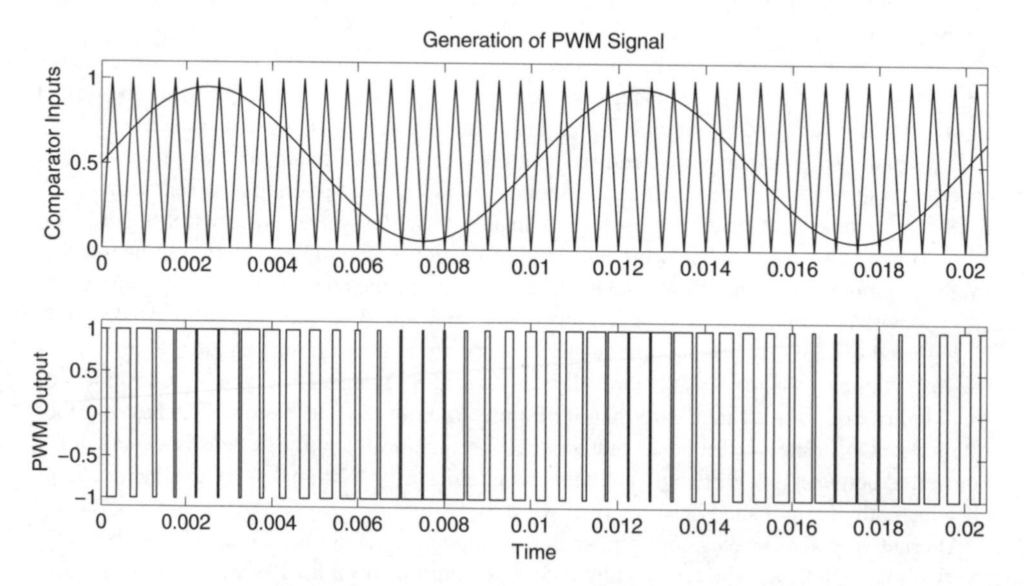

Figure 12.37 H-bridge PWM output, simple switching scheme

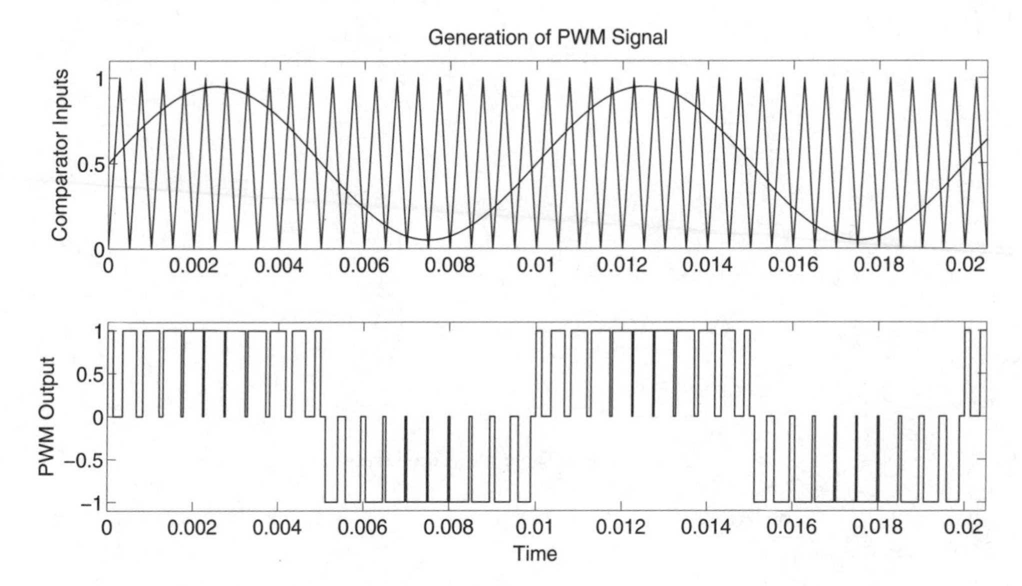

Figure 12.38 PWM modulation: single phase H bridge, top device modulation

12.5 Three-Phase Bridge Circuits

Three phase power electronic circuits share some of the advantages of other types of polyphase power circuits. With balanced currents comes the advantage of economy of use of components. A three-phase rectifier circuit, for example, requires only three stacks of components. High power rectifiers are usually implemented as three (or perhaps more) phase circuits, and motor drives are usually polyphase circuits. High-voltage DC transmission lines are implemented as line commutated three-phase bridges.

12.5.1 Rectifier Operation

Start by ignoring the effects of source inductance. The voltages on the AC side (left side of Figure 12.39 are represented by the phasor diagram shown in Figure 12.40. Note a time reference that makes the analysis convenient has been used.

$$v_a = \sqrt{2}V \cos\left(\omega t + \frac{\pi}{3}\right) \tag{12.1}$$

$$v_b = \sqrt{2}V \cos\left(\omega t - \frac{\pi}{3}\right) \tag{12.2}$$

$$v_c = \sqrt{2}V \cos(\omega t - \pi) \tag{12.3}$$

where V is the RMS line–neutral voltage. These are shown in Figure 12.41

The converter will be working from phase–phase voltages. For example, if switches 1 and 6 are ON, the voltage on the DC side of the switch array will be $v_a - v_c$. The two switches

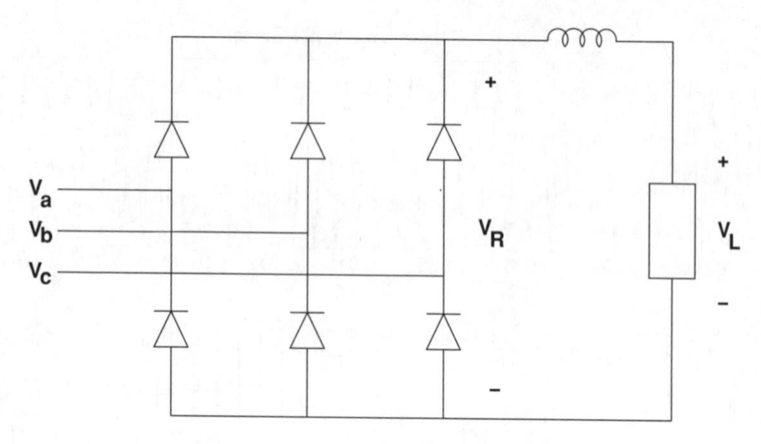

Figure 12.39 Three-phase diode rectifier

will both be ON only if this voltage is positive. For reference, the six possible line–line voltages are:

$$v_a - v_c = \sqrt{6}V\cos\left(\omega t + \frac{\pi}{6}\right) \tag{12.4}$$

$$v_b - v_c = \sqrt{6}V\cos\left(\omega t - \frac{\pi}{6}\right) \tag{12.5}$$

$$v_b - v_a = \sqrt{6}V\cos\left(\omega t - \frac{\pi}{2}\right) \tag{12.6}$$

$$v_c - v_a = \sqrt{6}V\cos\left(\omega t - \frac{5\pi}{6}\right) \tag{12.7}$$

$$v_c - v_b = \sqrt{6}V\cos\left(\omega t + \frac{5\pi}{6}\right) \tag{12.8}$$

$$v_a - v_b = \sqrt{6}V\cos\left(\omega t + \frac{\pi}{2}\right) \tag{12.9}$$

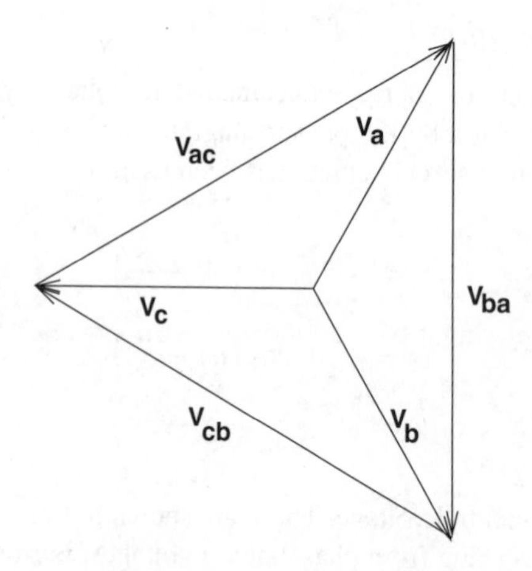

Figure 12.40 Voltage phasors

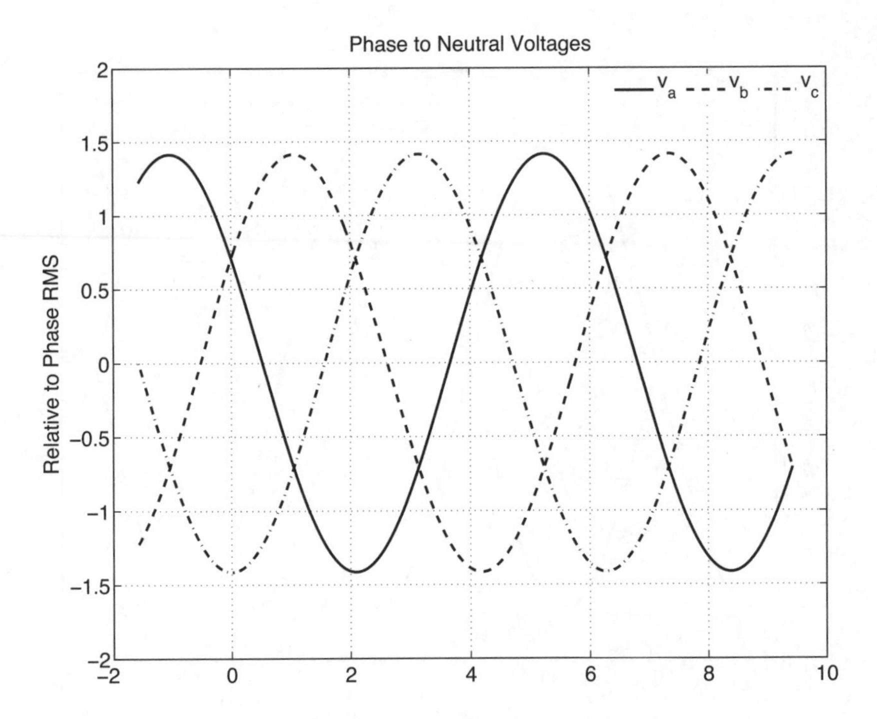

Figure 12.41 Line–neutral voltages

These are illustrated in Figure 12.42.

For this phasing, that is for the specific time origin chosen, v_b is the most positive of the three phases from time $\omega t = 0$ until $\omega t = \frac{2\pi}{3}$ and v_c is the most negative voltage from $\omega t = -\frac{\pi}{3}$ until $\omega t = \frac{\pi}{3}$. So $v_b - v_c$ is most positive over the interval $0 < \omega t < \frac{\pi}{3}$.

If the system were to use diode rectifiers, which means that the devices conduct whenever the voltage across the device would otherwise be positive, the average voltage on the DC line would be:

$$< V_{DC} >= \frac{3}{\pi} \int_0^{\frac{\pi}{3}} \sqrt{6} V \cos\left(\omega t - \frac{\pi}{6}\right) d\omega t = \frac{3}{\pi} \sqrt{6} V \qquad (12.10)$$

Since the system repeats this pattern of voltages identically for every interval of time of length $\omega t = \frac{\pi}{3}$, that is, every one sixth of a cycle, this is the average DC voltage. If the DC inductor is large enough, current will be about constant or varying very slowly. The voltage at the output of the rectifier (input of the inductor) will consist of this average value plus some 'ripple'.

12.5.2 Phase Control

If there is a large inductor in series with the DC side of the circuit, the DC current I_{DC} can be considered to be constant. Once two devices are ON and conducting, they can continue to

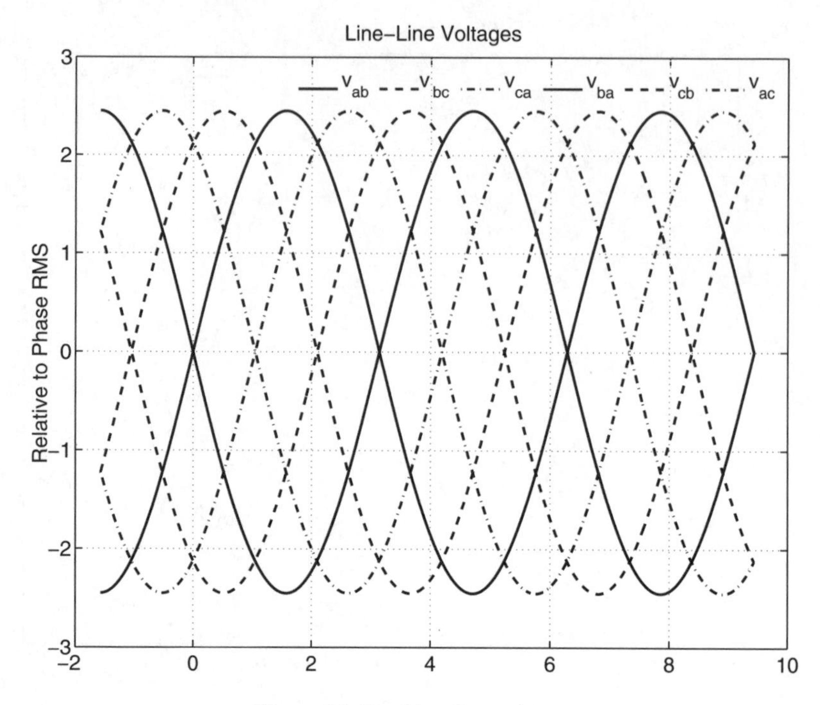

Figure 12.42 Line–line voltages

conduct as long as the voltage behind the 'positive' device is more positive than the voltage behind the 'negative' device. So if a regular firing pattern is established so that each of the 'positive' switches (Q1, Q3 and Q5) are fired at a time $\omega t = \alpha$ *after* their respective phase voltages become the most positive and do the same for the negative switches (Q2, Q4 and Q6), meaning firing of all switches is delayed by the phase angle α after they would turn on were they diodes, average voltage becomes:

$$< V_{DC} >= \frac{3}{\pi} \int_{\alpha}^{\alpha+\frac{\pi}{3}} \sqrt{6} V \cos\left(\omega t - \frac{\pi}{6}\right) d\omega t = \frac{3}{\pi} \sqrt{6} V \cos \alpha \tag{12.11}$$

This is how the phase-control rectifier can be used to control voltage. Note also that the voltage presented to the DC system can be *negative*, since α can be set to values greater than $\frac{\pi}{2}$, for which the cosine function is negative. In this case, the terminal is absorbing power from the DC side rather than supplying power. In a high voltage DC transmission line, this is called the 'receiving' terminal and serves as an inverter. It is important to note that a receiving terminal still absorbs reactive power from the AC system.

12.5.3 Commutation Overlap

The results of Equation 12.11 assume that the voltage at the input leads of the diodes are idealized sine waves. This is not usually the case because the source has series inductance,

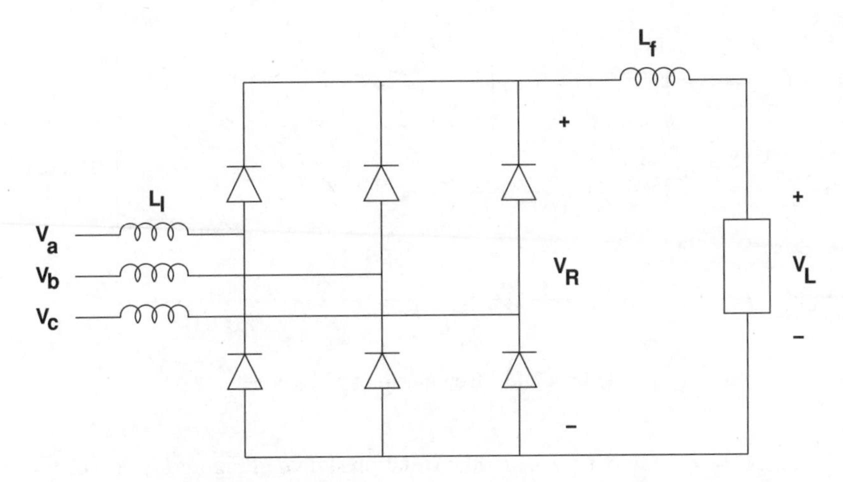

Figure 12.43 Diode rectifier with source inductance

shown in Figure 12.43, so that when one phase voltage exceeds the other there is a transient time while the current in one lead is driven down and the current in the other lead comes up. During this transient, the output voltage is not as high as it would otherwise be. The commutation side voltage is actually the average of the two phases undergoing commutation, as shown in Figure 12.44.

Commutation overlap can be shown to introduce a voltage drop in the DC side of a circuit that 'looks like' a resistance, but that does not dissipate power (it actually introduces a phase

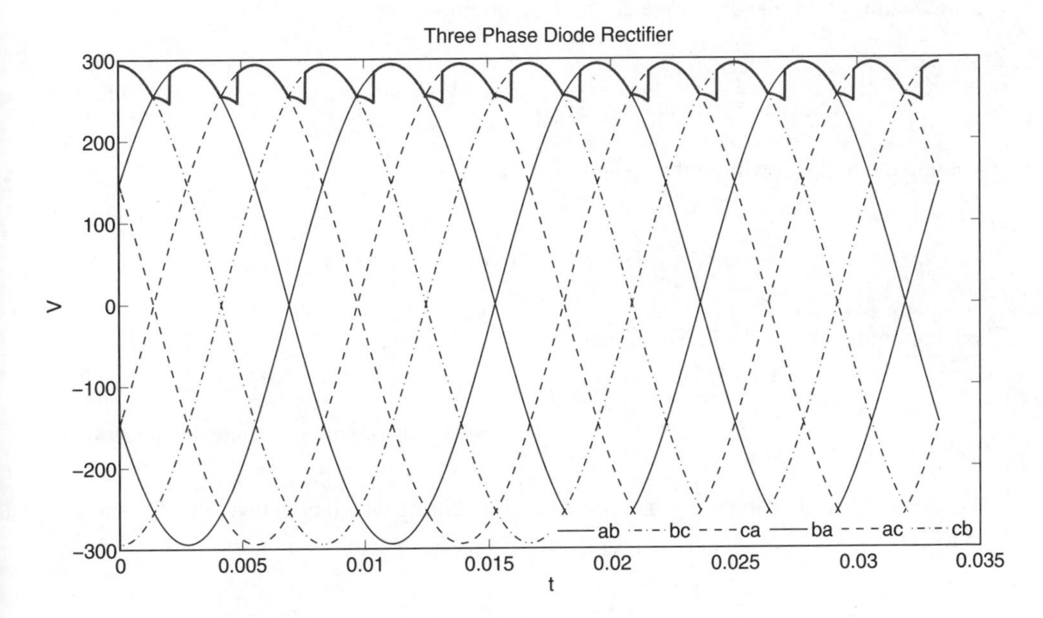

Figure 12.44 Diode rectifier voltage with input inductance

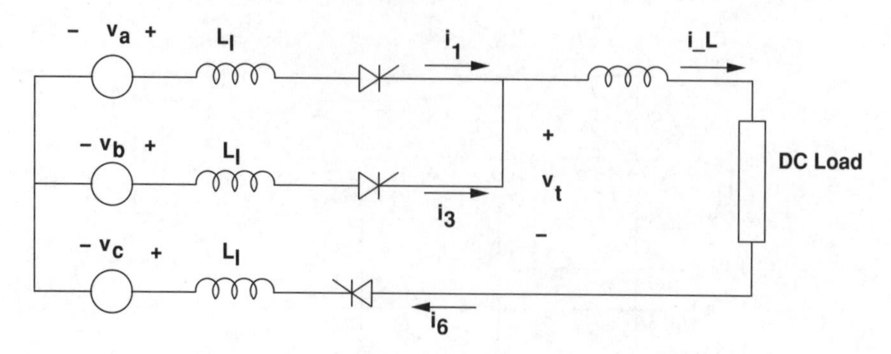

Figure 12.45 Commutation of a rectifier

shift and changes power factor). Consider the situation shown in Figure 12.45. The transfer of current from phase a (Q1) to phase b (Q3) can happen only when $v_b > v_a$.

During the transition between phase a and phase b, the voltage at the input to the filter inductor will be

$$v_t = \frac{v_a + v_b}{2} - v_c \tag{12.12}$$

During that transition, the sum of the two currents will be:

$$i_1 + i_3 = i_L \tag{12.13}$$

During the transition, the derivative of Phase b current will be:

$$\frac{di_3}{dt} = \frac{v_b - v_a}{2L_l} = \sqrt{6}V \sin \omega t \tag{12.14}$$

Assuming the initial condition that is $i_3 = 0$ at $t = 0$,

$$i_3 = \frac{\sqrt{6}V}{2\omega L_l}(\cos \alpha - \cos \omega t) \qquad \omega t > \alpha \tag{12.15}$$

Then, noting that the final value of i_3 must be the load current i_L,

$$i_L = \frac{\sqrt{6}V}{2\omega L_l}(\cos \alpha - \cos(\alpha + u)) \tag{12.16}$$

where u is the 'overlap angle', or that period of time during which both thyristors Q1 and Q2 are ON. That angle is:

$$u = \cos^{-1}\left(\cos \alpha - \frac{2\omega L_l i_L}{\sqrt{6}V}\right) - \alpha \tag{12.17}$$

Now, to find the average load voltage, break the time into two periods: the overlap and the non-overlap:

$$
\begin{aligned}
v_{\text{avg}} &= \frac{3}{\pi}\left(\int_{\alpha}^{\alpha+u}\left(\frac{v_{\text{a}}+v_{\text{b}}}{2}-v_{\text{c}}\right)\mathrm{d}\omega t+\int_{\alpha+u}^{\alpha+\frac{\pi}{3}}(v_{\text{b}}-v_{\text{c}})\,\mathrm{d}\omega t\right)\\
&= \frac{3}{\pi}\left(\int_{\alpha}^{\alpha+u}\left(\frac{v_{\text{a}}-v_{\text{c}}}{2}+\frac{v_{\text{b}}-v_{\text{c}}}{2}\right)\mathrm{d}\omega t+\int_{\alpha+u}^{\alpha+\frac{\pi}{3}}(v_{\text{b}}-v_{\text{c}})\,\mathrm{d}\omega t\right)\\
&= \frac{3}{\pi}\left(\int_{\alpha}^{\alpha+\frac{\pi}{3}}(v_{\text{b}}-v_{\text{c}})\,\mathrm{d}\omega t-\int_{\alpha}^{\alpha+u}\frac{v_{\text{b}}-v_{\text{a}}}{2}\mathrm{d}\omega t\right)\\
&= \frac{3\sqrt{6}V}{\pi}\cos\alpha-\frac{3}{\pi}\omega L_1 I_{\text{L}}
\end{aligned}
\tag{12.18}
$$

On the load side, the effect of leakage reactance is to produce load current dependent voltage drop. While it appears resistive from the DC side, this voltage drop does not involve any dissipation, as it is due solely to inductive elements.

12.5.4 AC Side Current Harmonics

If the inductor on the DC side is sufficiently large, the DC current will be nearly constant.Switching action will turn this current into what is shown in Figure 12.46.

With the assumption that DC side current is constant, phase A current is a pulse wave that is about constant for an interval of ωt that is $\frac{2\pi}{3}$ wide, and will be odd order periodic. See Figure 12.47, which is drawn with the assumption that the phase a current fundamental is a sine function. The actual waveform is time shifted from this, but the figure allows for ready calculation of the harmonic components.

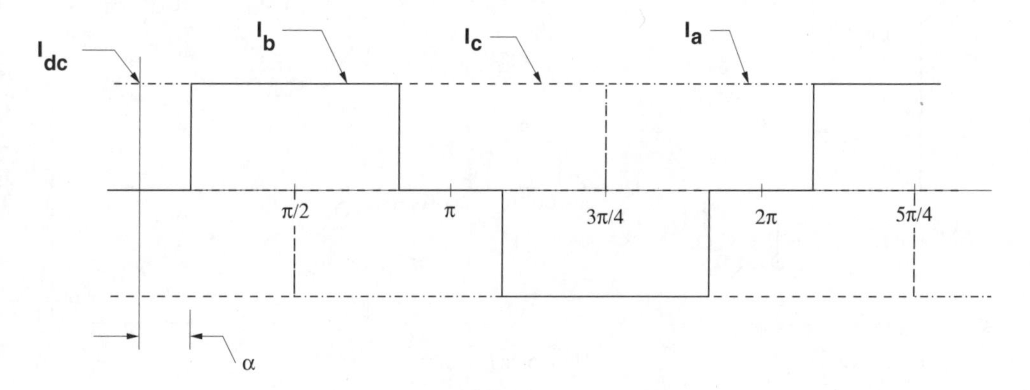

Figure 12.46 AC side currents

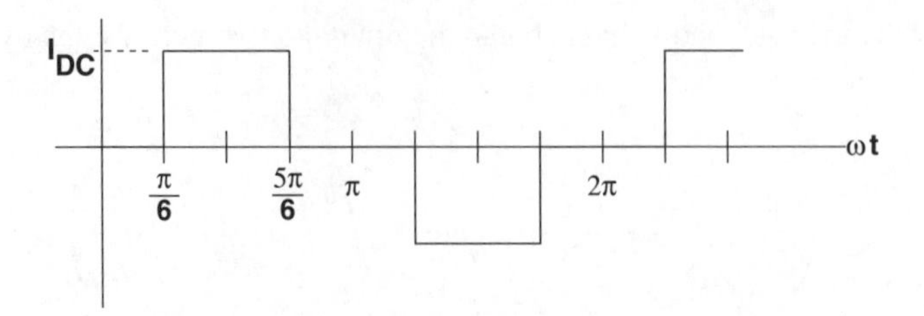

Figure 12.47 Calculation of time harmonics of AC current

The coefficients of the Fourier series for this can be found by:

$$I_n = \frac{2}{\pi}\int_{\frac{\pi}{2}-\frac{\pi}{3}}^{\frac{\pi}{2}+\frac{\pi}{3}} I_{DC}\sin n\omega t\, d\omega t = \frac{2I_{DC}}{n\pi}\left(\cos\left(\frac{\pi}{2}-\frac{\pi}{3}\right)-\cos\left(\frac{\pi}{2}+\frac{\pi}{3}\right)\right)$$
$$= \frac{4}{n\pi} I_{DC}\sin\left(n\frac{\pi}{2}\right)\sin\left(n\frac{\pi}{3}\right)$$

The coefficients are shown in Figure 12.48. Note that in the case of 120° conduction, all of the harmonics of triplen order (divisible by 3), have harmonic amplitudes of zero. Of course all even order harmonics have zero amplitude as well because the waveform has odd symmetry ($v(\phi + 180°) = -v(\phi)$).

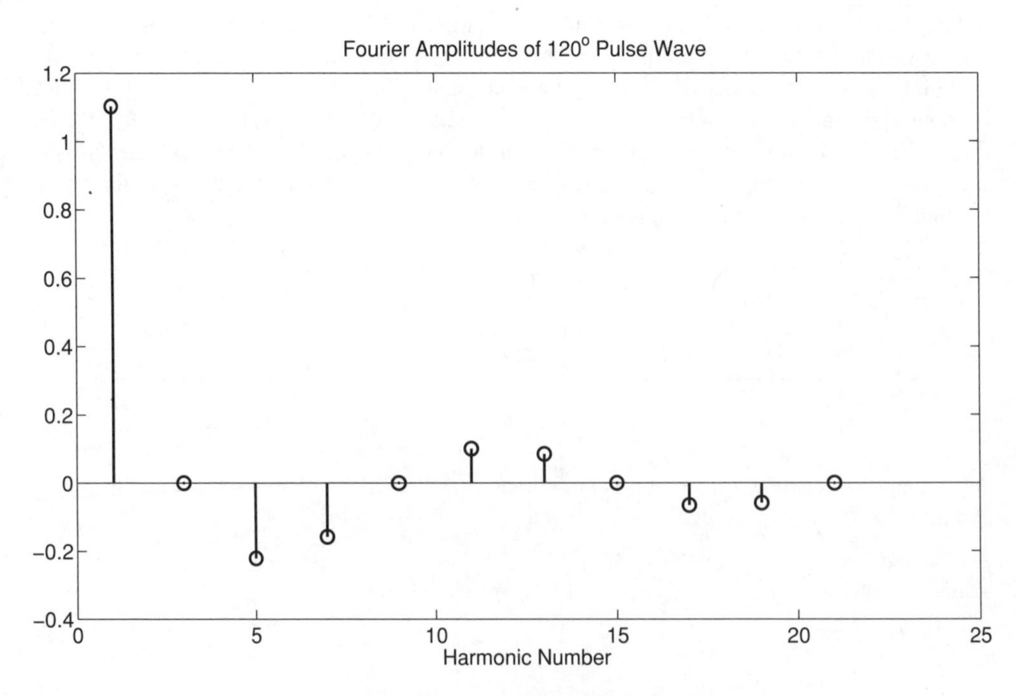

Figure 12.48 Time harmonic amplitudes for 120° conduction

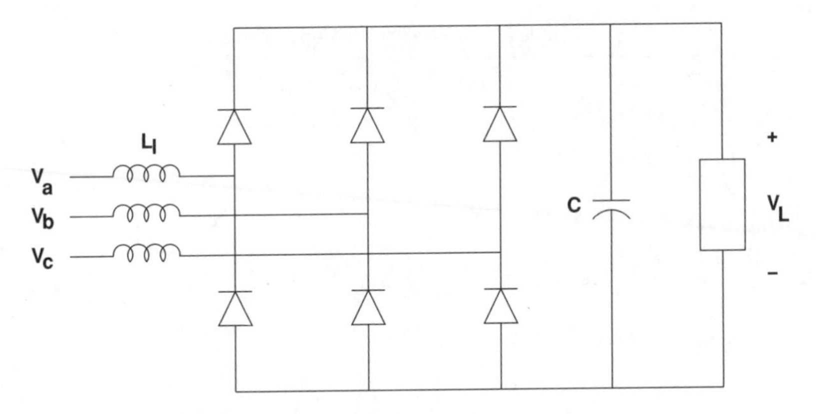

Figure 12.49 Three-phase diode bridge with filter capacitor

The phase currents are then:

$$i_a = \sum_{n \text{ odd}} I_n \cos\left(n\left(\omega t + \frac{\pi}{3} - \alpha\right)\right)$$
$$i_b = \sum_{n \text{ odd}} I_n \cos\left(n\left(\omega t - \frac{\pi}{3} - \alpha\right)\right)$$
$$i_c = \sum_{n \text{ odd}} I_n \cos\left(n\left(\omega t - \pi - \alpha\right)\right)$$

12.5.4.1 Power Supply Rectifiers

A common application for three-phase diode bridge rectifiers is to feed DC circuits that support roughly constant DC voltage using a filter capacitor such as the circuit as shown in Figure 12.49. As in single-phase bridge circuits with similar DC side configuration, this circuit draws currents that may be somewhat problematic to the AC side circuit. Phase current for this situation is shown in Figure 12.50. Here the phase current comes in two pulses, associated with the *other* two phases peaking. That is, the capacitor holds the DC side voltage roughly constant. When the phase a voltage becomes most positive, the rectifier voltage is below the capacitor voltage so no current flows. When the rectified voltage becomes more positive than the capacitor, current starts to flow, but then when the negative half phases commutate, the rectified voltage is lower and the phase current falls. But then when the next phase becomes more negative the rectified voltage increases and there is a second pulse. This ‘double pulse’ current pattern is characteristic of this kind of circuit. It results in relatively poor power factor and time harmonic distribution of current into this kind of power supply.

12.5.4.2 PWM Capable Switch Bridge

The circuit of Figure 12.51 is used in a wide variety of applications. It can be a motor drive circuit, it can be an active rectifier and it is also used as an inverter for connecting solar cells to the power system.

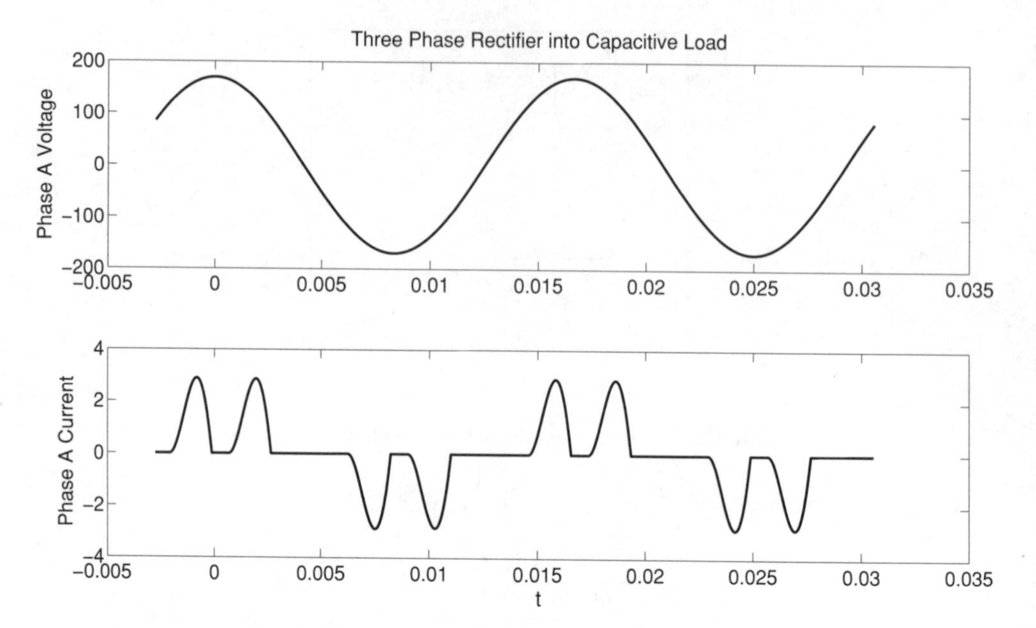

Figure 12.50 Voltage and current input with filter capacitor

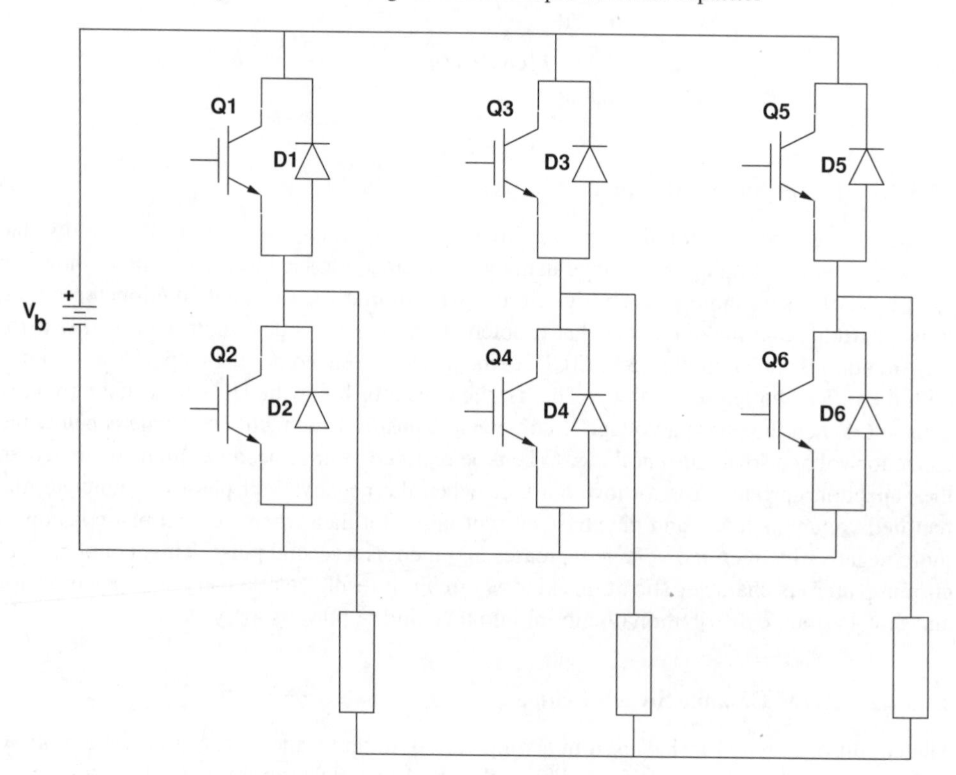

Figure 12.51 Three-phase IGBT bridge

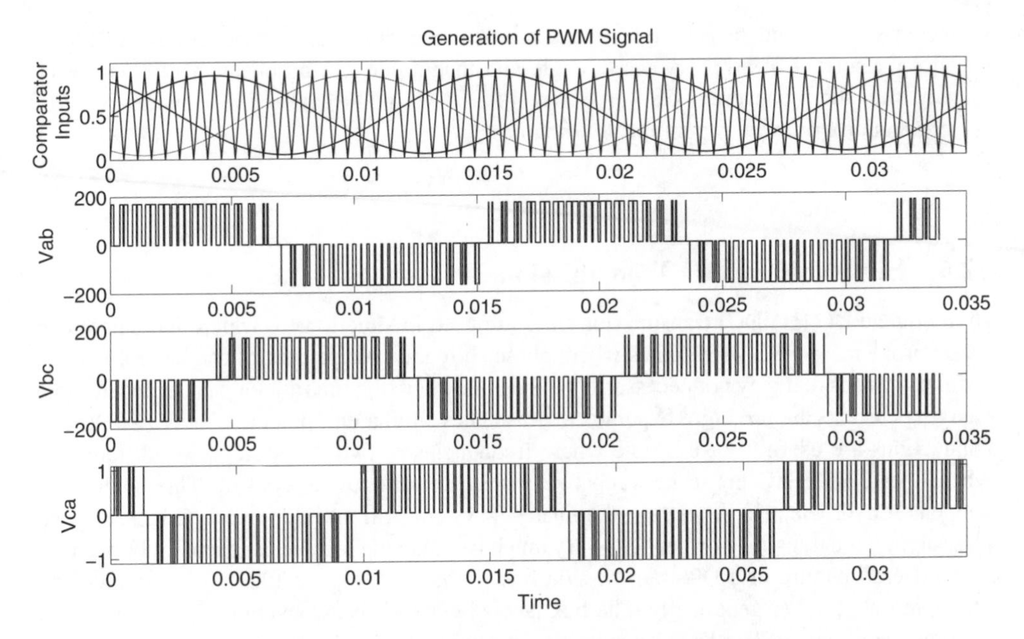

Figure 12.52 Three-phase PWM waveforms

The same system for generating PWM signals as was shown for single phase (H-bridge circuits) can be used for three-phase circuits. A sine wave is established for each of the three phases, and compared to a triangle wave, as shown in the top trace of Figure 12.52. Firing the switches in each totem pole of the bridge in complementary fashion (top switch ON and bottom switch OFF and vice versa), according to the comparison of the reference signal and triangle wave, yields line–line signals as shown in the lower three traces of Figure 12.52. When filtered by, say, the input inductance of a motor this gives a good approximation to a sine wave.

Three-phase PWM bridges can be controlled to serve as current sources. By adjusting the voltage produced by the bridge and its phase angle with respect to power system angle, the current from the bridge can be set (within limits related to DC bus voltage) to produce a current with any phase angle with respect to the power system. Inverters used in conjunction with solar cells and wind turbines are build this way and are often controlled to inject current into the power system at a fixed power factor, meaning the injected current has a fixed angle with respect to the system voltage. This is illustrated in Figure 12.53. By selecting the voltage

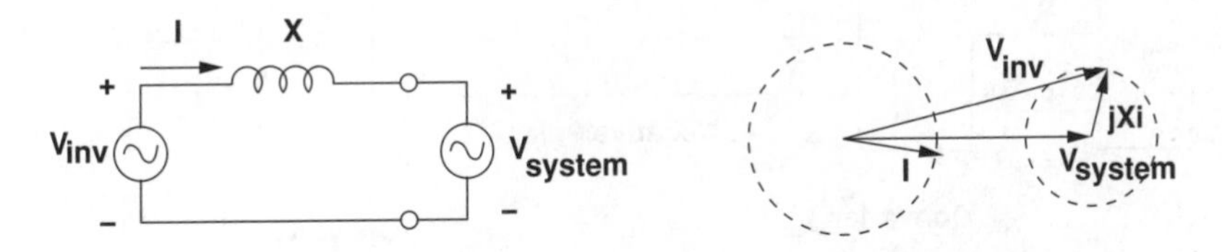

Figure 12.53 Voltage source inverter injecting current into power system

source magnitude and phase angle (the input to the comparator that generates the PWM waveform), current of arbitrary phase angle and magnitude can be injected into the power system. Thus these converters can serve as rectifiers (AC to DC conversion), inverters (DC to AC conversion), or even as pure reactive sources.

12.6 High-Voltage DC Transmission

High voltage DC (HVDC) transmission lines are used in situations in which line length can create problems: for very long lines where phase shift and voltage control can be problematic or for cable runs that may not necessarily be long but where the runs are long enough that cable charging currents are too high. In some circumstances similar equipment forming AC/DC/AC connections are useful. For example where frequencies of two AC systems are different or where the relative AC phase between two systems is difficult to control. This sometimes happens if hydroelectric generation dominates a system or if an inter-area link between two large areas has a transmission capacity very much less than the installed capacity of both areas.

An overall picture of a DC transmission line is shown in Figure 12.54. Two AC systems are represented rather generically. The box labeled generically 'Converter' will be discussed in some detail below. The DC transmission lines typically have positive and negative poles. Each pole has a high voltage relative to ground, and in most circumstances the currents in the two poles are approximately balanced so there is only small current in the ground. However, some HVDC systems are built so that ground return can be used if one pole is out of service.

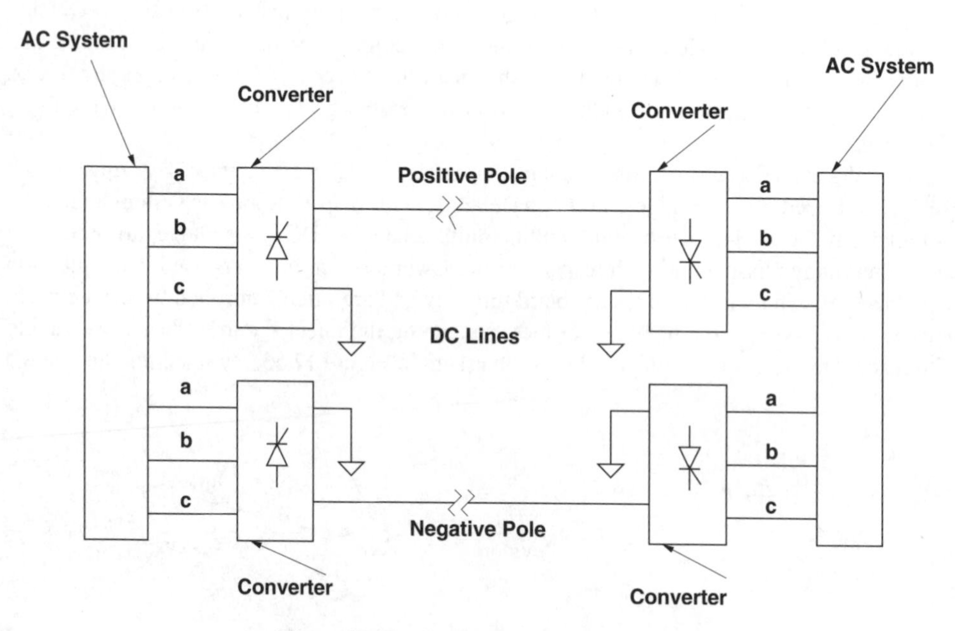

Figure 12.54 HVDC transmission system

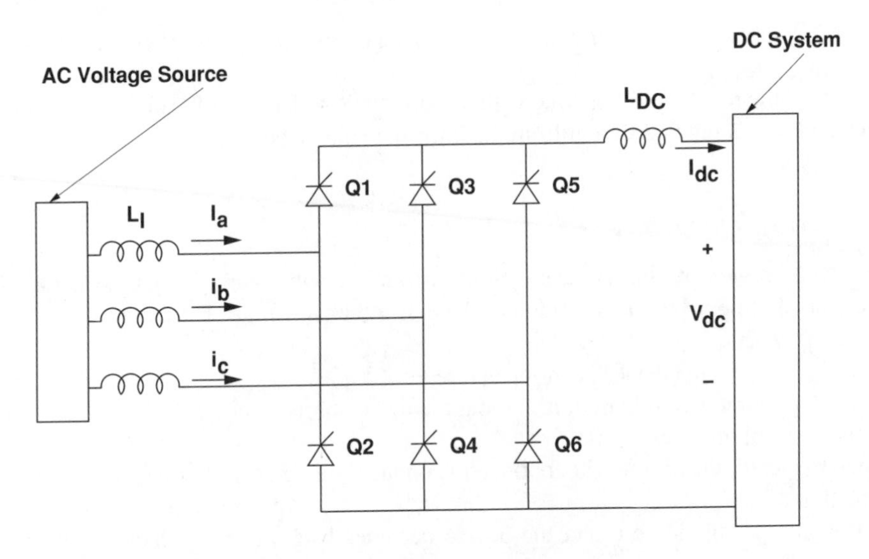

Figure 12.55 HVDC rectifier terminal

12.7 Basic Operation of a Converter Bridge

The basic setup of one of a rectifier converter is shown in Figure 12.55. This is the 'sending end' converter, if the voltage and current have the signs as shown in the picture. This is actually a fairly generic picture of a three-phase rectifier bridge, and the analysis presented earlier in this chapter applies. If the AC voltages have a line–line RMS voltage V, the peak line–line voltage is $\sqrt{2}\ V$. This is converted to DC by the action of the rectifier. The commutation interval and resulting voltage drop are represented by Figure 12.56.

It is straightforward to show that the time average DC voltage is:

$$V_{\mathrm{DC}} = \frac{3\sqrt{2}V}{\pi}\cos\alpha - \frac{3}{\pi}\omega L_1 I_{\mathrm{DC}}$$

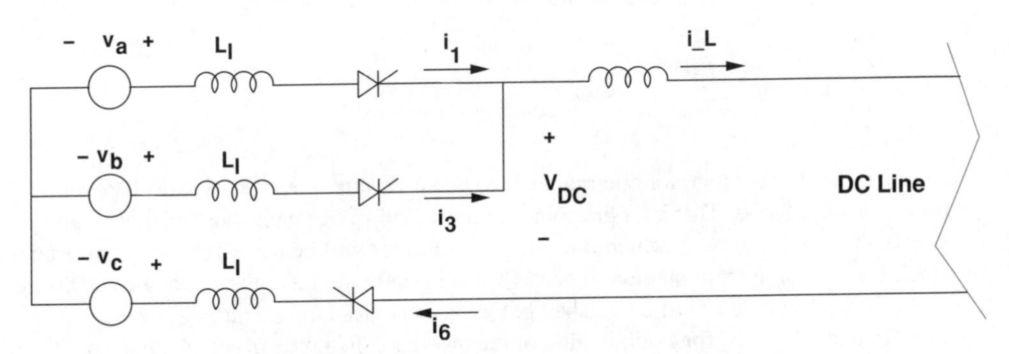

Figure 12.56 Commutation of the rectifier terminal

where L_l is the leakage, or commutation inductance of the transformers and the rest of the system.

Generally, the rectifier will operate with a phase angle α that is fairly small, so that the DC voltage will be not much different from the line–line voltage peak.

12.7.1 Turn-On Switch

Figure 12.55 represents the switches as thyristors. Actually, early HVDC systems used mercury arc rectifiers, but the operation of these switches is similar. They behave as follows:

1. Unless triggered, the device is an 'open circuit'.
2. If the device is forward biased, triggering the device turns it 'on', meaning it can conduct in the forward direction.
3. Once triggered, the device will continue to conduct as long as the current is in the forward direction.
4. If the current falls to zero and the device becomes back biased (voltage in the reverse direction), it stops conducting and remains 'open' until it is again forward biased and triggered.
5. Typically, triggering of the device is accomplished by injecting current into the 'gate'. In many applications, light triggered devices are used, so the trigger is a pulse of light through a fiber optic cable.

12.7.2 Inverter Terminal

At the 'receiving' end of the HVDC line, power must be turned back into AC. A schematic cartoon of the receiving end terminal is shown in Figure 12.57. Note that this is nearly the same as the rectifier terminal, but the switches are shown with their sense flipped. This is because the thyristor switches can carry current in only one direction. The voltage presented to the HVDC transmission line must be positive in this sense, so that the terminal absorbs power. To do this, the switches must operate with a phase angle α fairly close to π.

The commutation picture for the HVDC inverter terminal is as shown in Figure 12.58. This is very nearly the same as the picture for the rectifier, but with the sign of currents reversed.

It is straightforward to show that, for the inverter terminal, average DC voltage presented to the line is:

$$V_{DC} = \frac{3\sqrt{2}V}{\pi}\cos(\pi - \alpha) + \frac{3}{\pi}\omega L_l I_{DC}$$

It is important to note that, for commutation to be successful, the process must be over by the time voltages reverse. That is, if the point in time where v_b becomes more positive than v_a is $\omega t = 0$, the point in time at which this situation reverses will be $\omega t = \pi$. For the inverter terminal, commutation from phase a to phase b would normally take place at angles α close to π, but the commutation must be finished before v_a becomes larger than v_b at $\omega t = \pi$. The commutation overlap u is, for commutation of the inverter called the *extinction angle* and it is necessary that $\alpha + u < \pi$.

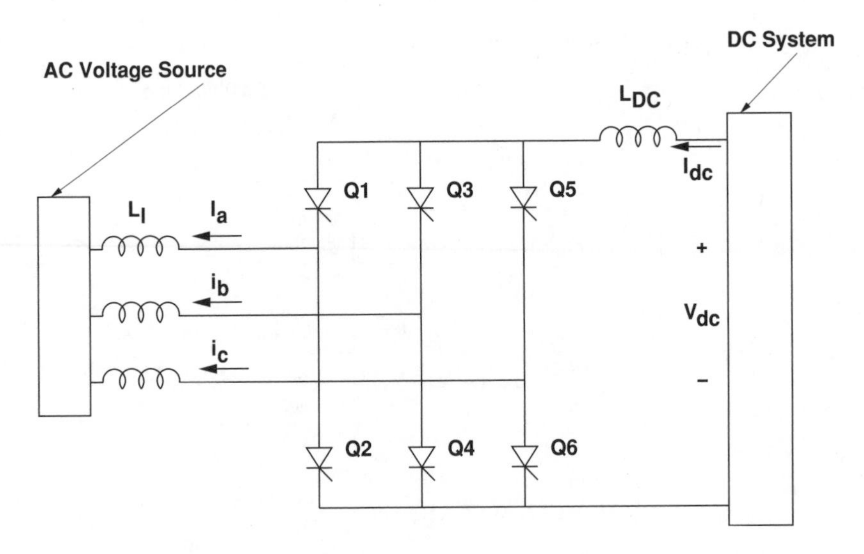

Figure 12.57 HVDC inverter terminal

12.8 Achieving High Voltage

HVDC systems are built with emphasis on *high*. They are often built with voltages on the order of ±400 kV or even higher. To achieve such voltages with existing devices it is necessary to employ relatively long strings of thyristors in series. In addition, whole converter strings are sometimes put in series, as is shown in Figure 12.59. The AC sides of a bank of transformers are connected in parallel, while the converters to which they are connected are connected in series. This does add to the insulation stress on the transformers, which must be specially built for this duty. However, another benefit of this kind of connection can be seen in the reduction of time harmonics. Both fifth and seventh harmonics are cancelled by this connection. To see this, note that the delta–wye connection shifts the voltage by 30°, which at fifth harmonic frequency is 150°. The time harmonic current is shifted by another 30° across the *delta–wye*, in the positive direction because fifth harmonic has reverse phase sequence. Thus there is a total of 180° of phase shift for a complete phase reversal. The seventh harmonic is similarly shifted, but it is shifted by 210° on the wye side of the transformer, but since it is positive

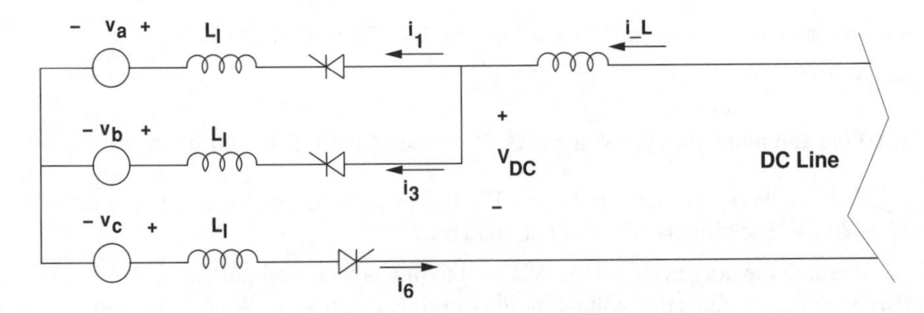

Figure 12.58 Commutation of the inverter terminal

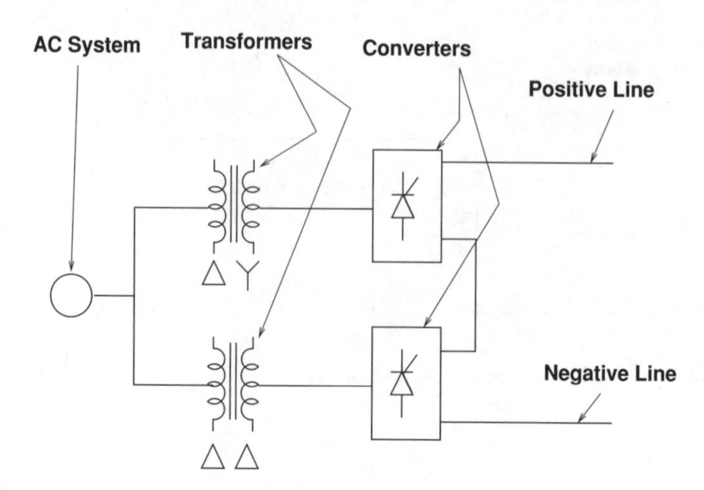

Figure 12.59 Twelve-pulse converter

sequence, the delta–wye shifts it *back* by 30°, achieving the same 180° shift. Some systems employ this technique with multiple stages and progressively smaller phase shifts across the transformers.

On the DC side of this system, the voltage ripple is made smaller because of the doubling of the 'pulse number'. A simple Graetz bridge (the six-switch bridge) achieves a pulse number of six. The stacked bridge with *delta–wye* and *delta–delta* transformers shown in Figure 12.59 gives a pulse number of 12. Higher pulse numbers result in smoother DC voltage and further current harmonic cancellation on the AC side.

12.9 Problems

1. Suppose you have a diode with reverse leakage current $I_0 = 2 \times 10^{-16}$ A. Assume the temperature is 299 K.
   (a) What is the forward voltage (ignoring resistance) if diode current is 1 A?
   (b) What is the forward voltage (ignoring resistance) if diode current is 10 A?
   (c) Plot, as best you can, forward voltage for current varying between 0 and 100 A.

2. For the same diode, carrying 10 A, remembering that 0° C is 273 K.
   (a) What is the forward voltage at 40° C?
   (b) What is the forward voltage at 0° C?
   (c) Find and plot forward voltage over the range of −40° C to 100° C.

3. A 'buck' converter is shown in Figure 12.60. The inductor is L = 6 mH. The capacitor is $C = 20\ \mu$F. The thing is switching at 10 kHz.
   (a) Assume the duty cycle is 0.5. What is the time average output voltage?
   (b) Assume the capacitor voltage is not varying very much. What is the current in the inductor? Sketch it.

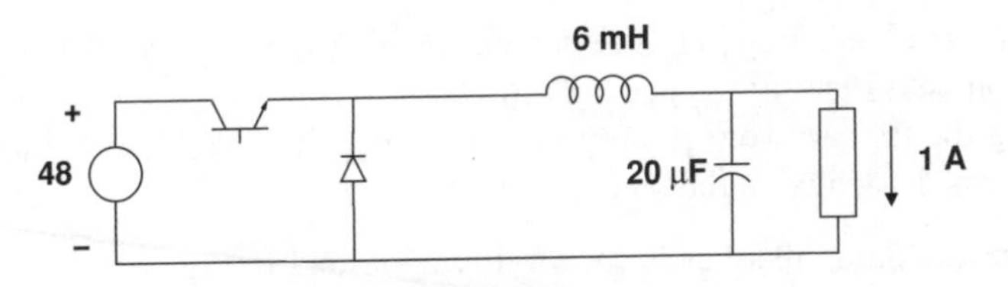

Figure 12.60 Buck converter

(c) Now, assuming the current variation you obtained in the previous part, what is the variation in capacitor voltage? Sketch it. How good was the assumption that capacitor voltage is nearly constant?

4. Using the buck converter of the last problem, calculate and plot ripple current in the inductor as a function of duty cycle D.

5. A buck converter, without a filter capacitor is driving a resistive load as in Figure 12.61.

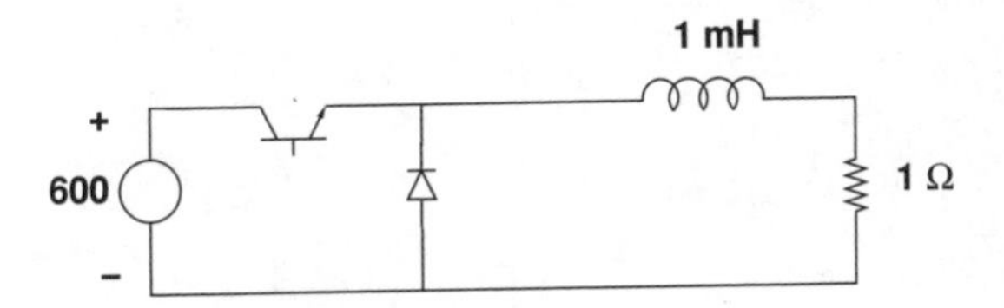

Figure 12.61 Buck converter without filter capacitor

(a) The device is in steady state with a duty cycle $D = 0.5$. The switching frequency is 2 kHz. What is average output voltage?
(b) Calculate the maximum and minimum output voltage. Sketch the waveform.

6. A single-phase rectifier is shown in Figure 12.62. It is driving a load that, for some reason, draws a constant current of 10 A. The source voltage is a sine wave of 120 V, RMS, 60 Hz.

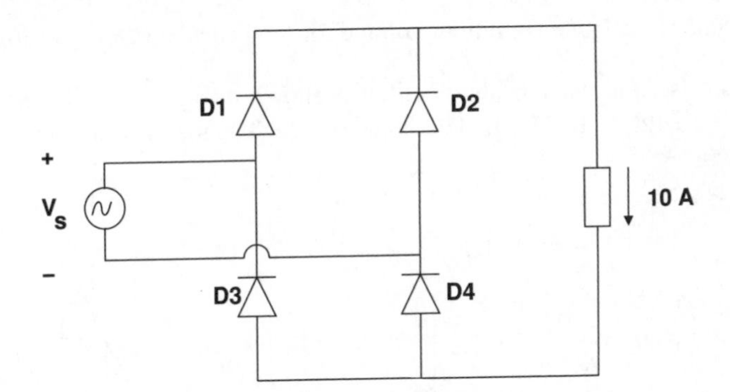

Figure 12.62 Single-phase rectifier

(a) Sketch the voltage across the load.
(b) Sketch current drawn by the source.

(c) Ignoring voltage drop in the diodes and any resistive loss, verify that the power out of the source is the same as power drawn by the (constant current) load.
(d) Using the forward voltage drop model developed in Problem 1, estimate how much power is dissipated in the diodes.

7. A three phase diode rectifier is shown driving a load in Figure 12.63. Assume that the three voltages on the left are sine waves forming a balanced three phase set and that the amplitude of the phase–phase voltage is 480 V, RMS.

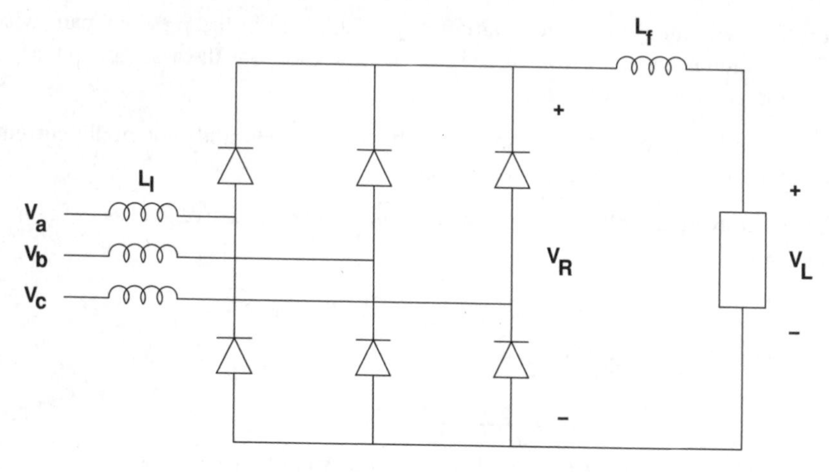

Figure 12.63 Three-phase Rectifier

(a) Assuming the leakage inductance L_l to be negligible and that the load current is a constant 10 A, what is average load voltage V_l?
(b) What is ripple voltage at V_R? Sketch and label output voltage.
(c) What is the current drawn from the phase a source?
(d) Now assume $L_l = 3$ mH. What is the average output voltage $< V_L >$ if load current $I_L = 10$ A?
(e) Find and sketch average output voltage for load current $0 < I_L < 20$ A.

8. A single-phase, full wave bridge rectifier is shown in Figure 12.64. Assume the voltage source V_s is a 120 V, RMS, 60 Hz voltage source. The load is drawing constant current.

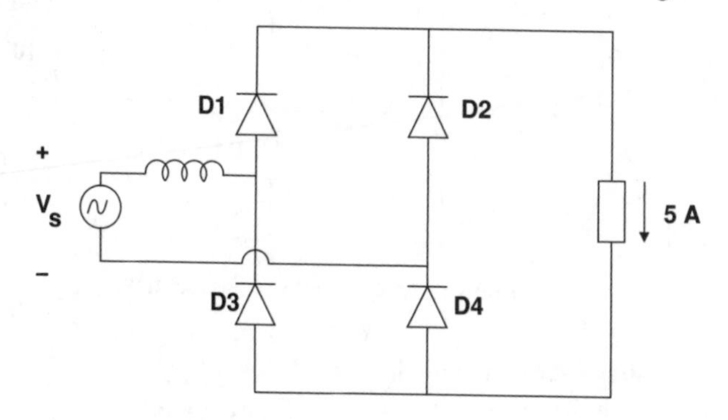

Figure 12.64 Full wave bridge rectifier

(a) Assuming the leakage inductance L_l to be negligible and that the load current is a constant 5 A, what is average load voltage V_l?
(b) What is ripple voltage at the load? Sketch and label output voltage.
(c) What is the current drawn from the source?
(d) Now assume $L_l = 10$ mH. What is the average output voltage $< V_o >$ if load current $I_L = 5$ A?
(e) Find and sketch average output voltage for load current $0 < I_L < 5$ A.

9. A boost converter is shown in Figure 12.65. The inductance is $L = 240$ μH and the capacitor is $C = 10$ μF. The switching frequency is 50 kHz. The source voltage $V_s = 12$ V.

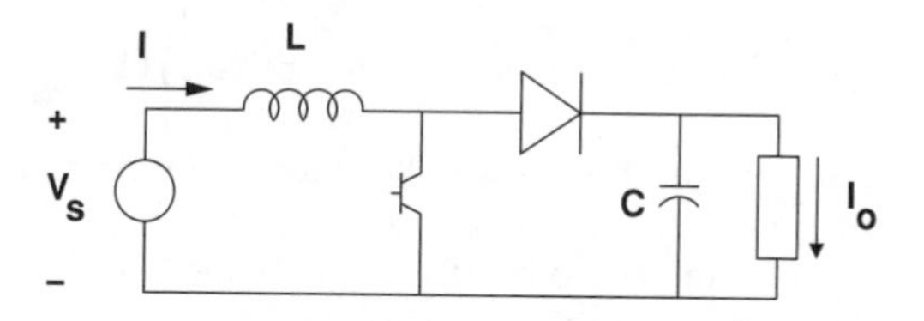

Figure 12.65 Boost converter

(a) Assuming that the inductor is in continuous conduction, what is the output voltage as a function of duty cycle D?
(b) Assume that duty cycle $D = 0.5$. What is the ripple current in the inductor?
(c) Assume that output current is a constant $I_o = 1$ A. What is the ripple voltage across the capacitor?

10. Another boost converter, similar to the one shown in Figure 12.65 is to be operated in most decidedly *dis*continuous conduction mode. The inductor $L = 72$ μH and the capacitor is $C = 100$ nF. Switching frequency is 100 kHz. The device is to be adjusted so that the capacitor voltage is 300 V, at which point the load is drawing 40 mA. This is a boost converter for a unity power factor supply such as one might use in a compact fluorescent lamp, so the voltage V_s is a full wave rectified version of a 120 V, RMS single phase line:

$$V_s = |\sqrt{2}\, 120 \sin 120\pi t|$$

(a) This sort of circuit should appear to the load as nearly the same as a linear resistor. What is the value of that resistor?
(b) Find the value of duty cycle required for this circuit to draw the right amount of power from the voltage source.

11. Here is a snippet of MATLAB code to generate a pulse width modulated waveform.

```
% generate a PWM waveform

f = 60;                    % sine wave of this frequency
T = 1/2000;                % pulse wave time
N_c = 2;                   % this many cycles
T_max = N_c/f;             % and this is for how long we run
N_pwm = ceil(T_max/T);     % need to go this many cycles
```

```
% now we generate the PWM reference triangle

N_s = 200;          % number of time steps per triangle half
dt = T/(2*N_s);     % this is the basic time step
t = 0:dt:T_max;     % and this is the basic time element

Tr1 = (1:N_s) ./ N_s;       % this is the first half
Tr2 = 1-((1:N_s) ./ N_s);% and this is the back side

Trf = [Tr1 Tr2];            % this should be the triangle
Trf0 = [0 Trf];             % leading edge needs a zero

tr = dt:dt:T;       % incremental time for triangles

% now put enough of these together to make a triangle

tt = 0:dt:T;        % these are the first triangle: time
Tr = Trf0;          % and triangle value

for k = 1:N_pwm   % now we concatenate on a bunch more
    Tr = [Tr Trf];
    tt = [tt k*T+tr];
end

% now make sine wave reference

vin = .5 + .45 .* sin(2*pi*f .* tt); %offset
pwm = zeros(size(tt));

for k = 1:length(tt)
    if vin(k) > Tr(k)
        pwm(k) = 1;
    else
        pwm(k) = 0;
    end
end
```

Using this code, or something like it, find the time harmonics produced in synthesizing a voltage source waveform, for frequencies between the time fundamental (60 Hz) and the PWM frequency (2 kHz).

12. A somewhat improbably HVDC line is shown in Figure 12.66. This line uses a single six pulse converter at each end. Line voltage is to be 400 kV and the line is carrying 5 kA. The AC power systems at either end are 330 kV, line–line, RMS at 60 Hz. The line is represented as an inductance, and you may assume this is big enough that line current is constant.

    (a) The transformers at each end have phase reactances of $X_t = 1.5\ \Omega$, and firing angles are set to fix DC line voltage at 400 kV. What value of firing angle is required at the sending end?

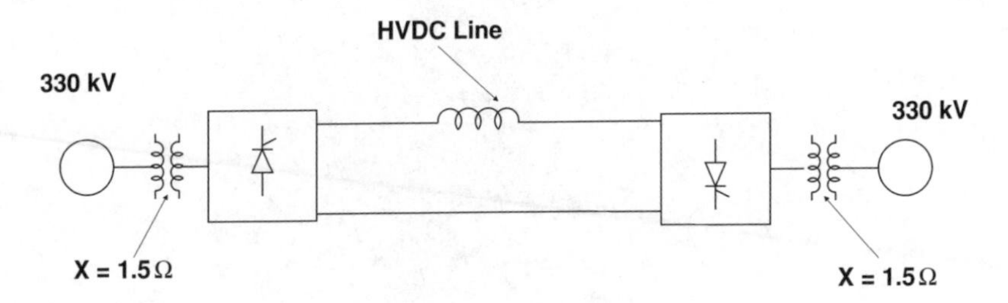

Figure 12.66 HVDC line

(b) What is the overlap angle u at the sending end?
(c) What is the voltage drop across the fictional resistor (due to overlap)?
(d) What is the firing angle at the receiving (inverter) end?
(e) What is the extinction angle?
(f) What are the first four time harmonics of current in the AC system?
(g) Suppose this were built as a twelve pulse system: what would be the first four time harmonics on the AC system?

Reference

Mohan, N., Undeland T.M. and Robbins, W. T. (1995) *Power Electronics Converters Applications and Design*, 2nd edition. Chichester: John Wiley & Sons, Ltd.

13

Induction Machines

13.1 Introduction

Induction machines are perhaps the most widely used of all electric motors. They are generally simple to build and rugged, offer reasonable asynchronous performance: a manageable torque–speed curve, stable operation under load, and generally satisfactory efficiency. Induction machines can also serve as generators, and induction machines, either wound rotor or squirrel cage, dominate as generators for wind turbine systems.

While induction motors have been around for quite a long time and are widespread in application, new uses for these machines are emerging: propulsion motors for transportation, high-speed motors for turbocompressors, compact machines for auxiliary drives in aircraft, and others.

Because they are so widespread and applications continue to grow, a general knowledge of induction machines is important. Space does not permit a thorough explication of all aspects if induction motor analysis, however, but it is hoped that this chapter will serve as a good introduction to the topic.

Figure 13.1 shows one reason why induction motors have such widespread application. This figure shows the torque vs. speed characteristic of an induction motor rated at 10 kW (about 15 horsepower), while the motor is connected to a 60 Hz, three-phase voltage source. With the load torque shown, the motor can start itself and the load: that is, the motor produces torque that is greater than load torque from zero speed to an equilibrium speed, which is, for this case, a bit less than 1,800 r.p.m. The equilibrium condition, where motor torque and load torque are equal, can be seen to be stable in the very important sense that small deviations from that equilibrium speed will create changes to torque that will tend to drive the machine/load system back toward that equilibrium.

- The most common type of induction motor is referred to as a 'squirrel-cage' machine. This motor is simple to build and tends to be quite rugged because it has no electrical connections to the rotor. A cutaway drawing of a squirrel-cage motor is shown in Figure 13.2.
- Wound rotor induction motors are less common. They have, as the name implies, windings on the rotor that are generally connected to collector rings that allow the rotor resistance to

Electric Power Principles: Sources, Conversion, Distribution and Use James L. Kirtley

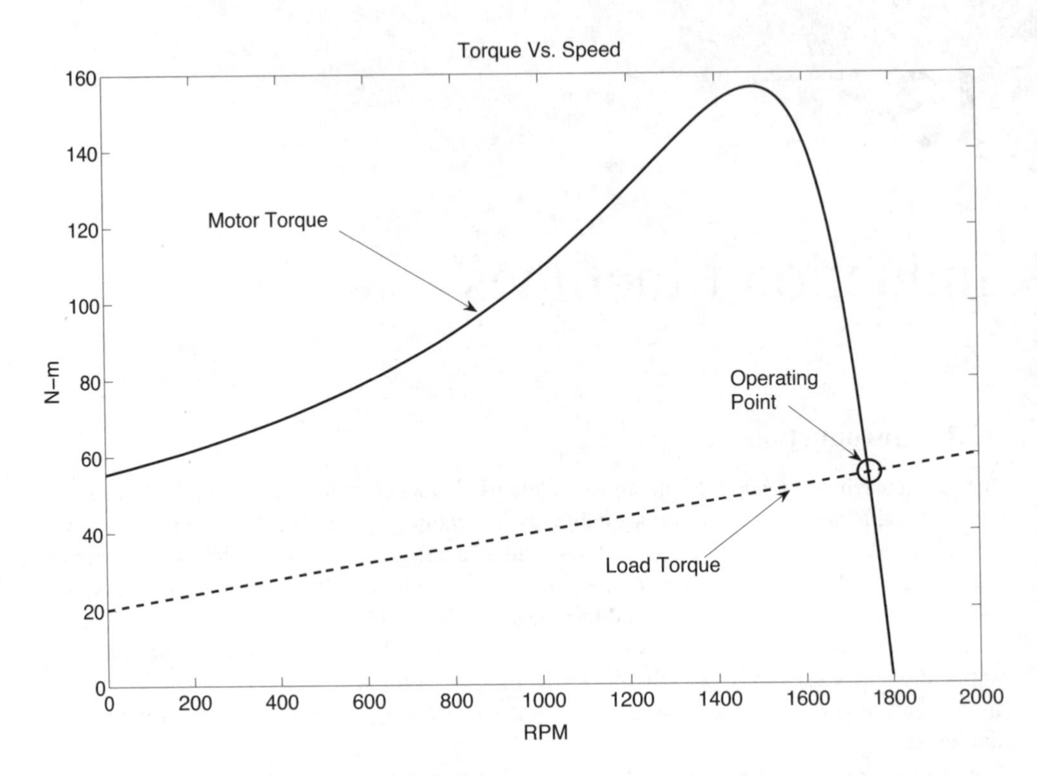

Figure 13.1 Torque-speed curve for an induction motor

be varied. There are some very specialized tasks that can employ such machines, but this kind of machine is more often used with power electronics to form a doubly fed machine.

- Doubly fed induction machines are wound rotor machines with power electronic converters connected, generally to the rotor. They allow for operation over a range of rotational speeds. They have come to be the dominant kind of generator in wind turbine systems.
- Polyphase induction motors are commonly used in most industrial applications, where three-phase power is available.
- Single-phase motors are used in smaller applications, generally where polyphase electric power is not available.

Operation of an induction motor can be explained in a qualitative fashion as follows: The polyphase winding of the machine produces a flux distribution that rotates, at synchronous speed, forming a 'flux wave'. The rotor, composed of conductors, is entrained by that flux wave. If there is a difference between the speed of the flux wave and the speed of the rotor (called the 'slip speed'), currents are induced in the rotor. Those currents interact with the flux distribution in such a way that they produce a force that tends to oppose the relative motion between the rotor and the flux distribution.

The analysis of this chapter starts with a description of the polyphase, wound rotor machine because this is the most straightforward machine to describe and to understand. Squirrel-cage machines and single phase machines follow.

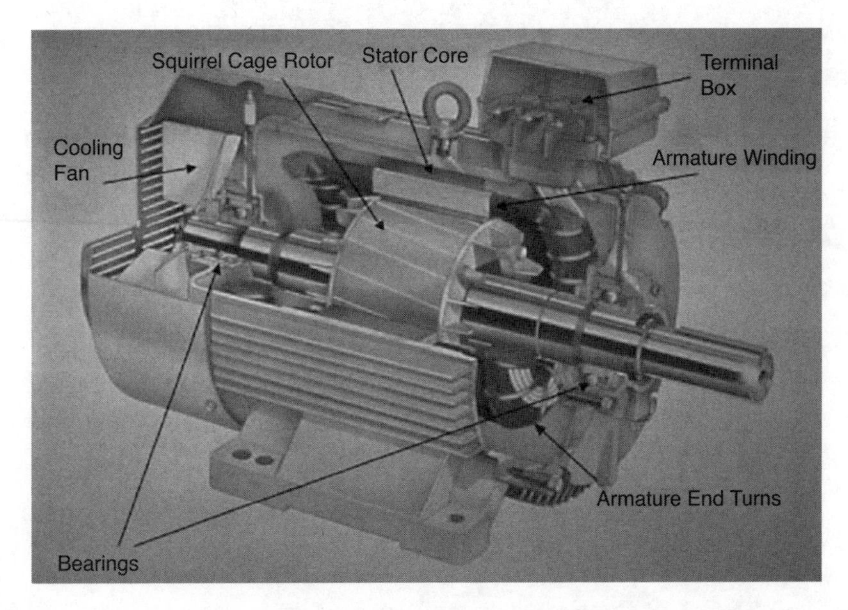

Figure 13.2 Squirrel-cage motor internal structure. Reproduced by permission of ABB Ltd., Annotation by the author

An elementary picture of the induction machine is shown in Figure 13.3, which is an axial view of a cut through the center of a machine. The rotor and stator are coaxial. The stator has a winding in slots. The rotor has either a winding or a cage, also in slots.

13.2 Induction Machine Transformer Model

The induction machine has two electrically active elements: a rotor and a stator. In normal operation, the stator is excited by polyphase voltage. The stator excitation creates a magnetic field in the form of a rotating wave, which induces currents in the circuits of the rotor. Those currents, in turn, interact with the traveling wave to produce torque.

To start the analysis of this machine, assume that both the rotor and the stator can be described by balanced, three-phase windings. Figure 13.4 shows the two winding sets. The rotor electrical angle is $p\theta$, where θ is the rotor physical angle. For the purpose of this analysis, the stator quantities will be subscripted with lower case letters; the rotor quantities with upper case letters.

The two sets of windings are coupled by mutual inductances which are dependent on rotor position. Stator fluxes are $(\lambda_a, \lambda_b, \lambda_c)$ and rotor fluxes are $(\lambda_A, \lambda_B, \lambda_C)$. The flux vs. current relationship is given by:

$$\begin{bmatrix} \lambda_a \\ \lambda_b \\ \lambda_c \\ \lambda_A \\ \lambda_B \\ \lambda_C \end{bmatrix} = \begin{bmatrix} \mathbf{L}_S & \mathbf{M}_{SR} \\ \mathbf{M}_{SR}^T & \mathbf{L}_R \end{bmatrix} \begin{bmatrix} i_a \\ i_b \\ i_c \\ i_A \\ i_B \\ i_C \end{bmatrix} \tag{13.1}$$

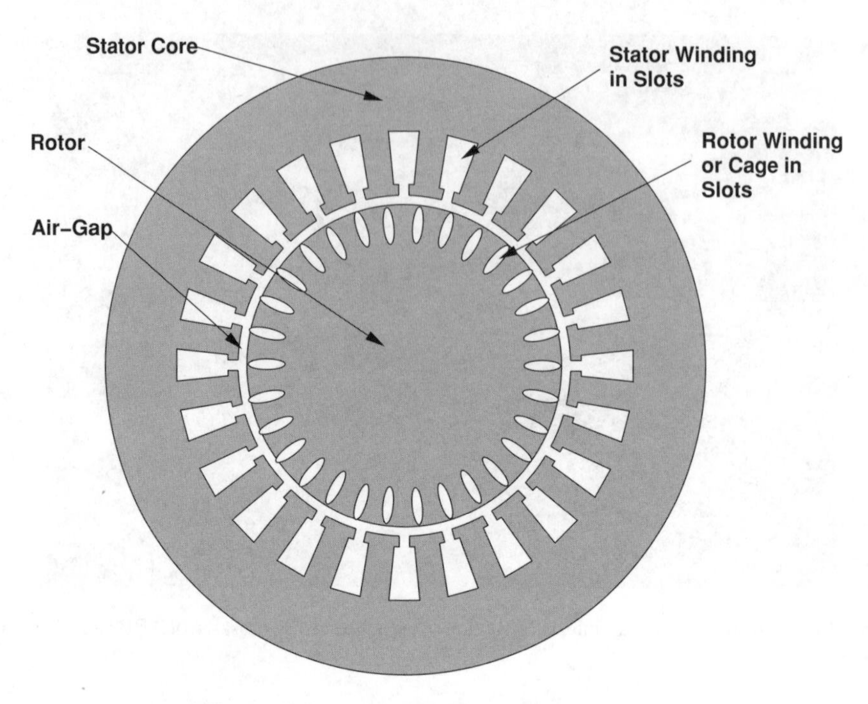

Figure 13.3 Axial view of an induction machine

where the self-inductance component matrices are:

$$\mathbf{L}_{\mathrm{S}} = \begin{bmatrix} L_{\mathrm{a}} & L_{\mathrm{ab}} & L_{\mathrm{ab}} \\ L_{\mathrm{ab}} & L_{\mathrm{a}} & L_{\mathrm{ab}} \\ L_{\mathrm{ab}} & L_{\mathrm{ab}} & L_{\mathrm{a}} \end{bmatrix} \qquad \mathbf{L}_{\mathrm{R}} = \begin{bmatrix} L_{\mathrm{A}} & L_{\mathrm{AB}} & L_{\mathrm{AB}} \\ L_{\mathrm{AB}} & L_{\mathrm{A}} & L_{\mathrm{AB}} \\ L_{\mathrm{AB}} & L_{\mathrm{AB}} & L_{\mathrm{A}} \end{bmatrix}$$

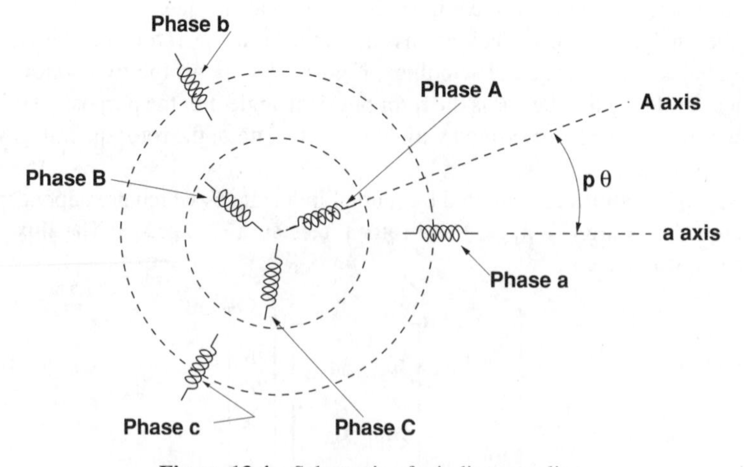

Figure 13.4 Schematic of winding coupling

The mutual inductance part of Equation 13.1 is:

$$\mathbf{M}_{\mathrm{SR}} = \begin{bmatrix} L_{\mathrm{aA}}\cos(p\theta) & L_{\mathrm{aA}}\cos(p\theta+\frac{2\pi}{3}) & L_{\mathrm{aA}}\cos(p\theta-\frac{2\pi}{3}) \\ L_{\mathrm{aA}}\cos(p\theta-\frac{2\pi}{3}) & L_{\mathrm{aA}}\cos(p\theta) & L_{\mathrm{aA}}\cos(p\theta+\frac{2\pi}{3}) \\ L_{\mathrm{aA}}\cos(p\theta+\frac{2\pi}{3}) & L_{\mathrm{aA}}\cos(p\theta-\frac{2\pi}{3}) & L_{\mathrm{aA}}\cos(p\theta) \end{bmatrix} \tag{13.2}$$

Note that when Equation 13.2 is transposed it makes Equation 13.1 reciprocal, as is necessary for an inductance matrix.

To carry the analysis further, it is necessary to make some assumptions regarding operation. To start, assume balanced currents in both the stator and rotor:

$$\begin{aligned} i_{\mathrm{a}} &= I_{\mathrm{S}}\cos(\omega t) \\ i_{\mathrm{b}} &= I_{\mathrm{S}}\cos(\omega t-\tfrac{2\pi}{3}) \\ i_{\mathrm{c}} &= I_{\mathrm{S}}\cos(\omega t+\tfrac{2\pi}{3}) \\ i_{\mathrm{A}} &= I_{\mathrm{R}}\cos(\omega_{\mathrm{R}}t+\xi_{\mathrm{R}}) \\ i_{\mathrm{B}} &= I_{\mathrm{R}}\cos(\omega_{\mathrm{R}}t+\xi_{\mathrm{R}}-\tfrac{2\pi}{3}) \\ i_{\mathrm{C}} &= I_{\mathrm{R}}\cos(\omega_{\mathrm{R}}t+\xi_{\mathrm{R}}+\tfrac{2\pi}{3}) \end{aligned}$$

The rotor position θ can be described by

$$\theta = \omega_{\mathrm{m}}t + \theta_0 \tag{13.3}$$

Under these assumptions, the expression for λ_{a} is:

$$\begin{aligned} \lambda_{\mathrm{a}} = (L_{\mathrm{a}} - L_{\mathrm{ab}})I_{\mathrm{s}}\cos(\omega t) &+ L_{\mathrm{aA}}I_{\mathrm{R}}\{\cos(\omega_{\mathrm{R}}t+\xi_{\mathrm{R}})\cos p(\omega_{\mathrm{m}}+\theta_0) \\ &+ \cos(\omega_{\mathrm{R}}t+\xi_{\mathrm{R}}+\frac{2\pi}{3})\cos(p(\omega_{\mathrm{m}}t+\theta_0)-\frac{2\pi}{3}) \\ &+ \cos(\omega_{\mathrm{R}}t+\xi_{\mathrm{R}}-\frac{2\pi}{3})\cos[p(\omega_{\mathrm{m}}t+\theta_0)+\frac{2\pi}{3}]\} \end{aligned}$$

which, after reducing some of the trigonometric expressions, becomes:

$$\lambda_a = (L_{\mathrm{a}} - L_{\mathrm{ab}})I_{\mathrm{s}}\cos(\omega t) + \frac{3}{2}L_{\mathrm{aA}}I_{\mathrm{R}}\cos[(p\omega_{\mathrm{m}}+\omega_{\mathrm{R}})t+\xi_{\mathrm{R}}+p\theta_0] \tag{13.4}$$

Doing the same thing for the rotor phase A yields:

$$\begin{aligned} \lambda_{\mathrm{A}} = L_{\mathrm{aA}}I_{\mathrm{s}}\{&\cos(p(\omega_{\mathrm{m}}t+\theta_0)\cos\omega t + \cos[p(\omega_{\mathrm{m}}t+\theta_0)-\frac{2\pi}{3}]\cos(\omega t-\frac{2\pi}{3}) \\ &+ \cos[p(\omega_{\mathrm{m}}t+\theta_0)+\frac{2\pi}{3}]\cos(\omega t+\frac{2\pi}{3})\} + (L_{\mathrm{A}} - L_{\mathrm{AB}})I_{\mathrm{R}}\cos(\omega_{\mathrm{R}}t+\xi_{\mathrm{R}}) \end{aligned}$$

This last expression is, after reducing the trigonometry:

$$\lambda_{\mathrm{A}} = \frac{3}{2} L_{\mathrm{aA}} I_{\mathrm{s}} \cos[(\omega - p\omega_{\mathrm{m}})t - p\theta_0] + (L_{\mathrm{A}} - L_{\mathrm{AB}}) I_{\mathrm{R}} \cos(\omega_{\mathrm{R}} t + \xi_{\mathrm{R}}) \tag{13.5}$$

These two equations, 13.4 and 13.5 give expressions for fluxes in the armature and rotor windings in terms of currents in the same two windings, assuming that both current distributions are sinusoidal in time and space and represent balanced distributions. The next step is to make another assumption, that the stator and rotor frequencies match through rotor rotation. That is:

$$\omega - p\omega_{\mathrm{m}} = \omega_{\mathrm{R}}$$

It is important to keep straight the different frequencies here:

ω is stator electrical frequency
ω_{R} is rotor electrical frequency
ω_{m} is mechanical rotation speed

so that $p\omega_{\mathrm{m}}$ is *electrical* rotation speed.

To refer rotor quantities to the stator frame (i.e. non-rotating), and to work in complex amplitudes, the following definitions are made:

$$\lambda_{\mathrm{a}} = \mathrm{Re}\left\{(\mathbf{\Lambda}_a e^{j\omega t})\right\}$$
$$\lambda_{\mathrm{A}} = \mathrm{Re}\left\{(\mathbf{\Lambda}_A e^{j\omega_{\mathrm{R}} t})\right\}$$
$$i_{\mathrm{a}} = \mathrm{Re}\left\{(\mathbf{I}_a e^{j\omega t})\right\}$$
$$i_{\mathrm{A}} = \mathrm{Re}\left\{(\mathbf{I}_A e^{j\omega_{\mathrm{R}} t})\right\}$$

With these definitions, the complex amplitudes embodied in Equations 13.4 and 13.5 become:

$$\mathbf{\Lambda}_{\mathrm{a}} = L_{\mathrm{S}} \mathbf{I}_{\mathrm{a}} + \frac{3}{2} L_{\mathrm{aA}} \mathbf{I}_{\mathrm{A}} e^{j(\xi_{\mathrm{R}} + p\theta_0)} \tag{13.6}$$

$$\mathbf{\Lambda}_{\mathrm{A}} = \frac{3}{2} L_{\mathrm{aA}} \mathbf{I}_{\mathrm{a}} e^{-jp\theta_0} + L_{\mathrm{R}} \mathbf{I}_{\mathrm{A}} e^{j\xi_{\mathrm{R}}} \tag{13.7}$$

where $L_{\mathrm{S}} = L_{\mathrm{a}} - L_{\mathrm{ab}}$ and $L_{\mathrm{R}} = L_{\mathrm{A}} - L_{\mathrm{AB}}$.

There are two phase angles embedded in these expressions: θ_0 which describes the rotor physical phase angle with respect to stator current and ξ_{R} which describes phase angle of rotor currents with respect to stator currents. It is helpful to rotate the rotor flux and current consistently:

$$\mathbf{\Lambda}_{\mathrm{AR}} = \mathbf{\Lambda}_{\mathrm{A}} e^{jp\theta_0} \tag{13.8}$$

$$\mathbf{I}_{\mathrm{AR}} = \mathbf{I}_{\mathrm{A}} e^{j(p\theta_0 + \xi_{\mathrm{R}})} \tag{13.9}$$

These are rotor flux and current referred to armature phase angle. Note that $\mathbf{\Lambda}_{\mathrm{AR}}$ and $\mathbf{I}_{\mathrm{AR}}$ have the same phase relationship to each other as do $\mathbf{\Lambda}_{\mathrm{A}}$ and $\mathbf{I}_{\mathrm{A}}$. Using Equations 13.8

and 13.9 in Equations 13.6 and 13.7, the basic flux/current relationship for the induction machine becomes:

$$\begin{bmatrix} \mathbf{\Lambda}_{\mathrm{a}} \\ \mathbf{\Lambda}_{\mathrm{AR}} \end{bmatrix} = \begin{bmatrix} L_{\mathrm{S}} & \frac{3}{2}L_{\mathrm{aA}} \\ \frac{3}{2}L_{\mathrm{aA}} & L_{\mathrm{R}} \end{bmatrix} \begin{bmatrix} \mathbf{I}_{\mathrm{a}} \\ \mathbf{I}_{\mathrm{AR}} \end{bmatrix}$$

This is an equivalent single-phase statement, describing the flux/current relationships in phases a and A, assuming balanced operation. The same expression will describe the flux/current relationships in phases b and c.

Voltage at the terminals of the stator and rotor windings is, in the sinusoidal steady state estimated by multiplying flux by frequency and rotating by 90° and adding current multiplied by resistance:

$$\mathbf{V}_{\mathrm{a}} = j\omega\mathbf{\Lambda}_{\mathrm{a}} + R_{\mathrm{S}}\mathbf{I}_{\mathrm{a}}$$
$$\mathbf{V}_{\mathrm{AR}} = j\omega_{\mathrm{R}}\mathbf{\Lambda}_{\mathrm{AR}} + R_{\mathrm{R}}\mathbf{I}_{\mathrm{AR}}$$

Substituting for flux:

$$\mathbf{V}_{\mathrm{a}} = j\omega L_{\mathrm{S}}\mathbf{I}_{\mathrm{a}} + j\omega\frac{3}{2}L_{\mathrm{aA}}\mathbf{I}_{\mathrm{AR}} + R_{\mathrm{S}}\mathbf{I}_{\mathrm{a}} \tag{13.10}$$

$$\mathbf{V}_{\mathrm{AR}} = j\omega_{\mathrm{R}}\frac{3}{2}L_{\mathrm{aA}}\mathbf{I}_{\mathrm{a}} + j\omega_{\mathrm{R}}L_{\mathrm{R}}\mathbf{I}_{\mathrm{AR}} + R_{\mathrm{R}}\mathbf{I}_{\mathrm{AR}} \tag{13.11}$$

To carry this further, it is necessary to go a little deeper into the machine's parameters. Note that L_{S} and L_{R} are synchronous inductances for the stator and rotor. Expressions for inductances were developed in the appendix to Chapter 9. The stator and rotor inductances may be separated into space fundamental and 'leakage' components as follows:

$$L_{\mathrm{S}} = L_{\mathrm{a}} - L_{\mathrm{ab}} = \frac{3}{2}\frac{4}{\pi}\frac{\mu_0 R l N_{\mathrm{S}}^2 k_{\mathrm{S}}^2}{p^2 g} + L_{\mathrm{Sl}}$$
$$L_{\mathrm{R}} = L_{\mathrm{A}} - L_{\mathrm{AB}} = \frac{3}{2}\frac{4}{\pi}\frac{\mu_0 R l N_{\mathrm{R}}^2 k_{\mathrm{R}}^2}{p^2 g} + L_{\mathrm{Rl}}$$

Where the normal set of machine parameters holds:

R is rotor radius
l is active length
g is the effective air-gap
p is the number of pole-pairs
N represents number of turns
k represents the winding factor
S as a subscript refers to the stator
R as a subscript refers to the rotor
L_{l} is 'leakage' inductance

The two leakage terms L_{Sl} and L_{Rl} contain higher order harmonic stator and rotor inductances, slot inductances, end-winding inductances and, if necessary, a provision for rotor skew. Essentially, they are used to represent all flux in the rotor and stator that is not mutually coupled.

In the same terms, the stator-to-rotor mutual inductance, which is taken to comprise *only* a space fundamental term, is:

$$L_{aA} = \frac{4}{\pi}\frac{\mu_0 R l N_S N_R k_S k_R}{p^2 g}$$

Note that there may be space harmonic mutual flux linkages. These are not of much interest with wound rotor machines but are responsible for 'stray' losses in squirrel-cage machines.

Air-gap permeance is defined as:

$$\wp_{ag} = \frac{4}{\pi}\frac{\mu_0 R l}{p^2 g}$$

so that the inductances are:

$$\begin{aligned} L_S &= \frac{3}{2}\wp_{ag} k_S^2 N_S^2 + L_{Sl} \\ L_R &= \frac{3}{2}\wp_{ag} k_R^2 N_R^2 + L_{Rl} \\ L_{aA} &= \wp_{ag} N_S N_R k_S k_R \end{aligned}$$

The voltage balance equations (13.10 and 13.11) become:

$$\begin{aligned} \mathbf{V}_a &= j\omega\left(\frac{3}{2}\wp_{ag} k_S^2 N_S^2 + L_{Sl}\right)\mathbf{I}_a + j\omega\frac{3}{2}\wp_{ag} N_S N_R k_S k_R \mathbf{I}_{AR} + R_S \mathbf{I}_a \\ \mathbf{V}_{AR} &= js\omega\frac{3}{2}\wp_{ag} N_S N_R k_S k_R \mathbf{I}_a + js\omega\left(\frac{3}{2}\wp_{ag} k_R^2 N_R^2 + L_{Rl}\right)\mathbf{I}_{AR} + R_R \mathbf{I}_{AR} \end{aligned}$$

Where 'slip' s is defined by:

$$\omega_R = s\omega \qquad (13.12)$$

so that

$$s = 1 - \frac{p\omega_m}{\omega} \qquad (13.13)$$

And the electrical equivalent of mechanical speed is:

$$p\omega_m = \omega(1 - s)$$

At this point, it is necessary to define rotor current referred to the stator. This is accomplished with a transformation that refers rotor currents across the transformer effective turns ratio between the rotor and stator. This transformation defines an equivalent stator current to produce the same fundamental MMF as a given rotor current:

$$\mathbf{I}_2 = \frac{N_R k_R}{N_S k_S} \mathbf{I}_{AR}$$

Note that this definition of I_2 in terms of I_{AR} is simply the equivalent of an ideal transformer between the rotor and stator of the induction machine. This is why this analysis is often referred to as the 'transformer model' for the induction machine.

Given this transformer turns ratio, the space fundamental inductance, referred to the stator, is:

$$L_m = \frac{3}{2} \wp_{ag} k_S^2 N_S^2$$

The voltage balance equations become:

$$\mathbf{V}_a = j(X_m + X_1)\mathbf{I}_a + jX_m \mathbf{I}_2 + R_S \mathbf{I}_a \qquad (13.14)$$

$$\mathbf{V}_2 = jX_m \mathbf{I}_a + j(X_m + X_2)\mathbf{I}_2 + \frac{R_2}{s} \mathbf{I}_2 \qquad (13.15)$$

where the following definitions have been made:

$$X_m = \omega L_m = \frac{3}{2} \omega \wp_{ag} N_S^2 k_S^2$$

$$X_1 = \omega L_{Sl}$$

$$X_2 = \omega L_{Rl} \left(\frac{N_S k_S}{N_R k_R} \right)^2$$

$$R_1 = R_S$$

$$R_2 = R_R \left(\frac{N_S k_S}{N_R k_R} \right)^2$$

Normally, the rotor of an induction motor is simply shorted, so that $\mathbf{V}_2 = 0$. If this is done, Equations 13.14 and 13.15 are loop equations that describe the equivalent circuit of Figure 13.5.

13.2.1 Operation: Energy Balance

Figure 13.5 represents one phase of a polyphase system that is operated under balanced conditions. 'Balanced conditions' means that each phase has the same terminal voltage magnitude and that the phase difference between phases is uniform. Assume the machine is connected to

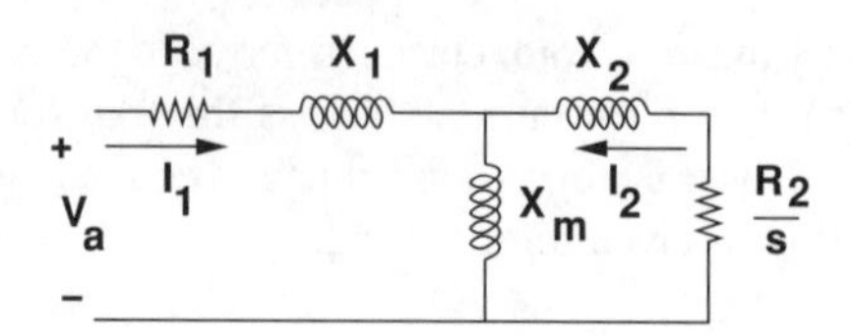

Figure 13.5 Equivalent circuit for an induction motor

a balanced polyphase voltage source. Power input to the stator of the machine (assuming three phases) is:

$$P_{\text{in}} = 3|I_1|^2 R_1 + 3|I_2|^2 \frac{R_2}{s}$$

where $|I|$ refers to the RMS amplitude of I.

This is the sum of power dissipated in the stator plus power dissipated in the rotor resistance as seen from the stator, or $\frac{R_2}{s}$.

'Air-gap' power (P_{ag}) is input power minus stator dissipation:

$$P_{\text{ag}} = 3|I_2|^2 \frac{R_2}{s} \tag{13.16}$$

This is real (time-average) power crossing the air-gap of the machine. Positive slip implies rotor speed less than synchronous and positive air-gap power (motor operation). Negative slip means rotor speed is higher than synchronous, negative air-gap power (from the rotor to the stator) and generator operation.

Note that this equivalent circuit represents a real physical structure, so it is possible to calculate power dissipated in the physical rotor resistance (P_{d}), and that is:

$$P_{\text{d}} = 3|I_2|^2 R_2 = s P_{\text{ag}}$$

(Note that, since both P_{ag} and s will always have the same sign, dissipated power is positive.)

The rest of this discussion is framed in terms of *motor* operation, but the conversion to *generator* operation is simple. The basic principle is conservation of energy (the first law of thermodynamics) and is illustrated in Figure 13.6. The principle is that power into the motor must match power out of the motor. Power into the motor is power at the armature terminals, and that is equal to power dissipated in the armature resistance plus power crossing the air-gap. The difference between power crossing the air-gap and power dissipated in the rotor resistance

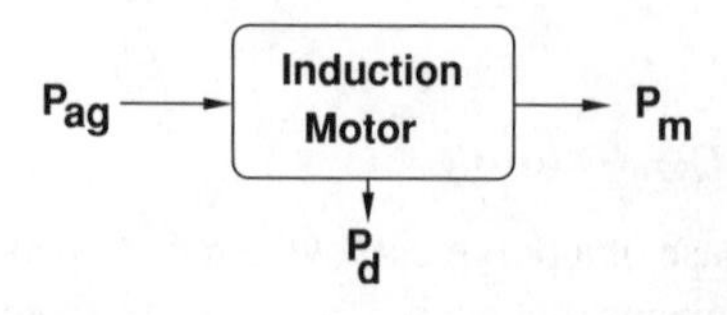

Figure 13.6 Schematic of power balance in the induction motor

must be converted into mechanical form P_m:

$$P_m = P_{ag} - P_d = P_{ag}(1 - s)$$

and *electrical input* power (P_{in}) is:

$$P_{in} = P_{ag} + P_a$$

where armature dissipation (P_a) is:

$$P_a = 3|I_1|^2 R_a$$

Output (mechanical) power is

$$P_{out} = P_m - P_w = P_{ag} - P_d - P_w$$

Where P_w describes friction, windage and certain stray losses.

Since power converted must be torque multiplied by mechanical speed:

$$P_m = P_{ag} - P_d = 3|I_2|^2 \frac{R_2}{s}(1 - s)$$

Converted mechanical power must also be

$$P_m = T_{em}\omega_m = T_{em}\frac{\omega(1 - s)}{p}$$

The torque produced by the motor must be:

$$T_{em} = \frac{p}{\omega} P_{ag} \tag{13.17}$$

The equivalent circuit of Figure 13.5 makes estimation of induction motor performance straightforward. The 'gap impedance' (Z_g), or the impedance looking to the right from the right-most terminal of X_1 is:

$$Z_g = jX_m || (jX_2 + \frac{R_2}{s})$$

A total, or terminal impedance is then

$$Z_t = jX_1 + R_1 + Z_g$$

and terminal current is

$$I_t = \frac{V_t}{Z_t}$$

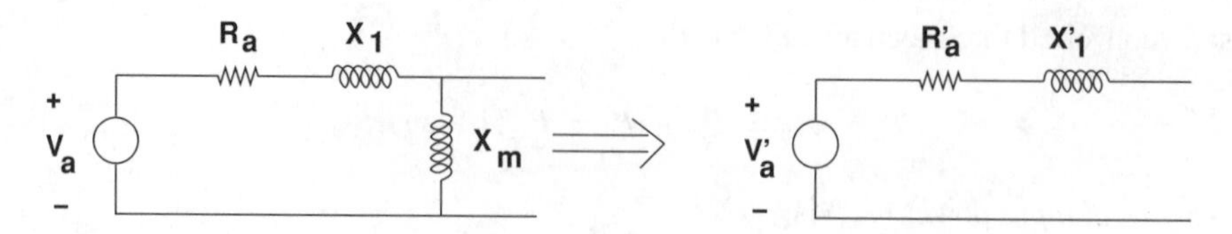

Figure 13.7 Thevenin equivalent of input and stator

Rotor current is found by using a current divider:

$$I_2 = I_t \frac{jX_m}{jX_m + jX_2 + \frac{R_2}{s}}$$

And, finally, once currents are determined, input and output power are calculated, efficiency and power factor are:

$$\eta = \frac{P_{out}}{P_{in}}$$
$$\cos\psi = \frac{P_{in}}{3V_t I_t}$$

13.2.1.1 Simplified Torque Estimation

To approach the torque–speed characteristic of the induction motor, consider this approach: The voltage source connected to the machine, stator resistance and leakage and magnetizing branch may be combined as shown in Figure 13.7. Some caution is required in using this circuit: it gives an accurate representation of rotor behavior, but hides the input to the motor.

The elements of the equivalent circuit are:

$$V_a' = |V_a| |\frac{X_m}{j(X_m + X_1) + R_1}|$$
$$R_1' = \mathrm{Re}\{jX_m || (R_1 + jX_a)\}$$
$$X_1' = \mathrm{Im}\{jX_m || (R_1 + jX_a)\}$$

With this simplification of the input to the machine, calculation of rotor circuit current is straightforward:

$$\mathbf{I}_2 = -\frac{V_a'}{j(X_1' + X_2) + (R_1' + \frac{R_2}{s})}$$

Then, according to Equation 13.17,

$$T_{em} = \frac{3|V_a'|^2 \frac{R_2}{s}}{(X_1' + X_2)^2 + (R_1' + \frac{R_2}{s})^2} \tag{13.18}$$

This may be interpreted in the following way:

- If slip s is *small*, as it will be in the vicinity of synchronous speed, the fraction $\frac{R_2}{s}$ will be large and will dominate the denominator of Equation 13.18, which will then approach:

$$T_{\mathrm{em}} \rightarrow 3|V_{\mathrm{a}}'|^2 \frac{s}{R_2}$$

That is, for slip sufficiently small enough, torque is linearly related to slip. This is often referred to as *resistance limited operation.*

- On the other hand, for slip that is sufficiently large, the ratio $\frac{R_2}{s}$ will be small enough that it will be negligible with respect to the other terms in the denominator. Torque approaches:

$$T_{\mathrm{em}} \rightarrow \frac{3|V_{\mathrm{a}}'|^2 \frac{R_2}{s}}{\left(X_1' + X_2\right)^2 + \left(R_1'\right)^2}$$

for which torque is inversely proportional to slip. This is referred to as 'inductance limited' operation.

A torque–speed curve for the motor described in Table 13.1 is shown in Figure 13.1.

13.2.1.2 Torque Summary

All this may be nicely summarized by:

$$P_{\mathrm{ag}} = 3|I_2|^2 \frac{R_2}{s}$$

$$P_{\mathrm{m}} = 3|I_2|^2 \frac{R_2}{s}(1 - s) = P_{\mathrm{ag}} \frac{\omega_{\mathrm{m}}}{\omega}$$

$$P_{\mathrm{R}} = 3|I_2|^2 R_2 = P_{\mathrm{ag}} \frac{\omega_{\mathrm{s}}}{\omega}$$

This implies that the torque generated by the machine is the same from all points of view: stator, rotor or mechanical, and power from each of those spots is torque multiplied by frequency at that spot.

Table 13.1 Parameters for a (roughly) 15 HP industrial motor

Number of pole pairs	2
RMS terminal voltage (line–line)	240
Frequency (Hz)	60
Stator resistance R_1	0.06 Ω
Rotor resistance R_2	0.15 Ω
Stator leakage X_1	0.44 Ω
Rotor leakage X_2	0.43 Ω
Magnetizing reactance X_{m}	12.6 Ω

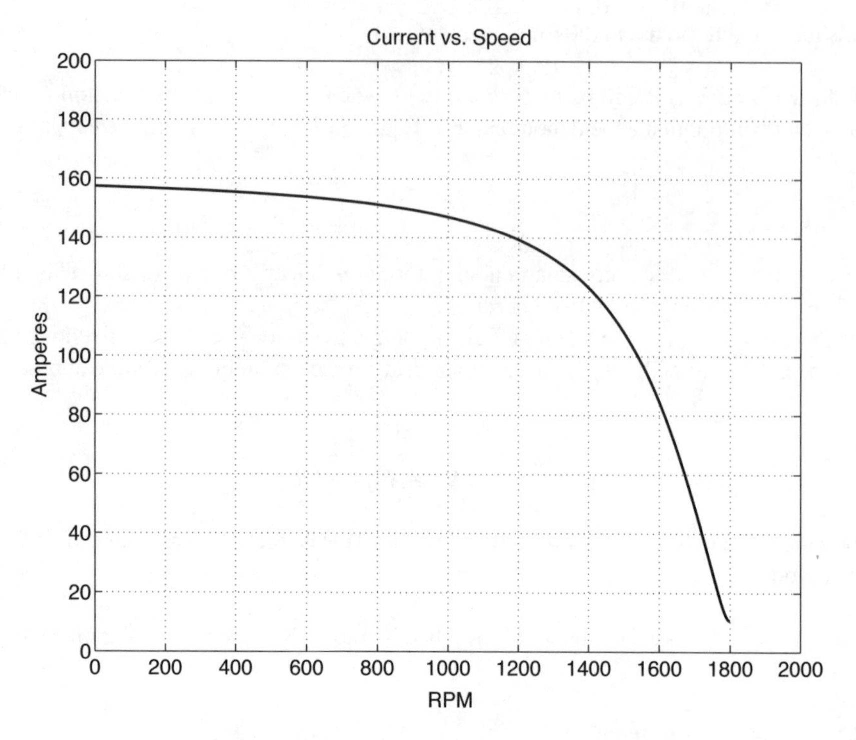

Figure 13.8 Current draw by example induction motor

13.2.2 Example of Operation

Circuit parameters for a standard motor are given in Table 13.1.

Estimates for torque, current drawn by the motor and power converted are shown in Figures 13.1, 13.8 and 13.9.

13.2.3 Motor Performance Requirements

There are a number of requirements for various induction motors set by national and international standards. For example, the National Electric Manufacturer's Association (NEMA) standard MG-1 NEMA (2004) specifies, for motors of specified class and rating, maximum value for current at 'locked rotor' (stall) conditions and minimum values for starting, 'pull up' and 'pull out' torque. Starting torque is torque at locked rotor conditions. 'pull-up' torque is the minimum value of torque in the range between starting and operating speeds. It can be lower than starting torque because of space harmonic and skin effect in rotor conductors. 'pull out' torque is the maximum torque achieved in the operating range. It is usually several times rated torque for a motor. Motor classes describe operational characteristics. Class A and B motors are general purpose motors generally intended for industrial operation, capable of being started 'across the line', or at full voltage and operating at relatively high efficiency, with low slip. Class C motors are also intended for low slip operation but have higher starting

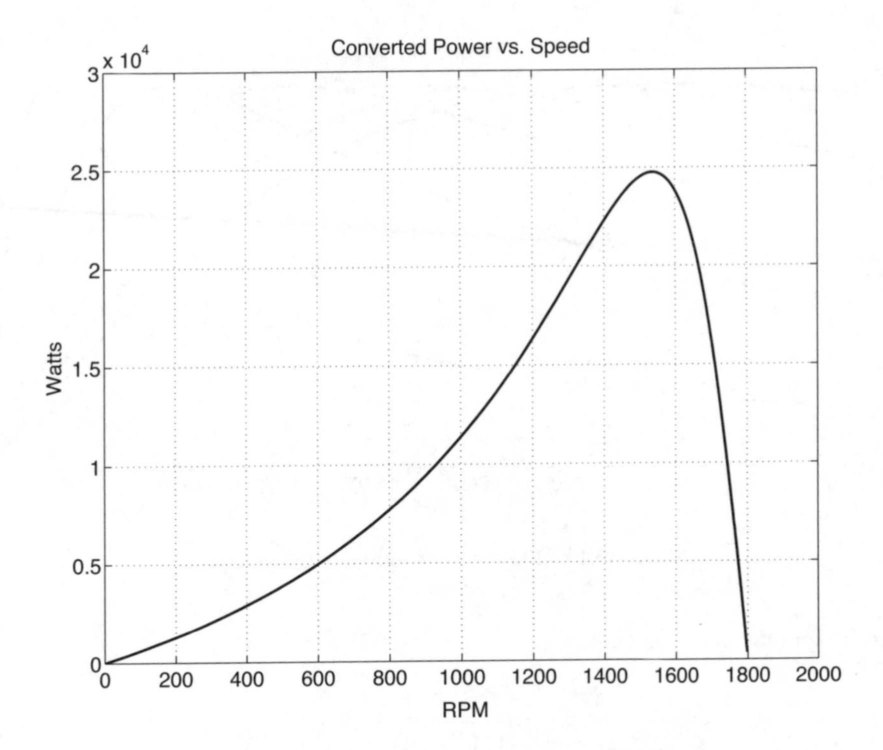

Figure 13.9 Power converted by example induction motor

torque. Such motors are intended for variable torque operation, as in driving trip hammers. Class D motors operate at higher slip and have even higher starting torque. They are intended for frequent starting in situations in which efficiency is not important, for example for opening and closing large valves.

13.2.3.1 Effect of Rotor Resistance

The torque vs. speed curve is affected by the value of rotor resistance. Figure 13.10 shows torque–speed curves plotted for different values of rotor resistance. It should be clear from the expressions above, that higher values of resistance will result in higher slip at rated conditions, meaning lower values of the slope of the torque–speed curve at the zero slip intercept. Clearly, high efficiency motors should have low values of rotor resistance. There is, of course, a tradeoff here, for low values of rotor resistance will generally result in lower values for starting, or stall, torque.

Note that peak, or pull-out torque, is the same for all values of rotor resistance. This is because torque is proportional to air-gap power, which is maximized when the quantity $\frac{R_2}{s}$ is equal to the source impedance. That is, torque is maximized when:

$$\left(\frac{R_2}{s}\right)_{\max} = \left|j\left(X_1' + X_2\right) + R_1'\right|$$

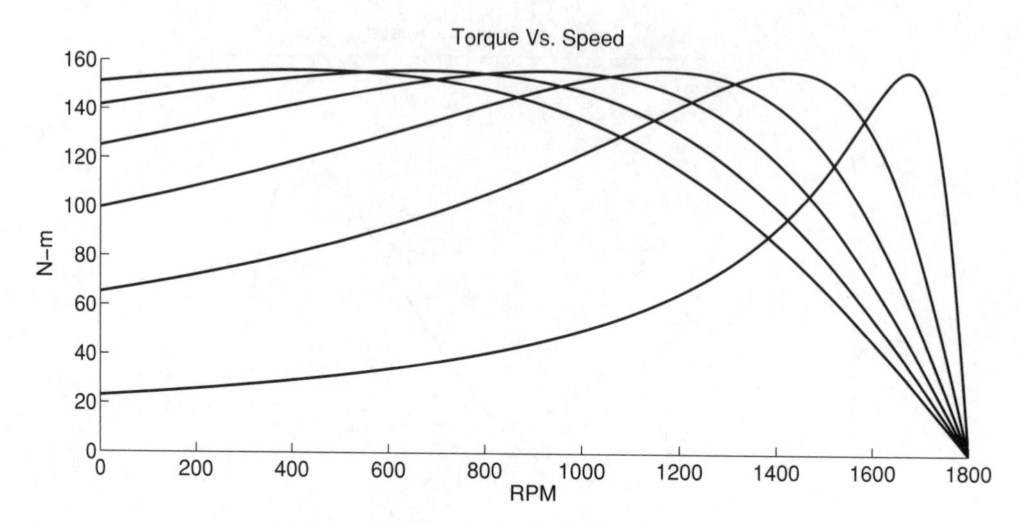

Figure 13.10 Illustration of rotor resistance effect

Then slip for maximum torque is:

$$s_{\max T} = \frac{R_2}{\left| j\left(X_1' + X_2\right) + R_1' \right|}$$

and

$$T_{\max} = \frac{3}{2}\frac{p}{\omega}\frac{|V_a'|^2}{R_1' + \sqrt{(R_1')^2 + \left(X_1' + X_2\right)^2}}$$

13.3 Squirrel-Cage Machines

While the ability to control rotor resistance is useful, wound rotor machines with slip rings and brushes are expensive and the brushes are a maintenance burden. for this reason, most induction motors are built with squirrel-cage rotors rather than discrete windings as was assumed in the derivation of motor characteristics in this chapter. The name 'squirrel cage' is evocative of what the rotor conductors would look like were the rotor iron to be removed. It is a bit like that circular cage a small animal might run around in, and so the name. It consists of a collection of conductors in regularly spaced slots, all conductors being shorted together by 'end rings' at each end. A picture of a squirrel-cage rotor and a squirrel cage is shown in Figure 13.11. To prepare the squirrel cage for this photograph, the iron around the aluminum cage was removed with an acid bath (using an acid that does not attack aluminum).

The behavior of a squirrel-cage motor is similar to that of a wound rotor machine, but there are a few differences. The armature winding of an induction machine produces not only the sinusoidal MMF described above, but also some space harmonics. A wound rotor can be made to not interact with those harmonics, but a squirrel cage cannot, so there are some interactions that

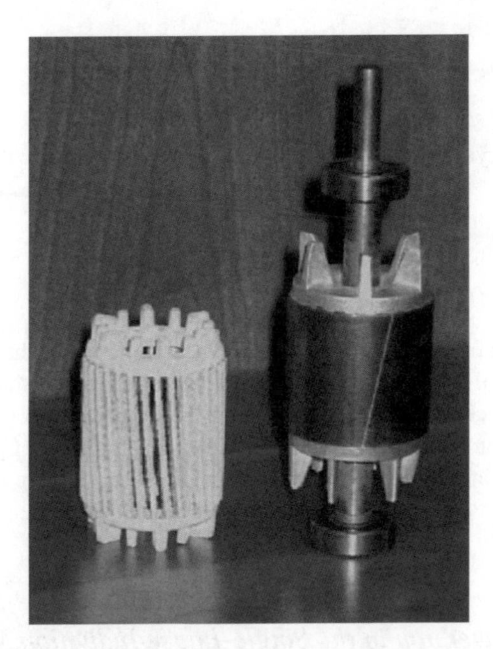

Figure 13.11 Squirrel-cage rotor (on right) and aluminum squirrel cage with rotor iron removed (on left). Reproduced by permission of Copper Development Association, Inc.

produce extra losses, sometimes described as 'stray' losses, that can amount to a few percent of power converted. On the other hand, a squirrel-cage motor can often be built, particularly in larger motors, to exhibit higher rotor resistance at starting because of skin effect (rotor frequency is higher at starting than it is at running speed). Both of these effects can make starting torque higher than torque at intermediate speed, hence the need to specify 'pullup' torque.

13.4 Single-Phase Induction Motors

Single-phase motors are required for service where polyphase electric power is not available, and are very widely used for service in smaller power ratings (typically below a few kilowatts). Single phase motors are a bit more difficult to analyze than polyphase machines and are actually a bit more complicated to build, primarily because special mechanisms are required to start them.

For light starting duty and very low power levels such as circulating fans, single-phase motors are often implemented as *shaded pole* motors. Such motors are quite inefficient, however, so for higher power levels or for more rigorous starting duty, *split phase* motors are used, often with capacitors to accomplish enough phase shift to provide high starting current.

13.4.1 Rotating Fields

To get started in understanding how single phase motors work, consider an ordinary single phase winding. If it is wound to produce air-gap flux that peaks at 90° (electrical), the space fundamental of air-gap flux density is:

$$B_r = \mu_0 \frac{4}{\pi} \frac{N k_w i}{2pg} \sin \theta$$

If the winding is carrying sinusoidal current $i = I \cos \omega t$, this flux density becomes:

$$B_r = \mu_0 \frac{4}{\pi} \frac{N k_w I}{2pg} \sin p\theta \cos \omega t = \mu_0 \frac{4}{\pi} \frac{N k_w I}{2pg} \left(\frac{1}{2} \sin (p\theta + \omega t) + \frac{1}{2} \sin (p\theta - \omega t) \right)$$

The key to operation of this machine should be becoming clear: the single phase winding, when carrying sinusoidal current produces a flux density distribution that has two components: one rotating in one direction and the other rotating in the other direction. If the machine has either a squirrel cage rotor or a wound rotor with the same number of poles as the stator, that rotor will interact separately with each of these components, yielding an equivalent circuit as shown in Figure 13.12. Parameters of this circuit are developed in Section 13.9.

13.4.2 Power Conversion in the Single-Phase Induction Machine

Following the same protocol as was used with polyphase induction machines, and assuming RMS amplitudes for currents, power input to the machine is:

$$P_{in} = I_a^2 R_a + I_F^2 \frac{R_2}{s} + I_R^2 \frac{R_2}{2 - s}$$

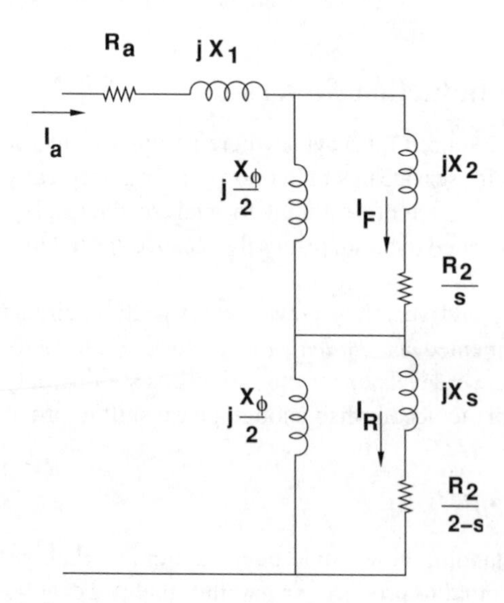

Figure 13.12 Equivalent circuit of single-phase induction motor

Note that slip of the reverse going flux wave is just $s_R = 2 - s$. Both values of slip are one at rotational speed of zero (stall) and slip for the reverse going wave is zero if the rotor is turning backwards at synchronous electrical speed.

Dissipation in the machine is:

$$P_d = I_a^2 R_a + I_F^2 R_2 + I_R^2 R_2$$

So, from conservation of energy, the difference must be power converted to mechanical form:

$$P_m = P_{in} - P_d = I_F^2 \frac{R_2}{s}(1-s) + I_R^2 \left(\frac{R_2}{2-s} - R_2 \right) = \left(I_F^2 \frac{R_2}{s} - I_R^2 \frac{R_2}{2-s} \right)(1-s)$$

Then torque is:

$$T_m = \frac{p}{\omega} \left(I_F^2 \frac{R_2}{s} - I_R^2 \frac{R_2}{2-s} \right)$$

Figure 13.13 shows a calculated torque vs. speed curve for a small (about 1/4 horsepower) single-phase motor. Note that the torque/speed curve is completely anti-symmetric, meaning the motor would turn 'backwards' just as well as it turns forwards. Note that there is no torque

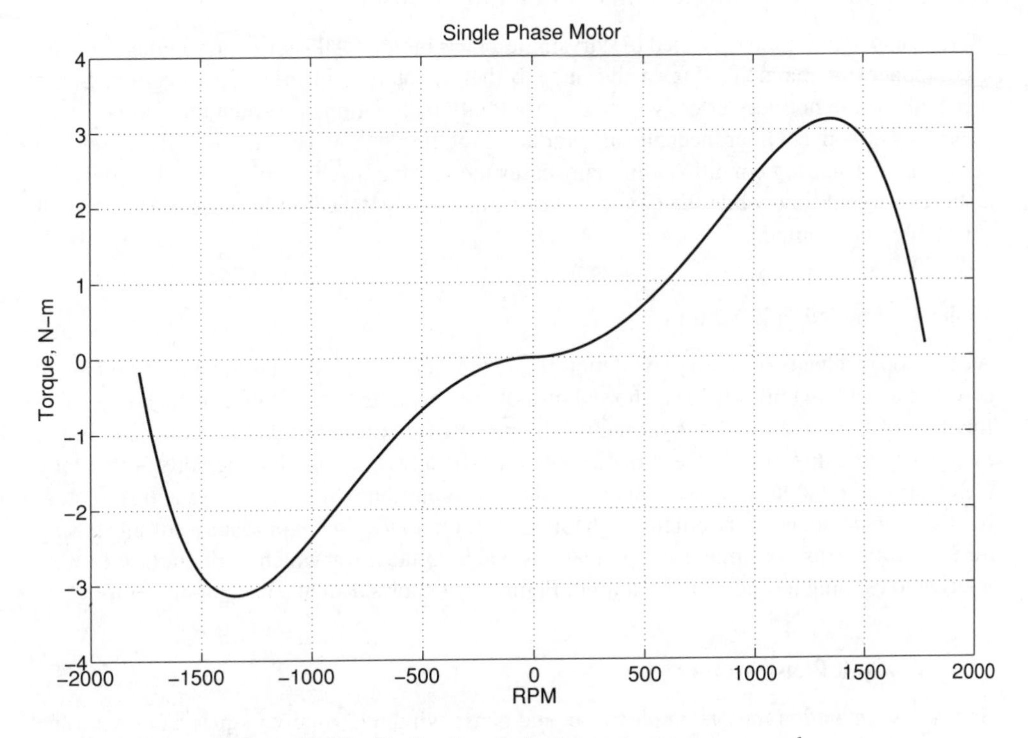

Figure 13.13 Single-phase induction motor torque vs. speed

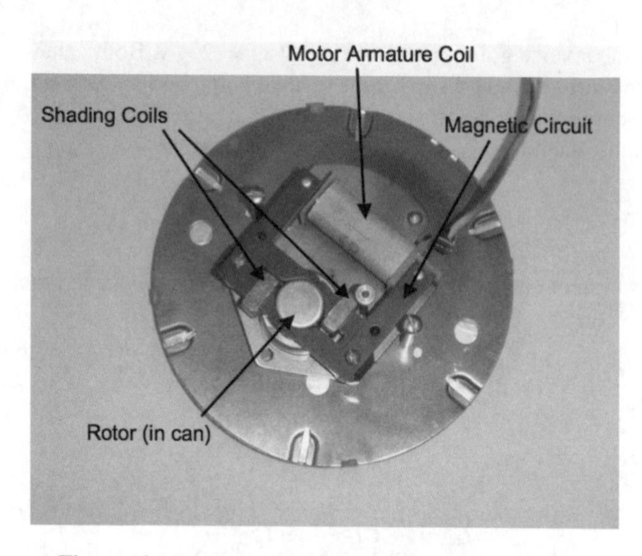

Figure 13.14 Shaded pole motor. Picture by the author

at stall (zero rotor speed), so that some special mechanism is required to cause the motor to start and to start in the right direction.

13.4.3 Starting of Single-Phase Induction Motors

More than one mechanism is used to start single-phase motors. All starting mechanisms create a component of magnetic flux in the air gap that is rotating in the right direction. Some mechanisms are not very effective and are used only in situations in which they do not need to be very effective. Other mechanisms produce a rotating flux wave with virtually no reverse component at starting conditions. In many cases the starting mechanism is turned on only for a short period of time while the motor is starting and are switched off because the motor will run stably once started.

13.4.3.1 Shaded Pole Motors

A common mechanism for very small motors and for motors that do not require high efficiency is to use a shading coil. This is a shorted turn of wire wrapped around part of each pole. The function of this shading coil is to modify the phase of flux through that part of the coil and thus generate a direction to the flux distribution. The advantage of shading coils is that they are cheap and reliable. The disadvantage is that they contribute to motor losses and so motors built with them are not very efficient. Shading coils, typically, do not produce a strong starting torque, so their use is limited to applications, such as fans, for which high starting torque, relative to running torque, is not required. Figure 3.14 shows a small shaded pole motor.

13.4.3.2 Split Phase Motors

Split phase induction motors employ a second phase winding, usually in quadrature with the main winding. A mechanism is used to cause current in the auxiliary winding to have a phase

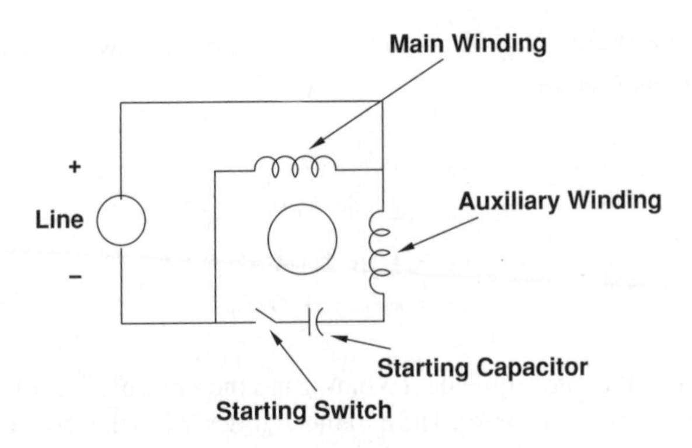

Figure 13.15 Split phase capacitor start motor

that is different from current in the main winding. There are many variations on split phase starting, but the scheme shown in Figure 13.15 is commonly used. Here, a capacitor is inserted in series with the auxiliary winding. The capacitor and auxiliary winding are left in the circuit only for a short period while the motor is starting and are removed from the circuit after the motor is started. That switch is often actuated by rotation of the rotor: when the rotor gets to a certain speed a centrifugally actuated mechanism opens the switch. Typically this is a mechanism that pushes on the switch with a circular sliding surface mounted on the rotor, holding it closed when the rotor is stationary or moving slowly. The mechanism withdraws when the rotor is turning at relatively high speed, so that while the machine is running there is no contact with the starting switch. There are many variations on the split phase motor scheme. Here are a few:

- In some cases, the starting switch mechanism is a timer rather than a centrifugally actuated mechanism.
- The starting mechanism may include a temperature-sensitive resistor with a resistance characteristic that reduces current in the starting winding when that resistor becomes hot, effectively removing it from the winding.
- Sometimes the main winding is energized through a capacitor and the auxiliary winding is connected directly to the line through the starting switch. This results in better running power factor.
- Sometimes the auxiliary winding is left in the circuit, but with a different 'run' capacitor to improve efficiency and power factor.
- Sometimes the auxiliary winding is not connected through a capacitor, but simply has a higher resistance than the main winding. This yields a smaller, but still effective phase shift.

13.4.4 Split Phase Operation

A 'split phase' single-phase induction motor has two, not necessarily identical, windings oriented in quadrature. The two windings are the 'main' or 'run' and 'auxiliary' windings. The 'main' winding is used for single phase run operation as described in the previous section, while the 'auxiliary' winding (sometimes called the 'starting' winding) is used to get the motor started.

The two stator phases are noted as a ('run') and b ('aux')and two equivalent rotor phases are A and B, flux linkages are:

$$\begin{aligned}\lambda_{\mathrm{a}} &= L_{\mathrm{a}} i_{\mathrm{a}} + L_{\phi} i_{\mathrm{A}} \\ \lambda_{\mathrm{b}} &= L_{\mathrm{b}} i_{\mathrm{b}} + \alpha L_{\phi} i_{\mathrm{B}} \\ \lambda_{\mathrm{A}} &= L_{\phi} i_{\mathrm{a}} + L_{\mathrm{A}} i_{\mathrm{A}} \\ \lambda_{\mathrm{B}} &= \alpha L_{\phi} i_{\mathrm{b}} + L_{\mathrm{A}} i_{\mathrm{B}}\end{aligned}$$

Here, it is assumed that the rotor equivalent winding has the same number of turns and winding factor as the 'run' winding of the stator. The parameter α describes the ratio between the mutual components of the main and auxiliary winding coupling to the rotor.

This is not a convenient set to use because the interaction of the rotor makes the equivalent phases A and B difficult to use. So a coordinate transformation is used to translate to equivalent forward- and reverse-going quantities. Assume that the equivalent quantities in the rotor coordinates are the sum of components rotating forward and backward:

$$\begin{bmatrix}\mathbf{I}_{\mathrm{A}} \\ \mathbf{I}_{\mathrm{B}}\end{bmatrix} = \frac{1}{2}\begin{bmatrix}1 & 1 \\ j & -j\end{bmatrix}\begin{bmatrix}\mathbf{I}_{\mathrm{F}} \\ \mathbf{I}_{\mathrm{R}}\end{bmatrix}$$

The inverse transformation is:

$$\begin{bmatrix}\mathbf{I}_{\mathrm{F}} \\ \mathbf{I}_{\mathrm{R}}\end{bmatrix} = \begin{bmatrix}1 & -j \\ 1 & j\end{bmatrix}\begin{bmatrix}\mathbf{I}_{\mathrm{A}} \\ \mathbf{I}_{\mathrm{B}}\end{bmatrix}$$

Then the complex amplitudes of flux linkages are:

$$\begin{aligned}\mathbf{\Lambda}_{\mathrm{a}} &= L_{\mathrm{a}}\mathbf{I}_{\mathrm{a}} + \frac{L_{\phi}}{2}\mathbf{I}_{F} + \frac{L_{\phi}}{2}\mathbf{I}_{\mathrm{R}} \\ \mathbf{\Lambda}_{\mathrm{b}} &= L_{\mathrm{b}}\mathbf{I}_{\mathrm{b}} + \frac{j\alpha L_{\phi}}{2}\mathbf{I}_{\mathrm{F}} - \frac{j\alpha L_{\phi}}{2}\mathbf{I}_{\mathrm{R}} \\ \mathbf{\Lambda}_{\mathrm{F}} &= L_{\phi}\mathbf{I}_{\mathrm{a}} - j\alpha L_{\phi}\mathbf{I}_{\mathrm{b}} + L_{\mathrm{A}}\mathbf{I}_{\mathrm{F}} \\ \mathbf{\Lambda}_{\mathrm{R}} &= L_{\phi}\mathbf{I}_{\mathrm{a}} + j\alpha L_{\phi}\mathbf{I}_{\mathrm{b}} + L_{\mathrm{A}}\mathbf{I}_{\mathrm{R}}\end{aligned}$$

Voltage equations are, in the stator coordinate system:

$$\begin{aligned}\mathbf{V}_{\mathrm{A}} &= (jX_{\mathrm{a}} + R_{\mathrm{a}})\mathbf{I}_{\mathrm{a}} + \frac{jX_{\phi}}{2}\mathbf{I}_{\mathrm{F}} + \frac{jX_{\phi}}{2}\mathbf{I}_{\mathrm{R}} \\ \mathbf{V}_{\mathrm{B}} &= (jX_{\mathrm{e}} + jX_{\mathrm{b}} + R_{\mathrm{b}})\mathbf{I}_{\mathrm{b}} - \frac{\alpha X_{\phi}}{2}\mathbf{I}_{\mathrm{F}} + \frac{\alpha X_{\phi}}{2}\mathbf{I}_{\mathrm{R}} \\ 0 &= \frac{jX_{\phi}}{2}\mathbf{I}_{\mathrm{a}} + \frac{\alpha X_{\phi}}{2}\mathbf{I}_{\mathrm{b}} + \left(jX_{\mathrm{A}} + \frac{R_2}{s}\right)\mathbf{I}_{\mathrm{F}} \\ 0 &= \frac{jX_{\phi}}{2}\mathbf{I}_{a} - \frac{\alpha X_{\phi}}{2}\mathbf{I}_{\mathrm{b}} + \left(jX_{\mathrm{A}} + \frac{R_2}{2-s}\right)\mathbf{I}_{\mathrm{R}}\end{aligned}$$

where jX_e is the impedance of some element placed in series with the auxiliary winding to shift the phase of current in that winding. This element might, for example, be a starting capacitor or simply some added resistance.

This set of four linear equations is readily solved for the four currents I_a, I_b, I_F and I_R. To find mechanical energy converted, see that air-gap power and power dissipated on the rotor are the same as were computed from the equivalent circuit.

$$T_m = \frac{p}{\omega(1-s)} P_m = \frac{p}{\omega}\left[|I_F|^2 \frac{R_2}{s} - |I_R|^2 \frac{R_2}{2-s}\right]$$

13.4.4.1 Example Motor

A split phase induction motor destined for service in a refrigerator application has these properties:

Stator voltage	V 120	V, RMS	
Electrical frequency	F	60	Hz
Magnetizing reactance	X_ϕ	169	Ω
Stator leakage	X_1	2.92	Ω
Rotor leakage	X_2	0.43	Ω
Stator resistance	R_1	4.13	Ω
Rotor resistance	R_2	2.07	Ω

For the purpose of this exercise, the auxiliary winding, which is oriented in quadrature with the main winding but has otherwise the same properties (distribution and number of turns), is connected to the same source as the primary winding, but in series with one of two additional circuit elements.

Consider two starting schemes:

1. The first of these is the capacitive start, in which for starting operation a 50 μF capacitor is put in series with the auxiliary winding.
2. The second starting scheme involves connecting the auxiliary winding in series with an additional resistance of 10 Ω.

In both of these starting schemes the auxiliary winding is connected in parallel with the main winding with the additional elements in series. The auxiliary winding is disconnected once the motor is turning. Torque vs. speed is shown in Figure 13.16.

13.5 Induction Generators

If a machine can be made to turn at a speed higher than synchronous, slip becomes negative. The value of $\frac{R_2}{s}$ also becomes negative, so that torque is negative for negative slip, corresponding to speed greater than synchronous, as shown in Figure 13.17. This implies, of course,

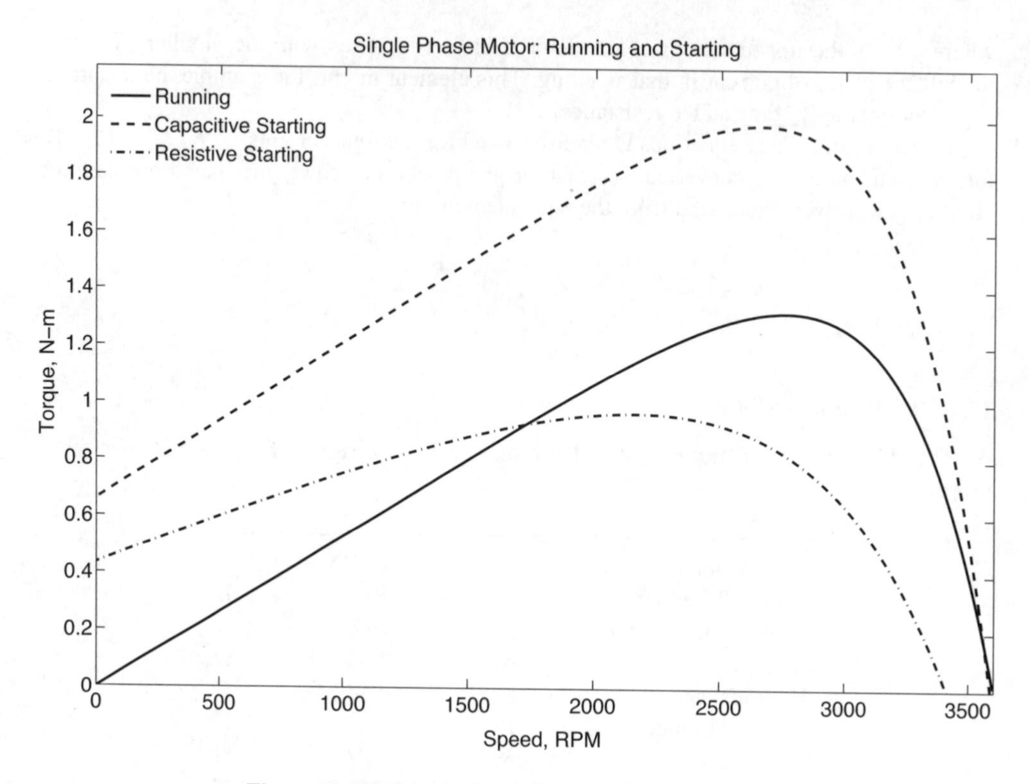

Figure 13.16 Single-phase motor: run and start torques

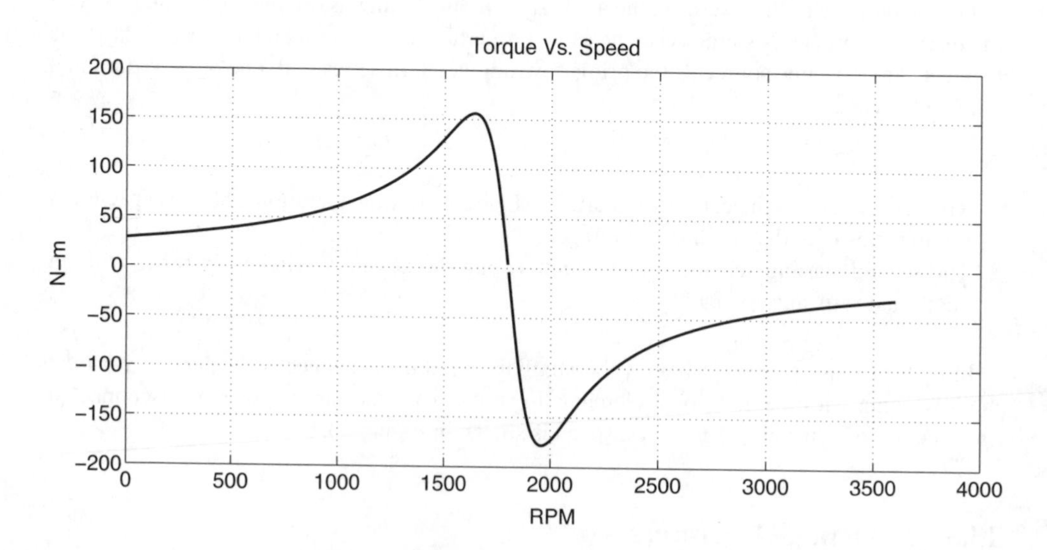

Figure 13.17 Torque vs. speed into the generation range

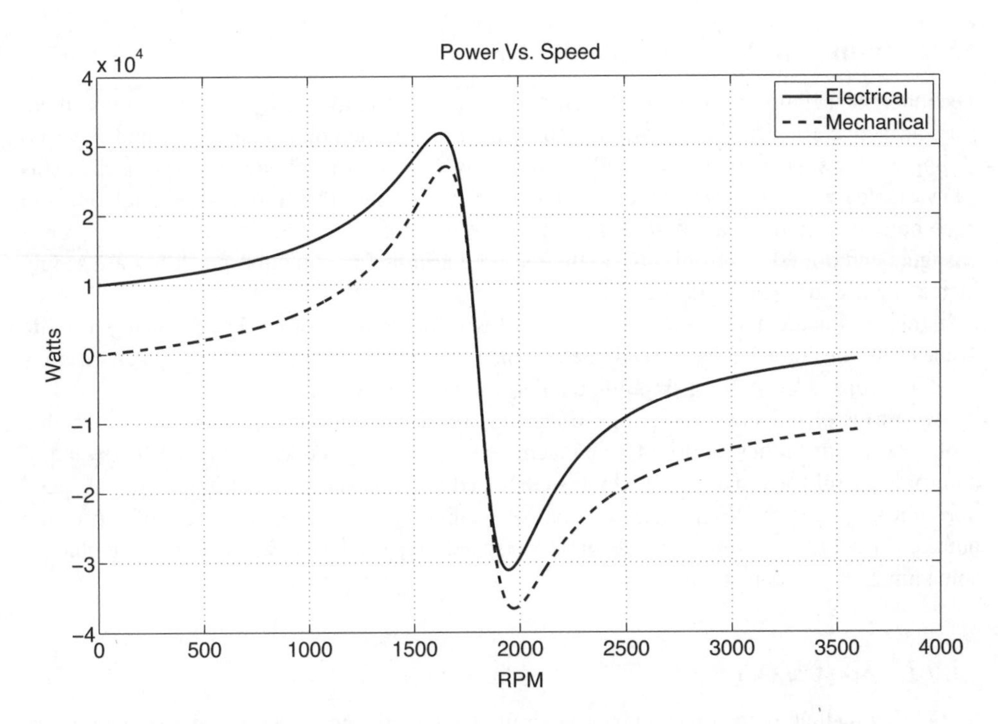

Figure 13.18 Electrical and mechanical power for an induction machine

operation as a generator. The machine absorbs mechanical power and produces electrical power. Figure 13.18 shows power vs. speed over a wider range than just motor operation. For speeds higher than synchronous, both mechanical power out of them machine and electrical power into the machine become negative. This sort of operation is often used when there is a mechanical system that produces power (such as a turboexpander in a refrigeration cycle).

Squirrel-cage induction generators have also been used with wind turbines of moderate power capacity. While there are many wind turbines with induction generators in service, this use of induction machines is not widely used any more for two reasons.

- Because squirrel-cage induction generators draw substantial amounts of reactive power. Further, the amount of reactive power drawn by an induction generator increases with real power, so that an induction generator draws increased amount of reactive power when the wind 'gusts'. Reactive power drawn by the machine affects (in the downward direction) system voltage, and this can be problematical, particularly on weak systems.
- The use of a squirrel-cage induction generator restricts the speed range of the turbine rotor. For aerodynamic reasons that are beyond the scope of this text, wind turbines work best if the 'tip speed ratio' is in a relatively narrow range, so that the rotor speed of the wind turbine must be able to vary if the wind turbine is to extract energy from the wind effectively. For this reason, variable speed machines are advantageous. Wound rotor induction machines controlled by power electronic cascades, known as 'doubly fed induction generators' have become the solution of choice for wind turbines. These are described in Section 13.7.

13.6 Induction Motor Control

The inherent attributes of induction machines, particularly squirrel-cage machines, make them very attractive for drive applications. They are rugged, economical to build and have no sliding contacts to wear. The difficulty with using induction machines in servomechanisms and variable speed drives is that they are 'hard to control', since their torque–speed relationship is complex and non-linear. With, however, modern power electronics to serve as frequency changers and digital electronics to do the required arithmetic, induction machines are seeing increasing use in drive applications.

In this section are developed models for control of induction motors. The derivation is quite brief for it relies on what has already been done for synchronous machines. This derivation is done in 'ordinary' variables, skipping the per-unit normalization.

Two forms of induction motor speed control are discussed in this section. What is called 'volts/Hz' control relies on driving the machine with variable frequency voltage. This kind of control is useful for situations in which shaft speed must be variable, but in which that speed does not vary quickly. 'Field oriented control', which comes in a wide variety of forms and names, is used for servomechanisms in which speed may need to be varied quickly, including some kinds of position control.

13.6.1 Volts/Hz Control

Remembering that induction machines generally tend to operate at relatively low *per-unit* slip, one might conclude that one way of building an adjustable speed drive would be to supply an induction motor with adjustable stator frequency, and this is, indeed, possible. One thing to remember is that voltage magnitude is the product of flux and frequency, so for a given voltage magnitude, flux is inversely proportional to frequency. To maintain constant flux one must make stator voltage proportional to frequency (hence the term 'constant volts/Hz'). However, voltage supplies are always limited, so that at some frequency it is necessary to switch to constant voltage control. The analogy to DC machines is fairly direct here: below some 'base' speed, the machine is controlled in constant flux ('volts/Hz') mode, while above the base speed, flux is inversely proportional to speed. It is easy to see that the maximum torque, then, varies inversely to the square of flux, or therefore inversely to the square of frequency.

To get a first-order picture of how an induction machine works at adjustable speed, refer back to the equivalent network that describes the machine, as shown in Figure 13.5.

Torque is:

$$T_e = 3\frac{p}{\omega}|I_2|^2\frac{R_2}{s}$$

where ω is the electrical frequency and p is the number of pole pairs. As an example, Figure 13.19 shows a series of torque/speed curves for an induction machine operated with a wide range of input frequencies, both below and above its 'base' frequency. The exemplar machine is the same one described in Table 13.1.

Strategy for operating the machine is to make terminal voltage magnitude proportional to frequency for input frequencies less than the 'base frequency', in this case 60 Hz, and to hold voltage constant for frequencies above the 'base frequency'.

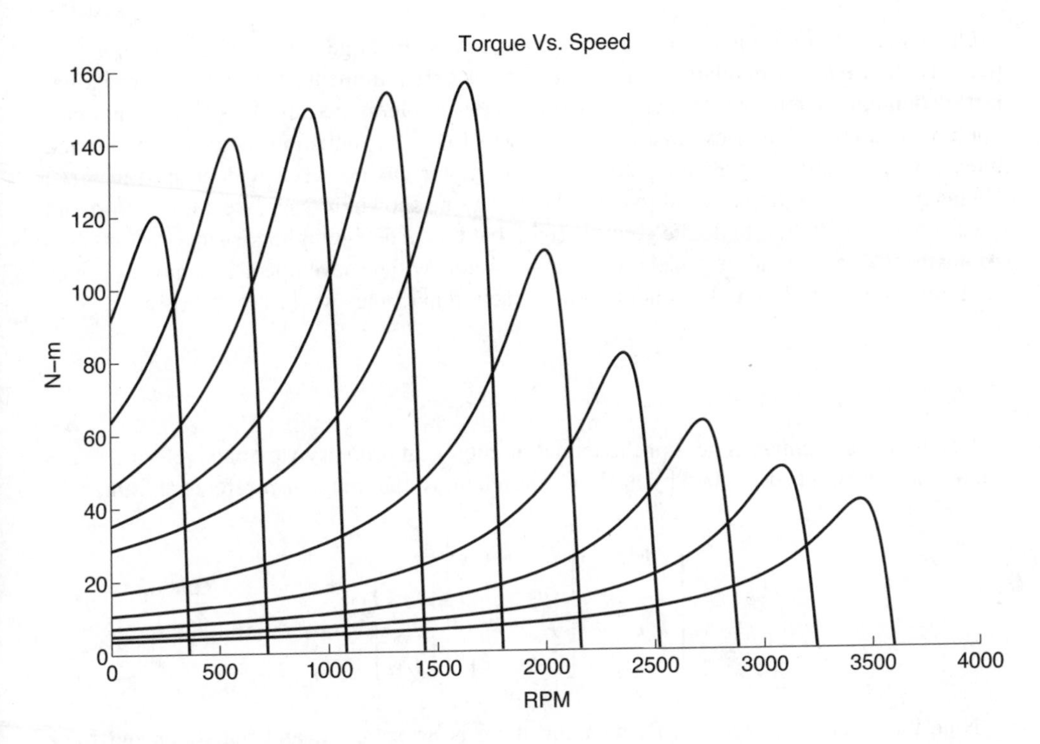

Figure 13.19 Induction machine torque–speed curves

For high frequencies, peak torque falls fairly rapidly with frequency (roughly proportional to the inverse of the square of frequency). It also falls with very low frequency because of the effects of terminal resistance. To understand why this is so, note that the reactive impedance elements have values that are proportional to frequency, while the stator resistance is constant. If terminal flux (volts/Hz) is held constant, the armature resistance becomes larger, relative to the reactances, as frequency is reduced. This can be compensated for by raising the volts/Hz ratio for very low frequencies.

13.6.2 Field Oriented Control

One of the more useful impacts of modern power electronics and control technology has enabled turning induction machines into high performance servomotors. The objective is to emulate the performance of a DC machine, in which torque is a simple function of applied current. For a machine with one field winding, this is simply:

$$T = G I_{\mathrm{f}} I_{\mathrm{a}}$$

This makes control of such a machine quite easy, for once the desired torque is known it is easy to translate that torque command into a current.

Of course DC (commutator) machines are, at least in large sizes, expensive, not particularly efficient, have relatively high maintenance requirements because of the sliding brush/commutator interface, provide environmental problems because of sparking and carbon dust and are themselves environmentally sensitive. The induction motor is simpler and more rugged. Until fairly recently the induction motor has not been widely used in servo applications because it was 'hard to control'. As will be shown, it does take a little effort and even some computation to do the controls right, but this is becoming increasingly affordable. Along the way, a simulation model for induction motors will be developed. This model 'comes for free' and is useful in understanding how induction machines work.

13.6.3 Elementary Model

Return to the elementary model of the induction motor. In ordinary variables, referred to the stator, the machine is described by flux–current relationships (in the d-q reference frame):

$$\begin{bmatrix} \lambda_{dS} \\ \lambda_{dR} \end{bmatrix} = \begin{bmatrix} L_S & L_m \\ L_m & L_R \end{bmatrix} \begin{bmatrix} i_{dS} \\ i_{dR} \end{bmatrix}$$
$$\begin{bmatrix} \lambda_{qS} \\ \lambda_{qR} \end{bmatrix} = \begin{bmatrix} L_S & L_m \\ L_m & L_R \end{bmatrix} \begin{bmatrix} i_{qS} \\ i_{qR} \end{bmatrix}$$

Note the machine is symmetric, meaning there is no saliency, and the stator and rotor self-inductances include leakage terms, which are the same for both axes. The meaning of the inductances is the same as for the ordinary squirrel cage machine (see Section 13.2).

$$L_S = L_m + L_1$$
$$L_R = L_m + L_2$$

The voltage equations are:

$$v_{dS} = \frac{d\lambda_{dS}}{dt} - \omega\lambda_{qS} + R_1 i_{dS}$$
$$v_{qS} = \frac{d\lambda_{qS}}{dt} + \omega\lambda_{dS} + R_1 i_{qS}$$
$$0 = \frac{d\lambda_{dR}}{dt} - \omega_s\lambda_{qR} + R_2 i_{dR}$$
$$0 = \frac{d\lambda_{qR}}{dt} + \omega_s\lambda_{dR} + R_2 i_{qR}$$

Note that both rotor and stator have 'speed' voltage terms since they are (or might be) both rotating with respect to the coordinate system. The speed of the rotating coordinate system is ω with respect to the stator, and $\omega_s = \omega - p\omega_m$ with respect to the rotor, where ω_m is the rotor mechanical speed. Note that this analysis does not require that the reference frame coordinate system speed ω be constant.

Torque was derived in Chapter 9 and is given by:

$$T_e = \frac{3}{2} p \left(\lambda_{dS} i_{qS} - \lambda_{qS} i_{dS} \right)$$

13.6.4 Simulation Model

As a first step in developing a simulation model, see that the inversion of the flux–current relationship is (The d-axis is shown here; the q-axis is identical):

$$\begin{aligned} i_{dS} &= \frac{L_R}{L_S L_R - L_m^2} \lambda_{dS} - \frac{L_m}{L_S L_R - L_m^2} \lambda_{dR} \\ i_{dR} &= \frac{L_m}{L_S L_R - L_m^2} \lambda_{dS} - \frac{L_S}{L_S L_R - L_m^2} \lambda_{dR} \end{aligned}$$

With the following definitions (the motivation for this should by now be obvious):

$$\begin{aligned} X_d &= \omega_0 L_S \\ X_{kd} &= \omega_0 L_R \\ X_{ad} &= \omega_0 L_m \\ X_d' &= \omega_0 \left(L_S - \frac{L_m^2}{L_R} \right) \end{aligned}$$

the currents become:

$$\begin{aligned} i_{dS} &= \frac{\omega_0}{X_d'} \lambda_{dS} - \frac{X_{ad}}{X_{kd}} \frac{\omega_0}{X_d'} \lambda_{dR} \\ i_{dR} &= -\frac{X_{ad}}{X_{kd}} \frac{\omega_0}{X_d'} \lambda_{dS} + \frac{X_d}{X_d'} \frac{\omega_0}{X_{kd}} \lambda_{dR} \\ i_{qS} &= \frac{\omega_0}{X_d'} \lambda_{qS} - \frac{X_{ad}}{X_{kd}} \frac{\omega_0}{X_d'} \lambda_{qR} \\ i_{qR} &= -\frac{X_{ad}}{X_{kd}} \frac{\omega_0}{X_d'} \lambda_{qS} + \frac{X_d}{X_d'} \frac{\omega_0}{X_{kd}} \lambda_{qR} \end{aligned}$$

Torque may be, with these calculations for current, written as:

$$T_e = \frac{3}{2} p \left(\lambda_{dS} i_{qS} - \lambda_{qS} i_{dS} \right) = -\frac{3}{2} p \frac{\omega_0 X_{ad}}{X_{kd} X_d'} \left(\lambda_{dS} \lambda_{qR} - \lambda_{qS} \lambda_{dR} \right)$$

Note that the foregoing expression was written assuming the variables are expressed as *peak* quantities. (Hence the factor of $3/2$.)

The simulation model is now quite straightforward. The state equations are:

$$\frac{d\lambda_{dS}}{dt} = V_{dS} + \omega\lambda_{qS} - R_1 i_{dS}$$
$$\frac{d\lambda_{qS}}{dt} = V_{qS} - \omega\lambda_{dS} - R_1 i_{qS}$$
$$\frac{d\lambda_{dR}}{dt} = \omega_s\lambda_{qR} - R_2 i_{dR}$$
$$\frac{d\lambda_{qR}}{dt} = -\omega_s\lambda_{dR} - R_2 i_{qR}$$
$$\frac{d\Omega_m}{dt} = \frac{1}{J}\left(T_e + T_m\right)$$

where the rotor frequency (slip frequency) is:

$$\omega_s = \omega - p\omega_m$$

For simple simulations and constant excitation frequency, the choice of coordinate systems is arbitrary, so we can choose something convenient. For example, we might choose to fix the coordinate system to a synchronously rotating frame, so that stator frequency $\omega = \omega_0$. In this case, the stator voltage could be aligned with one axis or another. A common choice is $V_d = 0$ and $V_q = V$.

13.6.5 Control Model

The key to turning the machine into a servomotor is to be a bit more sophisticated about the coordinate system. In general, the principle of field oriented control is much like emulating the function of a DC (commutator) machine. First, estimate where the flux is, then inject current to interact most directly with the flux.

As a first step, note that because the two stator flux linkages are the sum of air-gap and leakage flux,

$$\lambda_{dS} = \lambda_{agd} + L_{S\ell} i_{dS}$$
$$\lambda_{qS} = \lambda_{agq} + L_{S\ell} i_{qS}$$

Torque is:

$$T_e = \frac{3}{2} p \left(\lambda_{agd} i_{qS} - \lambda_{agq} i_{dS}\right)$$

Next, note that the rotor flux is, similarly, related to air-gap flux:

$$\lambda_{agd} = \lambda_{dR} - L_{R\ell} i_{dR}$$
$$\lambda_{agq} = \lambda_{qR} - L_{R\ell} i_{qR}$$

Torque now becomes:

$$T_e = \frac{3}{2}p\left(\lambda_{dR}i_{qS} - \lambda_{qR}i_{dS}\right) - \frac{3}{2}pL_{R\ell}\left(i_{dR}i_{qS} - i_{qR}i_{dS}\right)$$

Now, since the rotor currents could be written as:

$$i_{dR} = \frac{\lambda_{dR}}{L_R} - \frac{L_m}{L_R}i_{dS}$$
$$i_{qR} = \frac{\lambda_{qR}}{L_R} - \frac{L_m}{L_R}i_{qS}$$

That second term can be written as:

$$i_{dR}i_{qS} - i_{qR}i_{dS} = \frac{1}{L_R}\left(\lambda_{dR}i_{qS} - \lambda_{qR}i_{dS}\right)$$

So that torque is now:

$$T_e = \frac{3}{2}p\left(1 - \frac{L_{R\ell}}{L_R}\right)\left(\lambda_{dR}i_{qS} - \lambda_{qR}i_{dS}\right) = \frac{3}{2}p\frac{L_m}{L_R}\left(\lambda_{dR}i_{qS} - \lambda_{qR}i_{dS}\right)$$

13.6.6 Field-Oriented Strategy

What is done in field oriented control is to establish a rotor flux in a known position (usually this position is the d-axis of the transformation) and then put a current on the orthogonal axis (where it will be most effective in producing torque). That is, the fluxes are set to be:

$$\lambda_{dR} = \Lambda_0$$
$$\lambda_{qR} = 0$$

Then torque is produced by applying quadrature-axis current:

$$T_e = \frac{3}{2}p\frac{L_m}{L_R}\Lambda_0 i_{qS}$$

The process is almost that simple. There are a few details involved in determining where the quadrature axis is and how hard to drive the direct axis (magnetizing) current.

Suppose flux can be placed on the direct axis, so that $\lambda_{qR} = 0$, then the two rotor voltage equations are:

$$0 = \frac{\mathrm{d}\lambda_{dR}}{\mathrm{d}t} - \omega_s\lambda_{qR} + R_2 I_{dR}$$
$$0 = \frac{\mathrm{d}\lambda_{qR}}{\mathrm{d}t} + \omega_s\lambda_{dR} + R_2 I_{qR}$$

Now, since the rotor currents are:

$$i_{dR} = \frac{\lambda_{dR}}{L_R} - \frac{L_m}{L_R} i_{dS}$$

$$i_{qR} = \frac{\lambda_{qR}}{L_R} - \frac{L_m}{L_R} i_{qS}$$

The voltage expressions become, accounting for the fact that the rotor quadrature axis flux is zero:

$$0 = \frac{d\lambda_{dR}}{dt} + R_2 \left(\frac{\lambda_{dR}}{L_R} - \frac{L_m}{L_R} i_{dS} \right)$$

$$0 = \omega_s \lambda_{dR} - R_2 \frac{L_m}{L_R} i_{qS}$$

Noting that the rotor time constant is

$$T_R = \frac{L_R}{R_2}$$

the two operative expressions are:

$$T_R \frac{d\lambda_{dR}}{dt} + \lambda_{dR} = L_m i_{dS}$$

$$\omega_s = \frac{L_m}{T_R} \frac{i_{qS}}{\lambda_{dR}}$$

The first of these two expressions describes the behavior of the direct axis flux: as one would think, it has a simple first-order relationship with direct-axis stator current. The second expression, which describes slip as a function of quadrature axis current and direct axis flux, actually describes how fast to turn the rotating coordinate system with respect to the rotor to hold flux on the direct axis.

Now, a real machine application involves phase currents i_a, i_b and i_c, and these must be derived from the model currents i_{dS} and i_{qS}. This is done with, of course, a mathematical operation which uses a transformation angle θ. And that angle is derived from the rotor mechanical speed and computed slip:

$$\theta = \int (p\omega_m + \omega_s)\, dt$$

A generally good strategy to make this sort of system work is to measure the three phase currents and derive the direct- and quadrature-axis currents from them. A good estimate of direct-axis flux is made by running direct-axis flux through a first-order filter. The tricky operation involves dividing quadrature axis current by direct axis flux to get slip, but this is now easily done numerically (as are the trigonometric operations required for the rotating coordinate system transformation). An elementary block diagram of a (possibly) plausible scheme for this is shown in Figure 13.20.

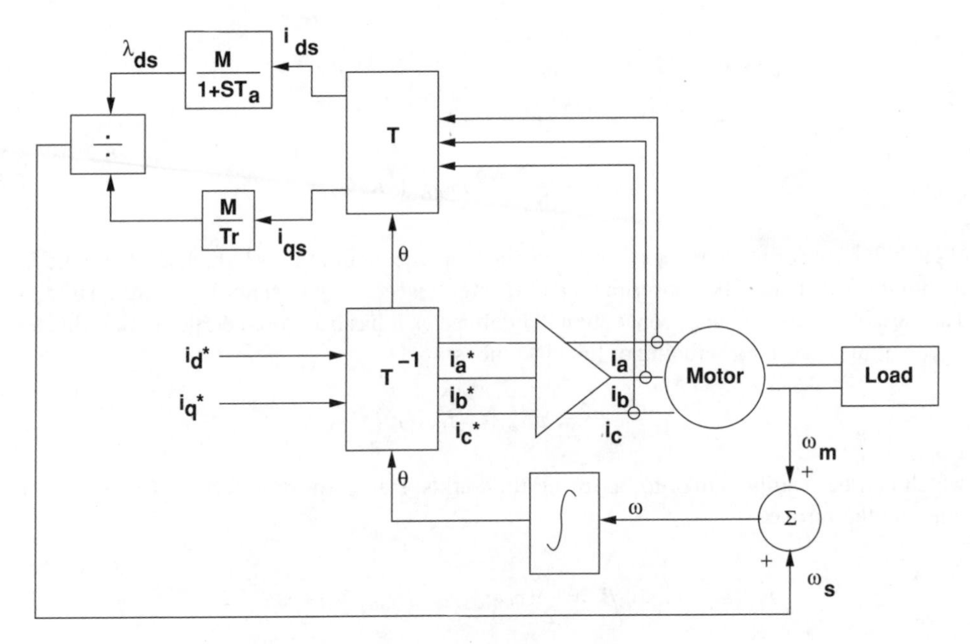

Figure 13.20 Field-oriented controller

This figure starts with commanded values of direct- and quadrature-axis currents, corresponding to flux and torque, respectively. These are translated through a rotating coordinate transformation into commanded phase currents. That transformation (simply the inverse Park's Transform) uses the angle θ derived as part of the scheme. That angle is the integral of a rotational velocity that is the sum of shaft speed ω_{m} and slip speed ω_{s}.

The commanded currents i_{a}^*, are shown as inputs to an 'Amplifier'. This might be implemented as a PWM voltage source inverter, for example, operating in a current feedback loop, and a tight loop here results in a high performance servo system.

13.7 Doubly Fed Induction Machines

The most common induction motors are squirrel-cage machines, in which the rotors are inherently shorted. Increasingly, however, induction machines with wound rotors have received application for a variety of uses. An important example is for variable speed, constant frequency generation in wind turbines. The basic principle involved is that, for relatively small variations in speed about and relative to a synchronous speed, power flow out of or in to the rotor are small relative to power flow out of or in to the stator of a machine.

To understand how a doubly fed machine works, consider voltages in a frame of reference frame (d-q) that is rotating at constant (e.g. synchronous) speed with frequency ω:

$$v_{d\mathrm{S}} = \frac{\mathrm{d}\lambda_{d\mathrm{S}}}{\mathrm{d}t} - \omega\lambda_{q\mathrm{S}} + R_{\mathrm{S}} i_{ds}$$

$$v_{qS} = \frac{d\lambda_{qS}}{dt} + \omega\lambda_{dS} + R_S i_{qs}$$
$$v_{dR} = \frac{d\lambda_{dR}}{dt} - s\omega\lambda_{qR} + R_R i_{dr}$$
$$v_{qR} = \frac{d\lambda_{qR}}{dt} + s\omega\lambda_{dR} + R_R i_{qr}$$

Here, the subscripts d and q refer to the direct and quadrature axes, which are the same for both rotor and stator. The subscripts S and R refer, respectively to stator and rotor quantities. The variable s refers to slip, conventionally defined as it has been used earlier in this chapter.

Power into the stator winding of the machine is:

$$P_S = v_a i_a + v_b i_b + v_c i_c$$

which can be readily shown to be, using the Park's Transform described in Chapter 8 and ignoring the zero axis:

$$P_s = \frac{3}{2}\left(v_{dS} i_{dS} + v_{qS} i_{qS}\right)$$

Substituting for voltage, this is:

$$P_S = \frac{3}{2}\left(i_{dS}\frac{d\lambda_{dS}}{dt} + i_{qS}\frac{d\lambda_{qS}}{dt}\right) + \frac{3}{2}\omega\left(\lambda_{dS} i_{qS} - \lambda_{qS} i_{dS}\right) + R_S\left(|i_{dS}|^2 + |i_{qS}|^2\right) \quad (13.19)$$

The first term of Equation 13.19 is related to changes in stored magnetic energy. The last term is clearly dissipation in the stator winding, while the middle, being proportional to electrical frequency, must be power converted from electrical to mechanical form:

$$P_{em}^S = \frac{3}{2}\omega\left(\lambda_{dS} i_{qS} - \lambda_{qS} i_{dS}\right) \quad (13.20)$$

A similar exercise done on the rotor terminals yields a similar expression for electromagnetic power:

$$P_{em}^R = \frac{3}{2}s\omega\left(\lambda_{dR} i_{qR} - \lambda_{qR} i_{dR}\right) \quad (13.21)$$

The relationship between fluxes and currents for the doubly fed machine is described by the transformer model of Figure 13.21. The ideal transformer is there to reflect a possible effective turns ratio between the stator of the machine and the rotor. Since the ideal transformer neither produces nor consumes any real or reactive power, Equation 13.21 could be written as:

$$P_{em}^R = \frac{3}{2}s\omega\left(\lambda_{d2} i_{q2} - \lambda_{q2} i_{d2}\right) \quad (13.22)$$

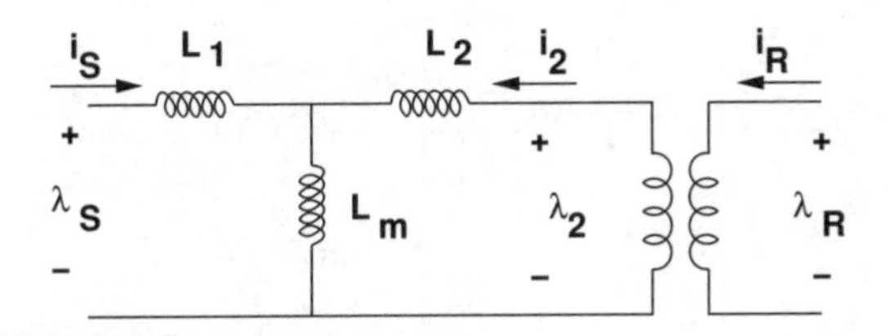

Figure 13.21 Doubly fed induction machine magnetic model

The flux-current relationship for the machine can be written as:

$$\begin{bmatrix} \lambda_S \\ \lambda_2 \end{bmatrix} = \begin{bmatrix} L_m + L_1 & L_m \\ L_m & L_m + L_2 \end{bmatrix} \begin{bmatrix} i_S \\ i_2 \end{bmatrix} \tag{13.23}$$

Note that the mutual inductance L_m is equal to $\frac{3}{2}$ times the space fundamental component of stator phase self inductance, L_1 is the leakage inductance of the stator winding and L_2 is the rotor winding leakage inductance referred across the ideal transformer as shown in Figure 13.21.

13.7.1 Steady State Operation

If the machine is operating in a steady state, it becomes convenient to employ complex notation. Let:

$$\begin{aligned}
\mathbf{V}_S &= v_{dS} + jv_{qS} \\
\mathbf{\Lambda}_S &= \lambda_{dS} + j\lambda_{qS} \\
\mathbf{I}_S &= i_{dS} + ji_{qS} \\
\mathbf{V}_2 &= v_{d2} + jv_{q2} \\
\mathbf{\Lambda}_2 &= \lambda_{d2} + j\lambda_{q2} \\
\mathbf{I}_2 &= i_{d2} + ji_{q2}
\end{aligned}$$

Then, in the sinusoidal steady state:

$$\begin{aligned}
\mathbf{V}_S &= j\omega\mathbf{\Lambda}_S + R_S\mathbf{I}_S \\
\mathbf{V}_2 &= js\omega\mathbf{\Lambda}_2 + R_2\mathbf{I}_2
\end{aligned}$$

Dividing the second of these expressions by slip yields:

$$\frac{\mathbf{V}_2}{s} = j\omega\mathbf{\Lambda}_2 + \frac{R_2}{s}$$

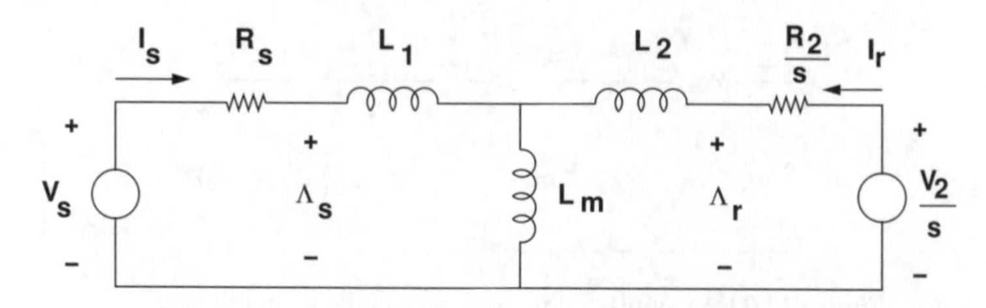

Figure 13.22 Equivalent circuit in sinusoidal steady state

and the magnetic model of Figure 13.21 yields the relationship between complex fluxes and currents:

$$\begin{bmatrix} \mathbf{\Lambda}_\mathrm{S} \\ \mathbf{\Lambda}_\mathrm{R} \end{bmatrix} = \begin{bmatrix} L_1 + L_\mathrm{m} & L_\mathrm{m} \\ L_\mathrm{m} & L_2 + L_\mathrm{m} \end{bmatrix} \begin{bmatrix} \mathbf{I}_\mathrm{S} \\ \mathbf{I}_\mathrm{R} \end{bmatrix}$$

These expressions describe the circuit shown in Figure 13.22.

Real power into the stator and rotor sides of the machine are, assuming that resistive dissipation is negligible:

$$P_\mathrm{em}^\mathrm{S} = \frac{3}{2}\mathrm{Re}\left\{ j\omega \mathbf{\Lambda}_\mathrm{S} \mathbf{I}_\mathrm{S}^* \right\}$$
$$P_\mathrm{em}^\mathrm{R} = \frac{3}{2}\mathrm{Re}\left\{ js\omega \mathbf{\Lambda}_\mathrm{R} \mathbf{I}_\mathrm{R}^* \right\}$$

As expected, these evaluate to:

$$P_\mathrm{em}^\mathrm{S} = \frac{3}{2}\omega \left(\lambda_{d\mathrm{S}} i_{q\mathrm{S}} - \lambda_{q\mathrm{S}} i_{d\mathrm{S}} \right)$$
$$P_\mathrm{em}^\mathrm{R} = \frac{3}{2}s\omega \left(\lambda_{d\mathrm{R}} i_{q\mathrm{R}} - \lambda_{q\mathrm{R}} i_{d\mathrm{R}} \right)$$

It is straightforward, using Equation 13.23 to show that:

$$\left(\lambda_{d\mathrm{R}} i_{q\mathrm{R}} - \lambda_{q\mathrm{R}} i_{d\mathrm{R}} \right) = -\left(\lambda_{d\mathrm{S}} i_{q\mathrm{S}} - \lambda_{q\mathrm{S}} i_{d\mathrm{S}} \right)$$

It becomes clear from this that the assertion made in Section 13.2.1, that torque is the same from all points of view (terminals and/or shafts) is meaningful and that:

$$P_\mathrm{em}^\mathrm{R} = -sP_\mathrm{em}^\mathrm{S}$$

This can be interpreted as follows:

- If slip is positive, meaning that the rotor is turning at a speed less than synchronous, real power coming out of the rotor has the same sign as power going into the stator. If the machine is a motor, both of these are positive: power goes into the stator and out of the rotor. (In a squirrel-cage machine, that power is dissipated in the rotor.)

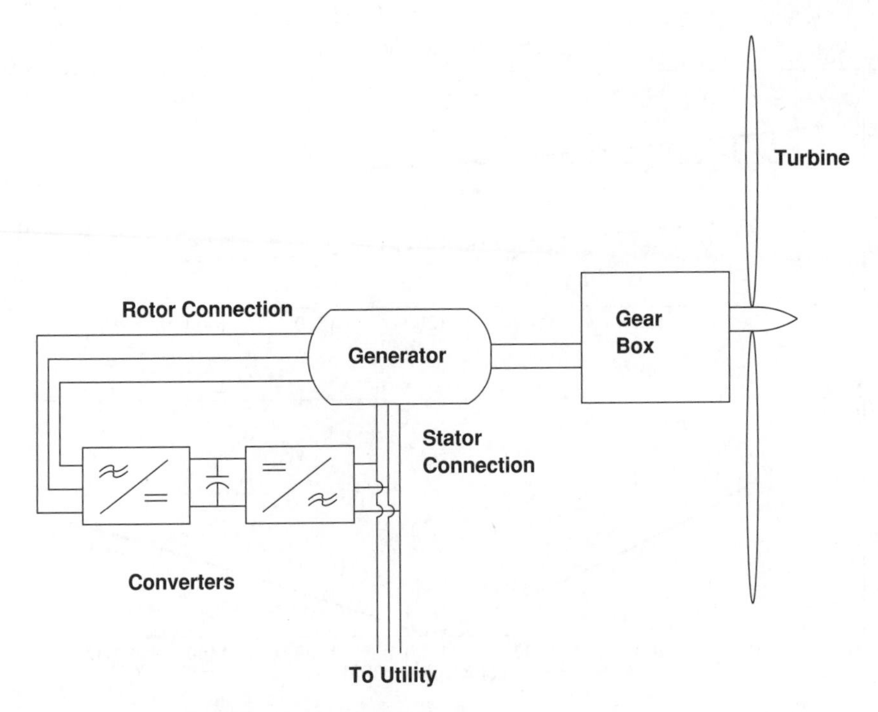

Figure 13.23 Modern wind turbine generator system

- If slip is negative, meaning that the rotor is turning at a speed greater than synchronous, real power going into the rotor terminals has the same sign as real power going into the stator. If the machine is a generator, real power comes out of the stator and real power comes out of the rotor. (This is similar to the case of a squirrel cage machine operating as a generator, in which the voltage $\mathbf{V}_2 = 0$, so all rotor power is dissipated in the rotor resistance.)

Figure 13.23 shows how a doubly fed induction generator is used in a wind turbine system. Generally, because wind turbines turn at a low speed, there is a gear box to increase shaft speed to the point where the doubly fed induction machine can operate efficiently. The stator of the machine is connected to the utility system that can receive generated power. Between the utility system and the rotor is a cascade of two AC/DC converters. One of these converts between the utility frequency (usually 50 or 60 Hz) and DC. The other converts between DC and the rotor frequency. Not shown in this figure are the controls that make the system work. When rotor speed is below synchronous, meaning 'slip' is positive, the cascade feeds low frequency current into the rotor so that real power flow is into the rotor. Stator output power is higher than wind input power by the amount of power flow into the rotor. Conversely, when rotor speed is higher than synchronous, so that slip is negative, currents fed by the power electronics cascade are such that real power comes out of the rotor. Stator power is less than wind input power by the amount of power coming out of the rotor. (Of course, these flows are modified in a minor way by losses in machine resistances.) The ratio of real power into the rotor to real power out of the stator is the per-unit slip, or the ratio of rotor frequency to stator frequency, as was shown in this section.

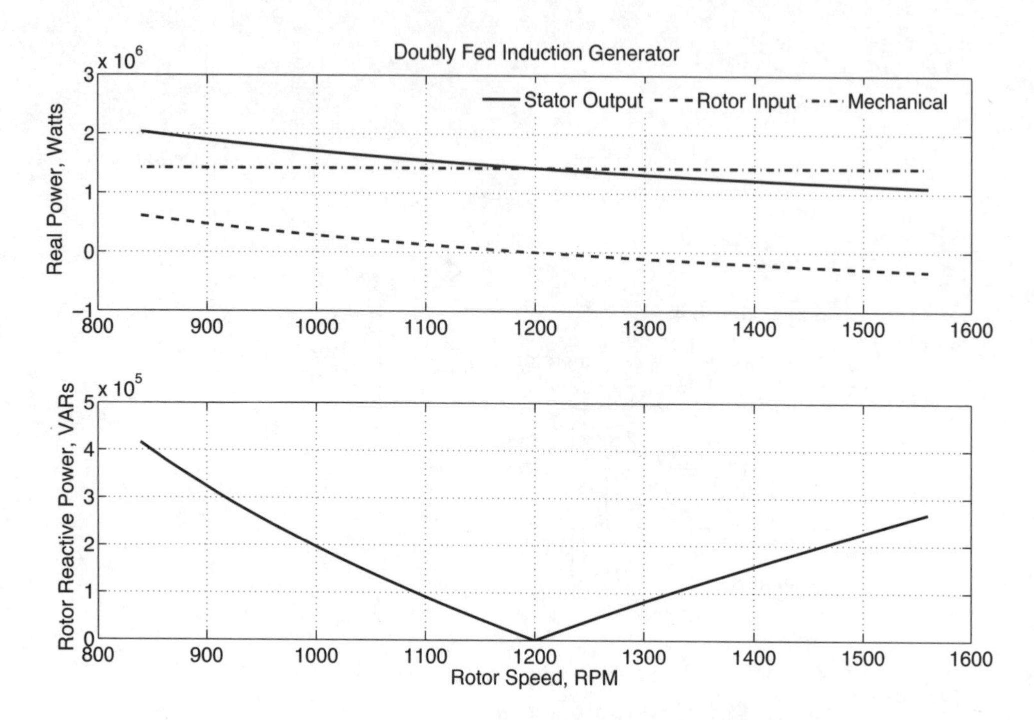

Figure 13.24 Operation of a doubly fed induction generator

Illustrated in Figure 13.24 are the real and reactive power flows in a doubly fed induction generator, assuming constant mechanical power, over a range of speeds of the generator. The machine and electronics combination are feeding 1.5 MVA to the system at a power factor of 0.95, so that $P + jQ = 1.425$ MW $+ j312.25$ kVAR. Actually, in real operation in a wind turbine, real power would be a function of speed, but this figure is intended to illustrate how the machine itself would operate. Rotor real power is stator power times slip, which goes through zero at synchronous speed (1,200 r.p.m). In this application, reactive power is supplied by and controlled by the rotor side converter. The reactive power supplied to the rotor includes reactive power consumed by the reactances of the machine itself as well as reactive power supplied to the system. It is proportional to the absolute value of slip because, for supersynchronous operation the phase (time) relationship between direct and quadrature axes reverses.

13.8 Appendix 1: Squirrel-Cage Machine Model

A circuit model for the squirrel-cage machine may be derived using field analytical techniques. The model consists of two major parts. The first of these is a description of stator flux in terms of stator and rotor currents. The second is a description of rotor current in terms of air- gap flux. The result of all of this is a set of expressions for the elements of the circuit model for the induction machine.

13.8.1 Rotor Currents and Induced Flux

To start, assume that the rotor is symmetrical and carries a surface current, the fundamental of which is:

$$\begin{aligned}\overline{K}_{\mathrm{R}} &= \overline{\imath}_z \mathrm{Re}\left(\mathbf{K}_{\mathrm{R}} e^{j(s\omega t - p\phi')}\right) \\ &= \overline{\imath}_z \mathrm{Re}\left(\mathbf{K}_{\mathrm{R}} e^{j(\omega t - p\phi)}\right)\end{aligned} \tag{13.24}$$

Note Equation 13.24 makes use of the simple transformation between rotor and stator coordinates:

$$\phi' = \phi - \omega_{\mathrm{m}} t \tag{13.25}$$

and that

$$p\omega_{\mathrm{m}} = \omega - \omega_{\mathrm{r}} = \omega(1 - s) \tag{13.26}$$

Here, the following symbols mean:

$\mathbf{K}_{\mathrm{R}}$	is complex amplitude of rotor surface current
s	is per-unit 'slip'
ω	is stator electrical frequency
ω_{r}	is rotor electrical frequency
ω_{m}	is rotational speed
$\mathbf{R}$	in rotor radius

The rotor current will produce an air gap flux density of the form:

$$B_{\mathrm{r}} = \mathrm{Re}\{\mathbf{B}_{\mathrm{r}} e^{j(\omega t - p\phi)}\} \tag{13.27}$$

where

$$\mathbf{B}_{\mathrm{r}} = -j\mu_0 \frac{R}{pg} \mathbf{K}_{\mathrm{R}} \tag{13.28}$$

Note that this describes only radial magnetic flux density produced by the space fundamental of *rotor* current. Flux linked by the armature winding due to this flux density is:

$$\lambda_{\mathrm{AR}} = l N_{\mathrm{S}} k_{\mathrm{S}} \int_{-\frac{\pi}{p}}^{0} B_{\mathrm{r}}(\phi) R d\phi$$

This yields a complex amplitude for λ_{AR}:

$$\lambda_{\mathrm{AR}} = \mathrm{Re}\{\mathbf{\Lambda}_{\mathrm{AR}} e^{j\omega t}\}$$

where

$$\mathbf{\Lambda}_{\mathrm{AR}} = \frac{2l\mu_0 R^2 N_{\mathrm{S}} k_{\mathrm{S}}}{p^2 g} \mathbf{K}_{\mathrm{R}}$$

Adding this to flux produced by the stator currents, yields an expression for total stator flux:

$$\mathbf{\Lambda}_{\mathrm{a}} = \left(\frac{3}{2} \frac{4}{\pi} \frac{\mu_0 N_{\mathrm{S}}^2 R l k_{\mathrm{S}}^2}{p^2 g} + L_{\mathrm{Sl}} \right) \mathbf{I}_{\mathrm{a}} + \frac{2l\mu_0 R^2 N_{\mathrm{S}} k_{\mathrm{S}}}{p^2 g} \mathbf{K}_{\mathrm{R}} \tag{13.29}$$

Expression 13.29 motivates a definition of an equivalent rotor current I_2 in terms of the space fundamental of rotor surface current density:

$$\mathbf{I}_2 = \frac{\pi}{3} \frac{R}{N_{\mathrm{S}} k_{\mathrm{S}}} \mathbf{K}_{\mathrm{R}} \tag{13.30}$$

Then the simple expression for stator flux is:

$$\mathbf{\Lambda}_{\mathrm{a}} = (L_{\mathrm{m}} + L_{\mathrm{Sl}}) \mathbf{I}_{\mathrm{a}} + L_{\mathrm{m}} \mathbf{I}_2 \tag{13.31}$$

where, as in the transformer model,

$$L_{\mathrm{m}} = \frac{3}{2} \frac{4}{\pi} \frac{\mu_0 N_{\mathrm{S}}^2 R l k_{\mathrm{S}}^2}{p^2 g}$$

13.8.2 Squirrel-Cage Currents

The second part of this derivation is the equivalent of finding a relationship between rotor flux and I_2. However, since this machine has no discrete windings, it is necessary to focus on the individual rotor bars.

Assume that there are N_{R} slots in the rotor. Each of these slots is carrying some current. If the machine is symmetrical and operating with balanced currents, current in the kth slot is:

$$i_{\mathrm{k}} = \mathrm{Re}\{\mathbf{I}_k e^{js\omega t}\}$$

where

$$\mathbf{I}_{\mathrm{k}} = \mathbf{I} e^{-j \frac{2\pi p}{N_{\mathrm{R}}}} \tag{13.32}$$

and $\mathbf{I}$ is the complex amplitude of current in slot number zero. Equation 13.32 shows a uniform progression of rotor current *phase* about the rotor. All rotor slots carry the same magnitude of current, but that current is phase delayed from slot to slot because of relative rotation of the current wave at slip frequency.

The rotor current density is a sum of impulses in space:

$$K_{\mathrm{R}} = \mathrm{Re}\left\{\sum_{k=0}^{N_{\mathrm{R}}-1} \frac{1}{R}\mathbf{I}\mathrm{e}^{j(\omega_r t - k\frac{2\pi p}{N_{\mathrm{R}}})}\delta\left(\phi' - \frac{2\pi k}{N_{\mathrm{R}}}\right)\right\} \tag{13.33}$$

The unit impulse function $\delta()$ used to approximate the current in each rotor slot as an impulse. R is rotor radius.

This rotor surface current may be also expressed as a Fourier Series of traveling waves:

$$K_{\mathrm{R}} = \mathrm{Re}\left\{\sum_{n=-\infty}^{\infty} \mathbf{K}_n \mathrm{e}^{j(\omega_r t - np\phi')}\right\} \tag{13.34}$$

Note that Equation 13.34, allows for negative values of the space harmonic index n to allow for reverse- rotating waves.

It is straightforward, but tedious, to show that the terms in the Fourier series are:

$$\begin{aligned} \mathbf{K}_n &= \frac{N_{\mathrm{R}}\mathbf{I}}{2\pi R} && \text{if} \quad (n-1)\frac{p}{N_{\mathrm{R}}} = \text{integer} \\ &= 0 && \text{otherwise} \end{aligned} \tag{13.35}$$

As it turns out, only the first three of the terms, for which the integer is zero, plus and minus one are important, because these produce the largest magnetic fields and therefore fluxes. These are:

$$\begin{aligned} (n-1)\frac{p}{N_{\mathrm{R}}} &= -1 \quad \text{or } n = -\frac{N_{\mathrm{R}} - p}{p} \\ &= 0 \quad \text{or } n = 1 \\ &= 1 \quad \text{or } n = \frac{N_{\mathrm{R}} + p}{p} \end{aligned} \tag{13.36}$$

Note that Equation 13.36 appears to produce space harmonic orders that may be of non-integer order. This is not really true: it is necessary that np be an integer, and Equation 13.36 will always satisfy that condition.

So, the harmonic orders of interest are one and

$$n_+ = \frac{N_{\mathrm{R}}}{p} + 1 \tag{13.37}$$

$$n_- = -\left(\frac{N_{\mathrm{R}}}{p} - 1\right) \tag{13.38}$$

Each of the space harmonics of the squirrel-cage current will produce radial flux density. A surface current of the form:

$$K_n = \mathrm{Re}\left\{\frac{N_{\mathrm{R}}\mathbf{I}}{2\pi R}\mathrm{e}^{j(\omega_r t - np\phi')}\right\} \tag{13.39}$$

produces radial magnetic flux density:

$$B_{rn} = \mathrm{Re}\{\mathbf{B}_{rn} e^{j(\omega_r t - np\phi')}\} \tag{13.40}$$

where

$$\mathbf{B}_{rn} = -j \frac{\mu_0 N_R \mathbf{I}}{2\pi npg} \tag{13.41}$$

In turn, each of the components of radial flux density will produce a component of induced voltage. To calculate that, use Faraday's Law:

$$\nabla \times \overline{E} = -\frac{\partial \overline{B}}{\partial t} \tag{13.42}$$

The radial component of Equation 13.42, assuming that the fields do not vary with z, is:

$$\frac{1}{R}\frac{\partial}{\partial \phi} E_z = -\frac{\partial B_r}{\partial t} \tag{13.43}$$

Or, assuming an electric field component of the form:

$$E_{zn} = \mathrm{Re}\left\{\mathbf{E}_n e^{j(\omega_r t - np\phi)}\right\} \tag{13.44}$$

Using Equations 13.41 and 13.44 in Equation 13.43, an expression for electric field induced by components of air-gap flux is found:

$$\mathbf{E}_n = \frac{\omega_r R}{np} \mathbf{B}_n \tag{13.45}$$

$$\mathbf{E}_n = -j \frac{\mu_0 N_R \omega_r R}{2\pi g (np)^2} \mathbf{I} \tag{13.46}$$

The total voltage induced in a slot pushes current through the conductors in that slot. This is:

$$\mathbf{E}_1 + \mathbf{E}_{n-} + \mathbf{E}_{n+} = \mathbf{Z}_{\mathrm{slot}} \mathbf{I} \tag{13.47}$$

In Equation 13.47, there are three components of air-gap field. E_1 is the space fundamental field, produced by the space fundamental of rotor current as well as by the space fundamental of stator current. The other two components on the left of Equation 13.47 are produced only by rotor currents and actually represent additional reactive impedance to the rotor. This is often called *rotor zigzag* leakage inductance. The parameter Z_{slot} represents impedance of the slot itself: resistance and reactance associated with cross-slot magnetic fields. Then Equation 13.47 can be re-written as:

$$\mathbf{E}_1 = \mathbf{Z}_{\text{slot}}\mathbf{I} + j\frac{\mu_0 N_{\text{R}}\omega_{\text{r}} R}{2\pi g}\left(\frac{1}{(pn_+)^2} + \frac{1}{(pn_-)^2}\right)\mathbf{I} \tag{13.48}$$

To finish this model, it is necessary to translate Equation 13.48 back to the stator. Note that Equations 13.30 and 13.35 make the link between $\mathbf{I}$ and $\mathbf{I}_2$:

$$\mathbf{I}_2 = \frac{N_{\text{R}}}{6N_{\text{S}}k_{\text{S}}}\mathbf{I}$$

Then the electric field at the surface of the rotor is:

$$\mathbf{E}_1 = \left[\frac{6N_{\text{S}}k_{\text{S}}}{N_{\text{R}}}\mathbf{Z}_{\text{slot}} + j\omega_r\frac{3}{\pi}\frac{\mu_0 N_{\text{S}}k_{\text{S}}R}{g}\left(\frac{1}{(pn_+)^2} + \frac{1}{(pn_-)^2}\right)\right]\mathbf{I}_2 \tag{13.49}$$

This must be translated into an equivalent stator voltage. To do so, use Equation 13.45 to translate Equation 13.49 into a statement of radial magnetic field, then find the flux liked and hence stator voltage from that. Magnetic flux density is:

$$\begin{aligned}\mathbf{B}_{\text{r}} &= \frac{p\mathbf{E}_1}{\omega_{\text{r}} R}\\ &= \left[\frac{6N_{\text{S}}k_{\text{S}}p}{N_{\text{R}}R}\left(\frac{R_{\text{slot}}}{\omega_{\text{r}}} + jL_{\text{slot}}\right)\right.\\ &\quad \left. + j\frac{3}{\pi}\frac{\mu_0 N_{\text{S}}k_{\text{S}}p}{g}\left(\frac{1}{(pn_+)^2} + \frac{1}{(pn_-)^2}\right)\right]\mathbf{I}_2\end{aligned} \tag{13.50}$$

where the slot impedance has been expressed by its real and imaginary parts:

$$\mathbf{Z}_{\text{slot}} = R_{\text{slot}} + j\omega_{\text{r}} L_{\text{slot}} \tag{13.51}$$

Flux linking the armature winding is:

$$\lambda_{\text{ag}} = N_{\text{S}}k_{\text{S}}lR\int_{-\frac{\pi}{2p}}^{0}\text{Re}\left\{\left(\mathbf{B}_r e^{j(\omega t - p\phi)}\right)\right\}\text{d}\phi \tag{13.52}$$

Which becomes:

$$\lambda_{\text{ag}} = \text{Re}\left(\mathbf{\Lambda}_{\text{ag}} e^{j\omega t}\right) \tag{13.53}$$

where:

$$\mathbf{\Lambda}_{\text{ag}} = j\frac{2N_{\text{S}}k_{\text{S}}lR}{p}\mathbf{B}_{\text{r}} \tag{13.54}$$

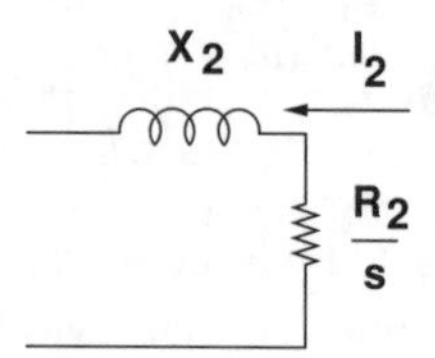

Figure 13.25 Rotor equivalent circuit

Then 'air-gap' voltage is:

$$\begin{aligned}\mathbf{V}_{\text{ag}} &= j\omega\mathbf{\Lambda}_{\text{ag}} = -\frac{2\omega N_S k_S l R}{p}\mathbf{B}_{\text{r}} \\ &= -\mathbf{I}_2\left[\frac{12 l N_S^2 k_S^2}{N_R}\left(j\omega L_{\text{slot}} + \frac{R_2}{s}\right)\right. \\ &\quad \left.+ j\omega\frac{6}{\pi}\frac{\mu_0 R l N_S^2 k_S^2}{g}\left(\frac{1}{(pn_+)^2} + \frac{1}{(pn_-)^2}\right)\right] \end{aligned} \quad (13.55)$$

Equation 13.55 describes the relationship between the space fundamental air-gap voltage $\mathbf{V}_{\text{ag}}$ and rotor current $\mathbf{I}_2$. This expression fits the equivalent circuit of Figure 13.25 if the definitions made below hold:

$$X_2 = \omega\frac{12 l N_S^2 k_S^2}{N_R}L_{\text{slot}} + \omega\frac{6}{\pi}\frac{\mu_0 R l N_S^2 k_S^2}{g}\left(\frac{1}{(N_R + p)^2} + \frac{1}{(N_R - p)^2}\right) \quad (13.56)$$

$$R_2 = \frac{12 l N_S^2 k_S^2}{N_R}R_{\text{slot}} \quad (13.57)$$

The first term in Equation 13.56 expresses slot leakage inductance for the rotor. Similarly, Equation 13.57 expresses rotor resistance in terms of slot resistance. Note that L_{slot} and R_{slot} are both expressed per unit length. The second term in Equation 13.56 expresses the 'zigzag' leakage inductance resulting from harmonics on the order of rotor slot pitch.

Next, see that armature flux is just equal to air- gap flux plus armature leakage flux. That is, Equation 13.31 could be written as:

$$\mathbf{\Lambda}_{\text{a}} = \mathbf{\Lambda}_{\text{ag}} + L_{\text{al}}\mathbf{I}_{\text{a}} \quad (13.58)$$

The rotor resistance typically must be corrected to account for dissipation in the end rings that short the conductor bars together. This is usually done by considering total current in the end rings, using that to estimate dissipation in the rings and then multiplying the bar resistance by a factor that gives the right total rotor dissipation. This factor is a strong function of the number of poles.

13.9 Appendix 2: Single-Phase Squirrel Cage Model

Derivation of the model elements R_s and X_2 for the single phase motor follows closely the derivation in Section 13.8. This derivation is carried out for the forward going flux wave, but the reverse going part is directly comparable. Currents in the rotor bars will are:

$$i_k = \mathrm{Re}\{\mathbf{I}_k e^{j\omega_r t}\}$$

where the complex amplitude in the ith bar is:

$$\mathbf{I}_k = \mathbf{I}_0 e^{-j\frac{2\pi pk}{N_R}}$$

This set of discrete currents is equivalent to a number of space harmonics of surface current:

$$K_R = \mathrm{Re}\left[\sum_n \frac{N_R I_0}{2\pi R} e^{j(\omega_r t \mp np\theta')}\right]$$

The values of n for which K_z is non-zero are

$$n = 1 \pm \text{integer} \times \frac{N_R}{p}$$

and they will produce magnetic flux across the air-gap:

$$\mathbf{B}_{rn} = -j\mu_0 \frac{N_R I_0}{2\pi npg}$$

Electric field induced by these fields is:

$$\mathbf{E}_n = \frac{\omega_r R}{np}\mathbf{B}_n = -j\frac{\mu_0 N_R \omega_r R}{2\pi g n^2 p^2}\mathbf{I}_0$$

Only the smallest order terms ($n = 1$, $n = 1 \pm \frac{N_R}{p}$) contribute substantially to the electric field driving current through the rotor conductor,

$$\mathbf{E}_1 + \mathbf{E}_{n+} + \mathbf{E}_{n-} = Z_{\mathrm{slot}}\mathbf{I}_0$$

Now it is necessary to put this back into the form of an equivalent circuit. Note that the space fundamental field is caused by currents in the stator as well as the rotor but that the higher order space harmonic voltage components are produced only by rotor currents. To refer the rotor current back to the stator, note that, if a current I_F were in the stator it would make a magnetic field:

$$\mathbf{B}_r = -\mu_0 \frac{4}{\pi}\frac{\mathrm{N_a}\mathbf{I}_F k_{wna}}{2pg}$$

then the correct turns ratio to refer rotor current to the stator is:

$$\mathbf{I}_0 = \frac{4N_\mathrm{a}k_{wna}}{N_\mathrm{R}}\mathbf{I}_\mathrm{F}$$

So now the space fundamental component of electric field seen from the rotor is:

$$\mathbf{E}_1 = \left(\frac{4N_\mathrm{a}k_\mathrm{w1a}}{N_\mathrm{R}} + j\omega_\mathrm{r}\frac{2}{\pi}\frac{\mu_0 N_\mathrm{a}k_{wna}R}{g}\left(\frac{1}{(N_{\mathrm{R+p}})^2} + \frac{1}{(N_{\mathrm{R}-p})^2}\right)\right)\mathbf{I}_\mathrm{F}$$

Finally, to relate stator air-gap voltage to rotor electric field, two things must be done: first, translate electric field according to relative frequencies between rotor and stator and integrate over the length of the winding. The result is:

$$\mathbf{V}_\mathrm{ag} = -2\ell N_\mathrm{a}k_\mathrm{w1a}\frac{\omega}{\omega_\mathrm{r}}\mathbf{E}_1$$

The equivalent circuit elements for the rotor are then just:

$$R_2 = \frac{4\ell N_\mathrm{a}^2 k_\mathrm{w1a}^2}{N_\mathrm{R}}R_\mathrm{slot}$$

$$X_2 = \frac{4\ell N_\mathrm{a}^2 k_\mathrm{w1a}^2}{N_\mathrm{R}}\omega L_\mathrm{slot} + \omega\frac{2}{\pi}\frac{\mu_0 N_\mathrm{a}^2 k_\mathrm{w1a}^2 R\ell}{g}\left(\frac{1}{(N_{\mathrm{R+p}})^2} + \frac{1}{(N_{\mathrm{R-p}})^2}\right)$$

Note that the slot resistance and reactance parameters may be frequency dependent and some care must be taken to compute those parameters at the right frequency. Note also that the same extension to space harmonics used for polyphase machines in Section 13.8 will be useful here. Note, however, that the triplen harmonics will, in general, be present and important in the single-phase machine, unlike three-phase motors. Thus the magnetizing inductance, slot leakage and rotor resistance for the higher order harmonic terms will be of the form:

$$L_{\mathrm{m},n} = \frac{4}{\pi}\frac{\mu_0 N_\mathrm{a}^2 k_{wna}^2 R\ell}{n^2p^2g}$$

$$X_{2,n} = \frac{4\ell N_\mathrm{a}^2 k_{wna}^2}{N_\mathrm{R}}\omega L_\mathrm{slot} + \omega\frac{2}{\pi}\frac{\mu_0 N_\mathrm{a}^2 k_{wna}^2 R\ell}{g}\left(\frac{1}{(N_{\mathrm{R}+np})^2} + \frac{1}{(N_\mathrm{R} - np)^2}\right)$$

$$R_{2,n} = \frac{4\ell N_\mathrm{a}^2 k_{wna}^2}{N_\mathrm{R}}R_\mathrm{slot}$$

Note that rotor resistance must be corrected for end ring effects as described in Alger (1969).

13.10 Appendix 3: Induction Machine Winding Schemes

Induction machines are typically wound in a fashion similar to synchronous machines, with their armature windings located in slots that are part of a stator *core*, which is itself part of the magnetic circuit of the induction machine. Typically, small and medium size induction

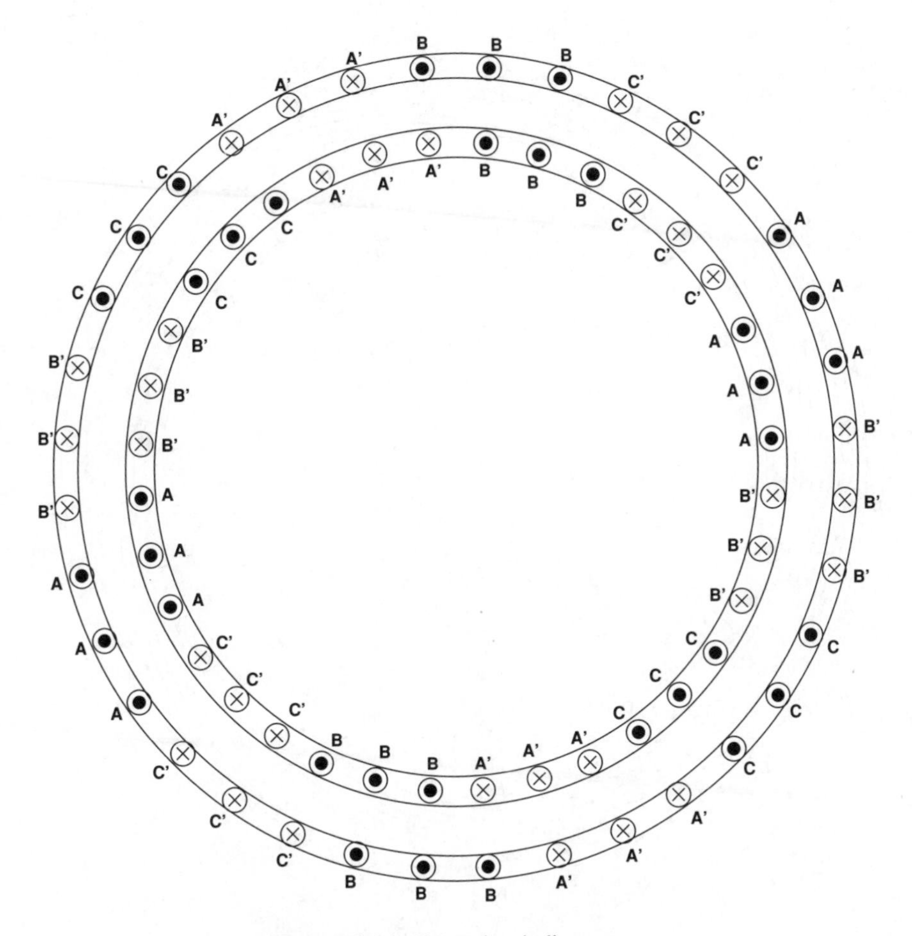

Figure 13.26 Example winding pattern

machines are wound with coils of magnet wire, while larger machines use 'form wound' coils, usually made with rectangular conductors. The coils and end windings of induction machines may take different forms. To illustrate, consider a four-pole, three-phase machine wound in a 36 slot stator. This is illustrated in very schematic form in Figure 13.26. Two layers of the winding are illustrated by two rings. The identities of the windings are denoted by letters. Conductors carrying current in one axial direction are noted by circles with a dot in the center, while conductors carrying current in the opposite axial direction are noted by circles with an 'X'.

A three-phase, four-pole winding in 36 slots has three slots per pole per phase. The winding pattern shown in Figure 13.26 is *short pitched* by one slot, and is referred to as an 8/9 pitch winding. A common way of connecting the turns on one side of each coil with the other side is called a 'lap' winding, illustrated in Figure 13.27. The end connections are spiral lines connecting the windings together, and it can be seen that they appear to nest together. This can be observed in machines with lap configuration windings.

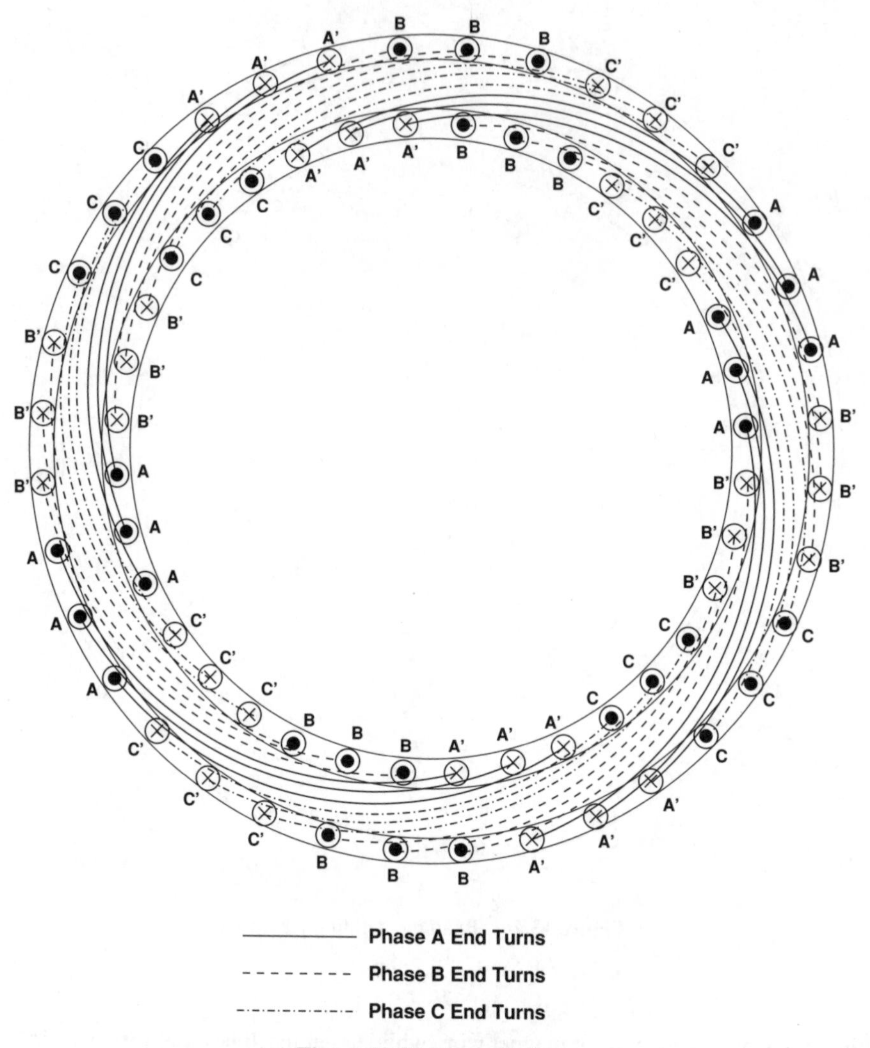

Figure 13.27 Lap winding

There is another way to configure an induction machine winding, however, called 'concentric winding'. Rather than connect the coils by overlapping end turns, the concentric winding does what the name implies: it connects the coils in a fashion that is symmetric about a centerline. Figure 13.28 illustrates such a winding. Note that this kind of winding is not always subdivided into two layers. The ellipses show upper and lower layers of some of the windings grouped together. In this winding scheme, some of the end turns are shorter than the others.

A variation on the concentric winding scheme is shown in Figure 13.29. This is called a 'consequent pole' winding because the end windings go around only two of the four poles. It has some rather long end windings and is used for economy in winding.

Note that all of the windings cited here have the same actual winding pattern. Concentric windings provide more flexibility in numbers of turns in each slot than lap windings, and

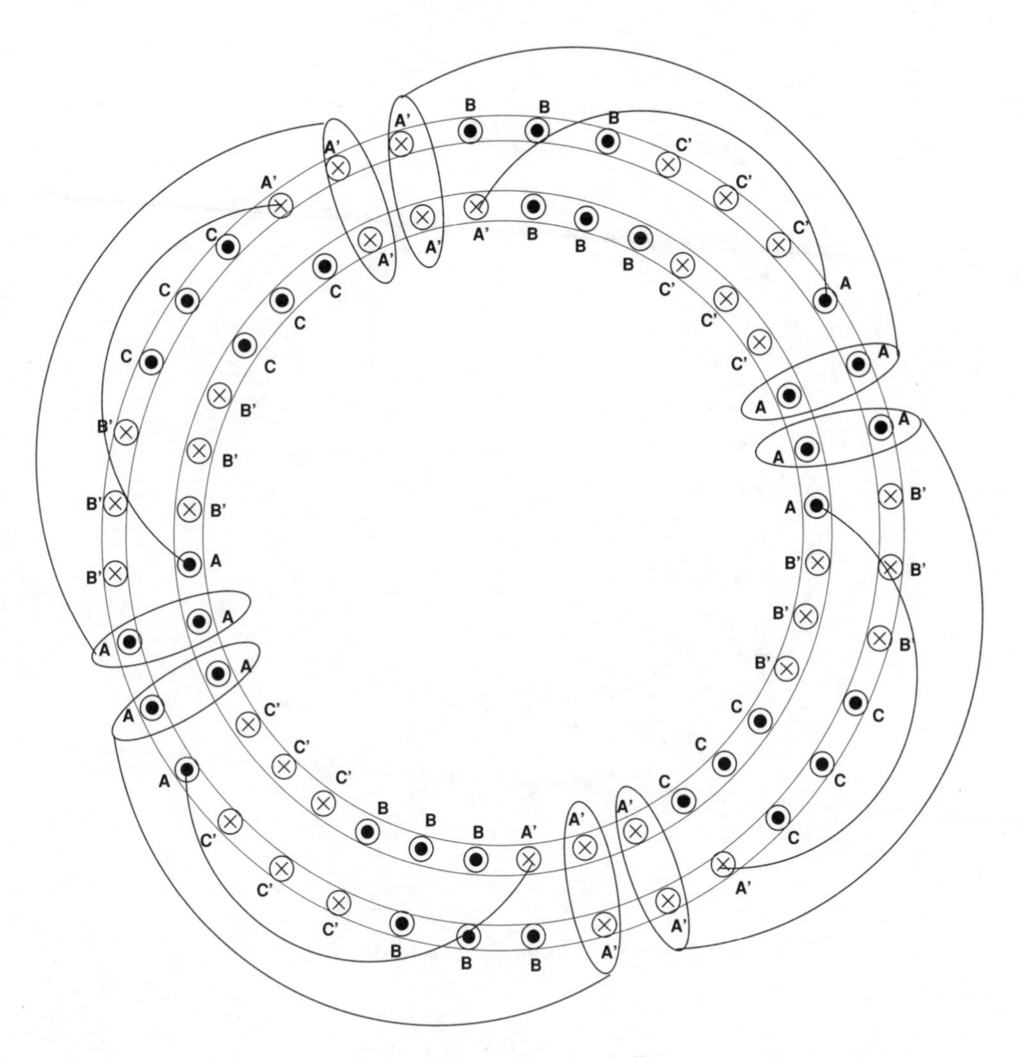

Figure 13.28 Concentric winding

can therefore be made to have more nearly sinusoidal winding patterns. All single-phase machines have concentric windings. While it is sometimes possible, as in the example case here, to recognize an equivalent lap winding and therefore to compute the winding factor of a concentric winding as the equivalent of a pitch factor times a breadth factor, this is not always possible.

13.10.1 Winding Factor for Concentric Windings

There is no 'breadth' factor in a concentric winding since all coils link flux with the same phase. Further, each of the coils will have a different 'pitch' factor. The actual winding factor can be computed as a turns-weighted average of all of the pitch factors.

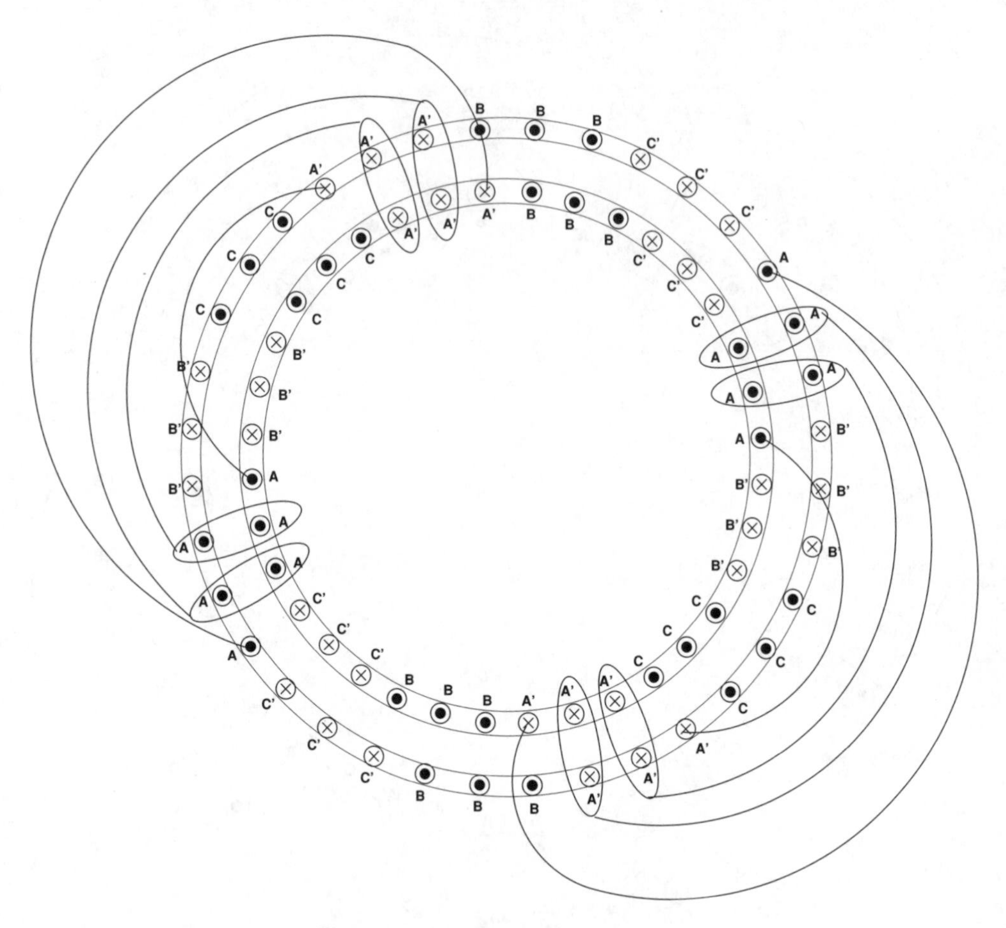

Figure 13.29 Consequent pole winding

If $N_s(k)$ is the number of turns in coil k and $N_c(k)$ as the coil throw, then the total number of turns is just the sum of all of the N_s's and the electrical span angle for coil k is

$$\phi_k = \frac{2p\pi N_c(k)}{N_S}$$

where N_S is the total number of slots in the stator. The total number of turns and consequent winding factor are then:

$$N_a = \sum_k N_c(k)$$

$$k_{wna} = \sum_k \frac{N_c(k)}{N_a} \sin \frac{n\phi_k}{2}$$

13.11 Problems

1. The single phase equivalent circuit of a three-phase, four-pole induction motor is shown in Figure 13.30. For the purposes of this problem, assume armature resistance is negligible. The machine is connected to a three-phase voltage source with line–neutral voltage of 240 V, RMS (416 V, line–line) and a frequency of 50 Hz.

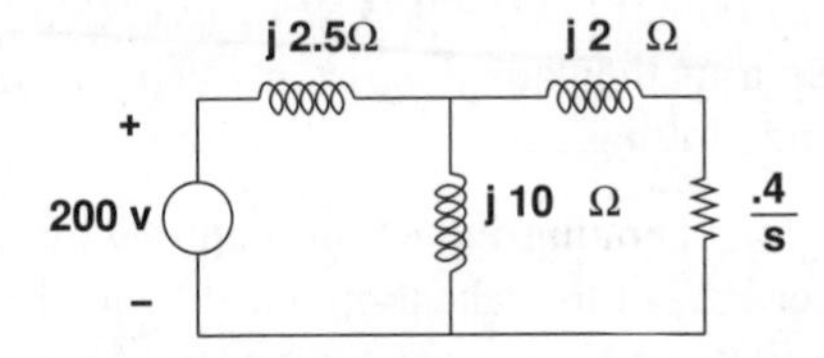

Figure 13.30 Induction motor equivalent circuit

(a) At what rotational speed does the motor achieve peak torque?
(b) What is that peak torque?

2. The next several problems deal with an induction motor with the following parameters. All of the impedance parameters are per phase, assuming the machine is connected in wye:

Rated output	25 Horsepower	18,650 W	
Terminal voltage	V_{ll}	480 V,	RMS, line-line
Number of pole pairs	p	2	(4-pole machine)
Supply frequency		60 Hz	
Armature resistance	R_1	0.2 Ω	
Rotor resistance	R_2	0.24 Ω	
Armature leakage inductance	X_1	1.0 Ω	
Rotor leakage inductance	X_2	1.2 Ω	
Magnetizing inductance	X_ϕ	50 Ω	
Core loss at rated voltage		400 W	
Friction and windage loss at rated speed		75 W	

Calculate and plot a torque–speed curve for this machine as a motor running with a power supply of rated voltage and frequency. This is 60 Hz, 480 V, RMS, line–line or 277 V, RMS per phase. In this part you should neglect windage but approximate the effects of core loss as a linear resistance.

3. Calculate and plot the magnitude of terminal current drawn as a function of speed.

4. What is 'breakdown' or maximum running torque for this motor with rated terminal voltage? What are the current and power factor when the motor is at breakdown?

5. In a test of this motor, the machine is run 'light', (without a mechanical load aside from friction and windage). If this test is carried out at rated terminal voltage, what are the real and reactive power drawn?

6. A similar test is called 'blocked rotor': the rotor of the machine is prevented from turning and the stator driven with a reduced voltage. What voltage should be used to result in a terminal current of 22.4 A, RMS? What are the real and reactive power drawn during such a test? What torque is produced?
7. Calculate and plot curves for efficiency and power factor vs. load power, from 5 to 100% percent of rated output power (933 W to 18,650 W). For this problem, you will probably want to use a mathematical assistant such as MATLAB or Maple to do the actual calculations and plotting.
8. A common strategy for operating induction motors with adjustable speed drives is to use 'constant volts per Hz', or to make terminal voltage proportional to drive frequency, for frequencies less than *base* or rated frequency. Above rated frequency voltage is held constant. To see how this would work, plot a family of torque-speed curves for this motor operating with terminal frequencies of 10, 20, ... 100 Hz. Be sure to use voltage proportional to frequency below 60 Hz and uniform voltage for 60 Hz and above.
9. Continuing with the volts/Hz operation of the previous problem, assume the machine is driving a load that has a constant torque characteristic. The load torque is 75 N- m. Find and plot against motor speed:

   (a) Input power
   (b) Output power
   (c) Power factor
   (d) Efficiency

   You will probably find it appropriate to calculate, using a larger number of frequencies than in the earlier problem, and cross-plotting to generate the curves asked for in this problem.
10. Here is a description of a 100 horsepower (75,600 W) induction motor:

Voltage	V	600	Line–line, RMS
Frequency	F	60	Hz
Number of pole pairs	p	2	
Stator leakage reactance	X_1	0.180	Ω
Stator resistance	R_1	0.080	Ω
Rotor leakage reactance	X_2	0.530	Ω
Rotor resistance	R_2	0.047	Ω
Magnetizing reactance	X_m	15.7	Ω

   For this motor, assume core loss at rated frequency and voltage is 2 kW and proportional to air-gap voltage.

   Assume also that friction and windage loss is proportional to the cube of rotor speed and is 3 kW at rated speed.

   Finally, assume that 'stray' loss is 2% of output power (1492 Watts) and is directly proportional to output power.

(a) Ignoring all of those loss elements, generate and plot a torque-speed curve for this motor, at rated terminal voltage and frequency.
(b) Estimate and plot motor efficiency and terminal power factor, while the motor is operated at rated terminal voltage and frequency for mechanical loads between 20 and 120 horsepower (14,920 to 89,520 Watts).

11. The very same motor you analyzed in the previous problem is to be used in an adjustable speed drive application. For lower speeds, the motor is to be driven by a balanced voltage that is proportional to frequency ('constant volts per Hz'). For drive frequency greater than 60 Hz, the terminal voltage is fixed at 600 V, RMS, line-line.

(a) Plot torque–speed curves for this motor and drive combination, for frequencies of 20, 40, 60, 80, 100 and 120 Hz. Be sure to use the right terminal voltages.
(b) Calculate and plot efficiency and power factor for the motor operating at a power output of 50 horsepower (37,300 W) over a range of speeds from about 900 to 3,600 r.p.m. Note that you may find it convenient to fix electrical frequency and do a cross-plot, so your plot may not extend exactly between these two speeds.

12. This concerns a real, 10 horsepower induction motor. The rotor diameter is 145 mm and the active length is 152 mm. The air-gap dimension (rotor to stator spacing) is 1/2 mm. The stator has 48 slots. Here are some details on the winding:

- This is a four-pole motor but the winding is called 'consequent pole'. This means that each phase winding is wound around only two poles. Thus each phase winding has two groups, and in this case each group has six coils. The coils are 'concentric', so that if you consider the six groups, they all link flux in the same axis.
- The coils have, respectively, 8, 9, 17, 17, 9 and 8 turns each, and span 17, 15, 13, 11, 9 and 7 slots.
- You will note that the winding scheme has some overlap from phase to phase: the coils with 8 and 9 turns each overlap with the adjacent phase with 9 and 8 turns, so the total number of wires in each slot is a uniform 17.

(a) Create a winding plan that shows how many turns are in each slot, phase half by phase half. Convince yourself that there are indeed 17 conductors in each slot.
(b) Compute the winding factor for the space fundamental and for the two 'belt' harmonics (order 5 and 7).
(c) Compute the magnetizing reactance for this machine, assuming it is to be operated at 60 Hz.
(d) If all coils of a phase are connected in series and if the three-phase winding is connected in 'star' (same as '*wye*'), and if the machine is operated with line-line voltage of 480 V, RMS, what is the *peak* value of flux density in the air-gap when the machine is running at no load? (Assume air-gap voltage is equal to terminal voltage.)

13. This is about a three-phase wound-rotor induction generator that might be used as a wind turbine generator. The stator and rotor windings are identical, except for the numbers of turns. It has characteristics as shown here:

Number of poles	2p	4
Armature phase self inductance	L_a	5.6 mH
Armature phase-to-phase mutual inductance	L_{ab}	−2.8 mH
Rotor phase self inductance	L_A	50.4 mH
Rotor phase-to-phase mutual inductance	L_{AB}	−25.2 mH
Rotor to stator (peak) mutual inductance	L_{aA}	16.59 mH
Effective transformer turns ratio	$\frac{N_r}{N_s}$	3
Synchronous rotational speed		1800 r.p.m.
Terminal voltage (RMS, line–line)	V_a	690 V
Rated power		1500 kVA
Frequency		60 Hz

The rotor windings are connected to a set of slip rings and so can be driven by an inverter. Machines such as these are used for adjustable speed drives as well as windmill generators.

Suppose this machine is operating as a generator at some speed other than synchronous. This will cause the rotor to have an electrical frequency different from the stator. The stator is supplying rated volt-amperes at a power factor of 0.8 (so that the stator is supplying VARs). What is the complex power required into the rotor terminals if the machine speed is:

(a) 70% of synchronous?
(b) 130% of synchronous?

14. A doubly fed induction machine is often used as the generator in wind turbine generation schemes, as shown in Figure 13.31. Here the slip rings are fed through a bidirectional converter which can provide both real and reactive power to the rotor windings. Assume that the power electronics interacts with the machine (and power bus) terminals at unity power factor (that is, the reactive power either drawn or supplied by the right-hand end of the converter is zero).

The characteristics of the machine are:

Number of poles	2p	4
Armature phase self inductance	L_a	5.6 mH
Armature phase-to-phase mutual inductance	L_{ab}	−2.8 mH
Rotor phase self inductance	L_A	50.4 mH
Rotor phase-to-phase mutual inductance	L_{AB}	−25.2 mH
Rotor to stator (Peak) mutual inductance	L_{aA}	16.59 mH
Effective transformer turns ratio	$\frac{N_r}{N_s}$	3
Synchronous rotational speed		1800 r.p.m.
Terminal voltage (RMS, line–line)	V_a	690 V
Rated power		1500 kVA
Frequency		60 Hz

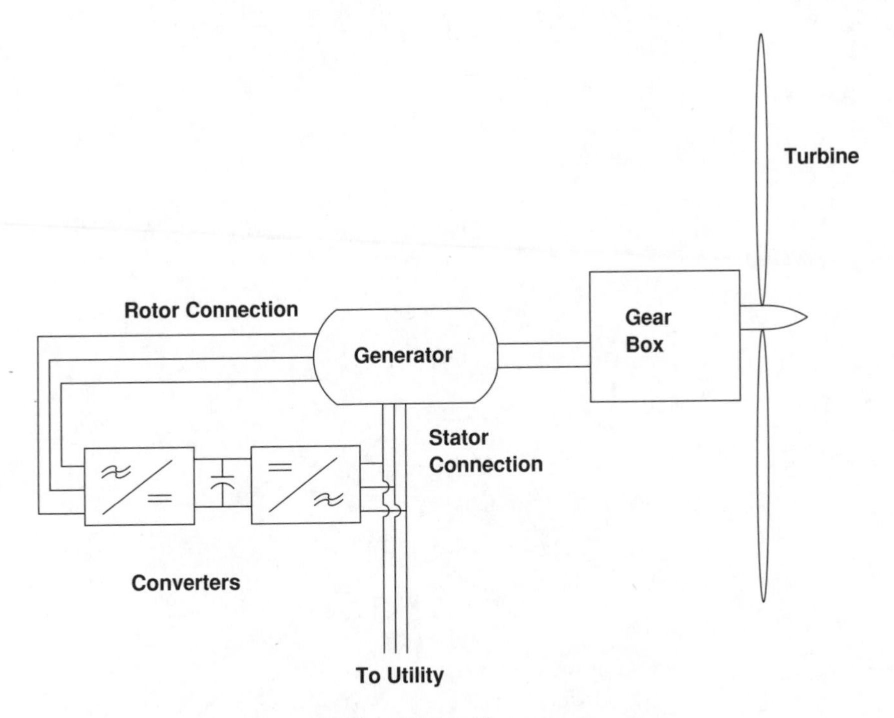

Figure 13.31 Wind turbine generator setup

Assume that the load is drawing $\rho = 750$ kW, $Q = 0$. Ignoring losses in the system, find and plot the following quantities over a speed range of between 70% and 130% of synchronous:

(a) Power *out of* the stator winding
(b) Power *in to* the slip rings (and rotor winding)
(c) Power delivered by the wind turbine

References

Alger, P.L. (1969) Induction Machines. Newark, NJ: Gordon and Breach.
NEMA (2004) Standard Publication MG-1. Motors and Generators. Rosslyn, VA: National Elecrical Manufacturer's Association.

14

DC (Commutator) Machines

Virtually all electric machines, and all practical electric machines employ some form of rotating or alternating field/current system to produce torque. While it is possible to produce a 'true DC' machine (e.g. the 'Faraday Disk'), for practical reasons such machines have not reached application and are not likely to. In the machines examined so far the machine is operated from an alternating voltage source. Indeed, this is one of the principal reasons for employing AC in power systems.

Historically, the first electric machines employed a mechanical switch, in the form of a carbon brush/commutator system, to produce a rotating field. That is, a field rotating with respect to the armature, which is on the rotor of these machines. That field is stationary with respect to the stator. The commutator can be seen at the near end of the rotor in Figure 14.1. While the widespread use of power electronics is making 'brushless' motors (which are really just synchronous machines) more popular and common, commutator machines are still economically very important. They are relatively cheap due to mass production, particularly in small sizes.

Commutator machines are used in a very wide range of applications. The starting motor in all automobiles is a commutator machine. Many of the other electric motors in automobiles, from the little motors that drive the outside rear-view mirrors to the motors that drive the windshield wipers are permanent magnet commutator machines. Most of the large traction motors that drive subway trains and diesel/electric locomotives are DC commutator machines (although induction machines are making some inroads here), and many common appliances use 'universal' motors: commutator motors adapted to AC.

14.1 Geometry

A schematic picture of a commutator type machine is shown in Figure 14.2. The armature of this machine is on the rotor. (The armature is the part that handles the electric power.) Current is fed to the armature through the brush/commutator system. The interaction magnetic field is provided (in this picture) by a field winding. Permanent magnets are sometimes used to provide this field, and as better permanent magnet materials are developed, this type of machine has achieved substantial application.

Electric Power Principles: Sources, Conversion, Distribution and Use James L. Kirtley

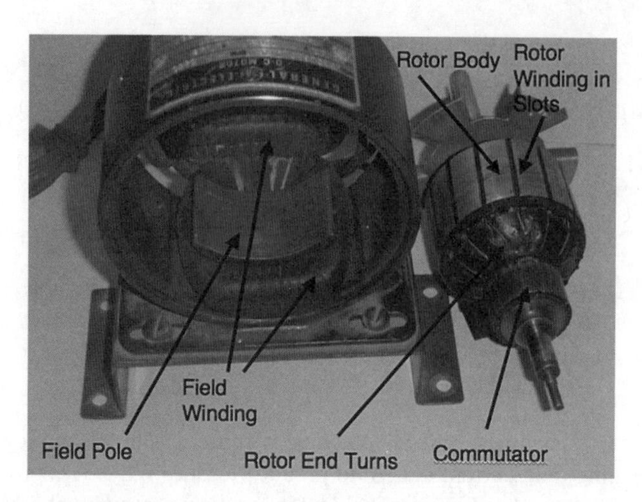

Figure 14.1 Wound field DC motor stator and rotor. Photo by author

14.2 Torque Production

The field winding will produce a radial magnetic flux density:

$$B_r = \mu_0 \frac{N_f I_f}{g}$$

where N_f is the number of field turns per pole, carrying a field current of I_f and g is the magnetic gap between the pole and the armature.

If there are C_a conductors underneath the poles at any one time, arranged in m parallel paths, then torque produced by the machine is:

$$T_e = \frac{C_a}{m} R \ell B_r I_a$$

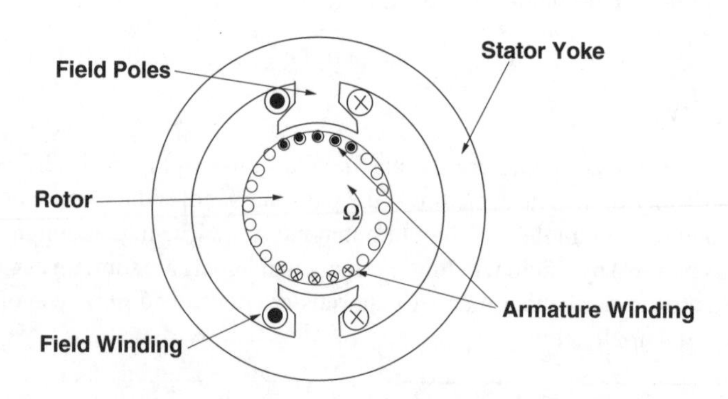

Figure 14.2 Wound-field DC machine geometry

where R and ℓ are rotor radius and length, respectively and I_a is armature terminal current. Note that C_a is not necessarily the total number of conductors, but rather the total number of *active* conductors (that is, conductors underneath the poles and therefore subject to the interaction field).

Torque may be expressed in terms of the two currents:

$$T_e = G I_a I_f$$

where G is the motor coefficient (units of N-m/ampere squared):

$$G = \mu_0 \frac{C_a}{m} \frac{N_f}{g} R\ell \tag{14.1}$$

14.3 Back Voltage

It is time to look at this from the point of view of voltage. Start with Faraday's Law:

$$\oint \vec{E} \cdot d\vec{\ell} = -\iint \frac{\partial \vec{B}}{\partial t}$$

This is a bit awkward to use, particularly in this case in which the edge of the contour is moving. This is made more convenient to use by noting:

$$\frac{d}{dt}\iint \vec{B} \cdot \vec{n}\, da = \iint \frac{\partial \vec{B}}{\partial t} \cdot \vec{n}\, da + \oint \vec{v} \times \vec{B} \cdot d\vec{\ell}$$

where $\vec{v}$ is the velocity of the contour. This yields a convenient way of noting the apparent electric field as viewed from a moving object (as in the conductors in a DC machine):

$$\vec{E}' = \vec{E} + \vec{v} \times \vec{B}$$

This yields the simple 'VLB' rule for voltage induced in a moving conductor: as illustrated in Figure 14.3 the voltage induced in a conductor of length L is simply minus the vector length

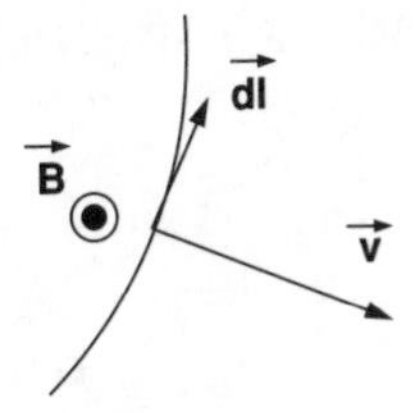

Figure 14.3 Motion of a contour through a magnetic field produces flux change and electric field in the moving contour

$\vec{L}$ times the cross product of velocity and flux density:

$$E = -\vec{L} \times \left(\vec{v} \times \vec{B}\right)$$

In the case of the DC machine, the vectors are conveniently mutually perpendicular. The armature conductors are moving through the magnetic field produced by the stator (field) poles, and one can ascribe to them an axially directed electric field:

$$E_z = -R\Omega B_\mathrm{r}$$

If the armature conductors are arranged as described above, with C_a active conductors in m parallel paths underneath the poles and with a mean active radial magnetic field of B_r, one can compute a voltage induced in the stator conductors:

$$E_\mathrm{b} = \frac{C_\mathrm{a}}{m} R\Omega B_\mathrm{r}$$

Note that this is only the voltage induced by motion of the armature conductors through the field and does not include brush or conductor resistance. Using (Equation 14.1, the back voltage is:

$$E_\mathrm{b} = G\Omega I_\mathrm{f}$$

which leads us to the conclusion that newton-meters per ampere squared equals volt seconds per ampere (that is, H). This stands to reason if one examines electric power into the interaction and mechanical power out:

$$P_\mathrm{em} = E_\mathrm{b} I_\mathrm{a} = T_\mathrm{c}\Omega$$

Figure 14.4 illustrates an equivalent circuit for a DC motor. Including the effects of armature, brush and lead resistance, in steady state operation, terminal voltage is:

$$V_\mathrm{a} = R_\mathrm{a} I_\mathrm{a} + G\Omega I_\mathrm{f}$$

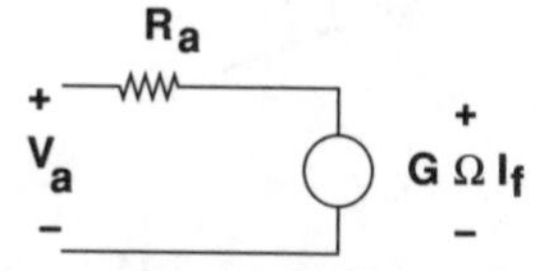

Figure 14.4 DC machine equivalent circuit

Now, consider this machine with its armature connected to a voltage source and its field operating at steady current, so that:

$$I_a = \frac{V_a - G\Omega I_f}{R_a}$$

14.4 Operation

Operation of a DC machine is illustrated by Figures 14.5 through 14.11. The first five of these assume a machine described by the parameters given in Table 14.1. This is a fairly large (about 50 kW motor with a normal speed of 1000 r.p.m.). It is assumed that the machine is driven by a 600 V DC source for these examples.

Torque, electric power in and mechanical power out are:

$$T_e = GI_f \frac{V_a - G\Omega I_f}{R_a}$$

$$P_e = V_a \frac{V_a - G\Omega I_f}{R_a}$$

$$P_m = G\Omega I_f \frac{V_a - G\Omega I_f}{R_a}$$

These expressions define three regimes defined by rotational speed. The two 'break points' are at zero speed and at the 'zero torque' speed:

$$\Omega_0 = \frac{V_a}{GI_f}$$

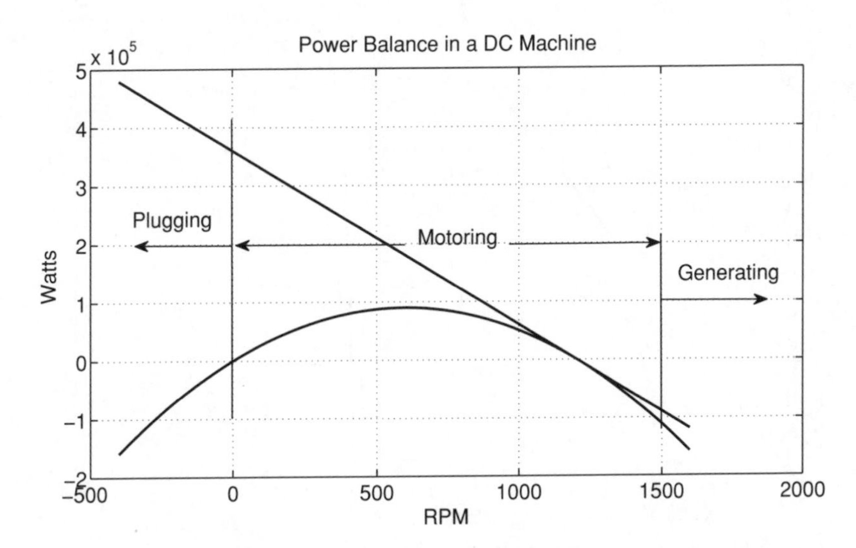

Figure 14.5 DC machine operating regimes

Table 14.1 Parameters of an example DC machine

Terminal voltage	600	V
Armature resistance	1	Ω
Field resistance	60	Ω
Motor constant G	0.476	N m/A^2

14.4.1 Shunt Operation

Figure 14.5 shows input electrical power and output mechanical power for the example motor driven by 600 V on both the armature and field. This is called 'shunt' operation as both windings could be driven by the same power supply. For $0 < \Omega < \Omega_0$, the machine is a motor: electric power in and mechanical power out are both positive. For higher speeds: $\Omega_0 < \Omega$, the machine is a generator, with electrical power in and mechanical power out being both negative. For speeds less than zero, electrical power in is positive and mechanical power out is negative. There are few needs to operate machines in this regime, short of some types of 'plugging' or emergency braking in traction systems.

Figure 14.6 shows the electrical efficiency of operation of this particular motor, neglecting important elements such as friction and windage. Efficiency becomes very poor when the machine is converting little power, because the field winding consumes a fixed amount of power. But the efficiency is also poor when the machine is operating well away from the zero

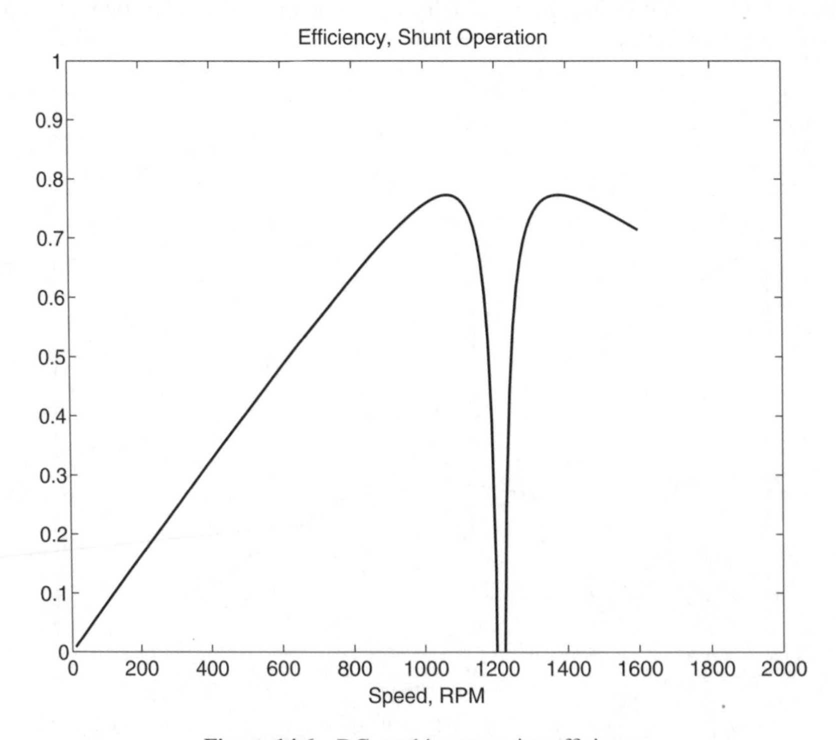

Figure 14.6 DC machine operating efficiency

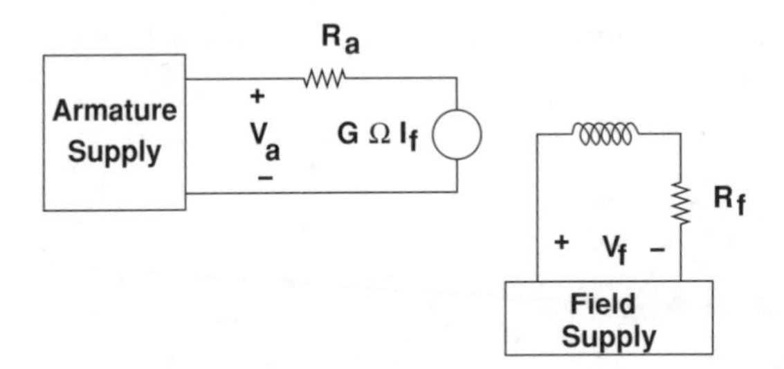

Figure 14.7 Separately excited machine hookup

torque speed, because power converted is torque times speed which is equal to back voltage times armature current, but armature loss is proportional to the square of armature current.

14.4.2 Separately Excited

'Shunt' operation is a subset of 'separately excited' operation in which both the armature and field source are the same source. If the two power supplies to the motor are independent of each other, speed can be controlled in a different ways. This mode of operation is used in some types of traction applications in which the flexibility it affords is useful. For example, some traction applications apply voltage control in the form of 'choppers' (buck converters) to separately excited machines. The connection is shown in Figure 14.7.

14.4.2.1 Armature Voltage Control

Note that the 'zero torque speed' is dependent on armature voltage and on field current. For high torque at low speed one would operate the machine with high field current and enough armature voltage to produce the requisite current. Figure 14.8 shows a portion of the torque–speed curves with armature voltage as a parameter. For a range of speeds consistent with the available armature voltage supply, this connection allows for very flexible control of motor speed. It is often used for traction drives.

14.4.2.2 Field Weakening Control

As speed increases so does back voltage, Another method for controlling motor speed is to change field current. This might be accomplished with an added field resistance, and this is why this mode of control is sometimes called 'field weakening'. Reducing field current increases the zero-torque speed but it also reduces the slope of the torque–speed curve, as is shown in Figure 14.9.

14.4.2.3 Dynamic Braking

A DC machine can serve as a motor or as a generator, and a separately excited machine can be reconnected to serve as a dynamic brake in traction applications. The connection is shown in

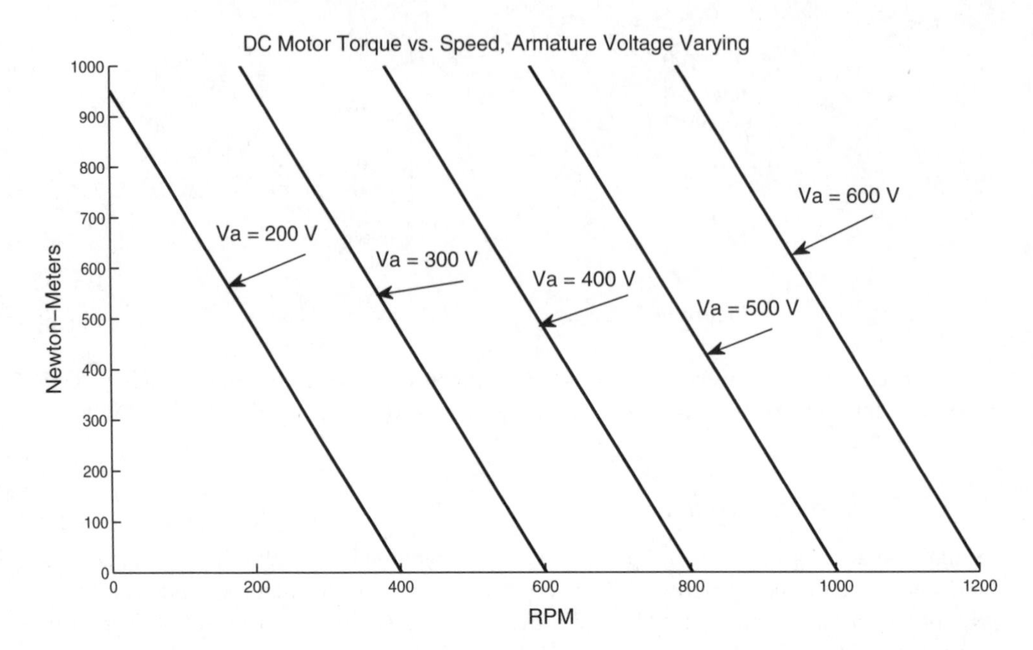

Figure 14.8 Armature voltage control of separately excited DC motor

Figure 14.10. The load resistor must be capable of absorbing the energy produced. In traction applications the resistor is often made of cast iron and is mounted where the heat produced can be easily dissipated (as on the roof of a transit car). Even with a fixed value of the resistor, the field winding can achieve arbitrary and smooth control of braking torque.

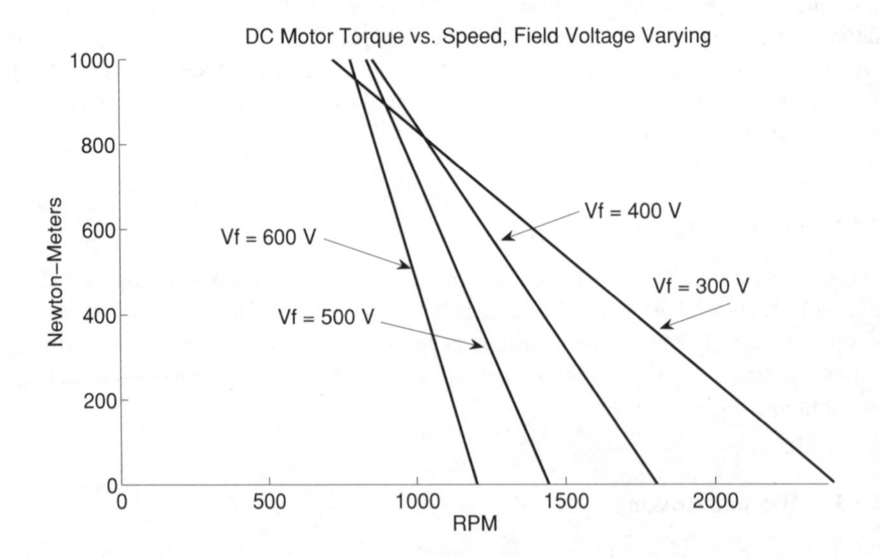

Figure 14.9 Field voltage control of separately excited DC motor

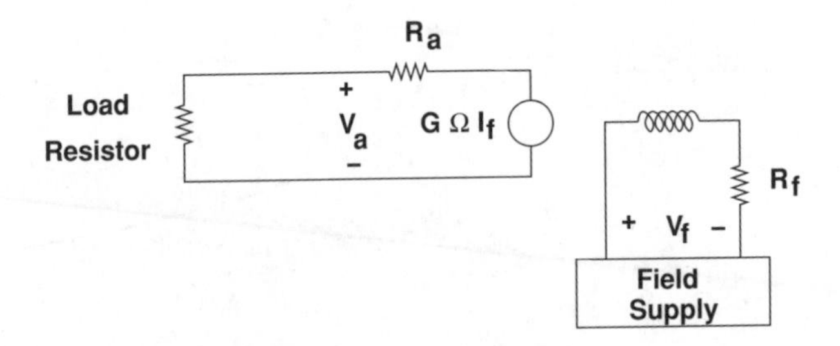

Figure 14.10 Dynamic braking connection

14.4.3 Machine Capability

The limits to operation of a DC machine could be framed in a number of ways, but it is clear that there will be limits to both armature and field current and on armature voltage.

1. Armature current will be limited by heating of the armature conductors.
2. Field current will be similarly limited by heating or by saturation of the magnetic circuit of the machine.
3. Armature voltage will be limited by risk of dielectric failure of the commutator or the air around it (flashover).

For low speeds, back voltage is not limiting so that the torque for the DC machine is limited by the two currents:

$$T_{\mathrm{max}} = G I_{\mathrm{alim}} I_{\mathrm{flim}}$$

For the low speed regime, converted power is limited to speed times torque.

For higher speeds, the main limit to machine operation is the voltage that can be carried by the commutator and by the armature current. To control that voltage, typically field current must be controlled (turned down as speed increases), so that the machine limit for high speed is roughly constant power:

$$P_{\mathrm{max}} \approx V_{\mathrm{alim}} I_{\mathrm{alim}}$$

Which means that field current must be controlled:

$$I_{\mathrm{f}} = \frac{V_{\mathrm{alim}}}{G\Omega}$$

so that torque is inversely proportional to speed:

$$T_{\mathrm{max}} = \frac{V_{\mathrm{alim}} I_{\mathrm{alim}}}{\Omega}$$

These limits are shown in Figure 14.11. The low speed part of this curve is referred to as the 'constant torque' region while the high speed part is referred to as 'constant power'.

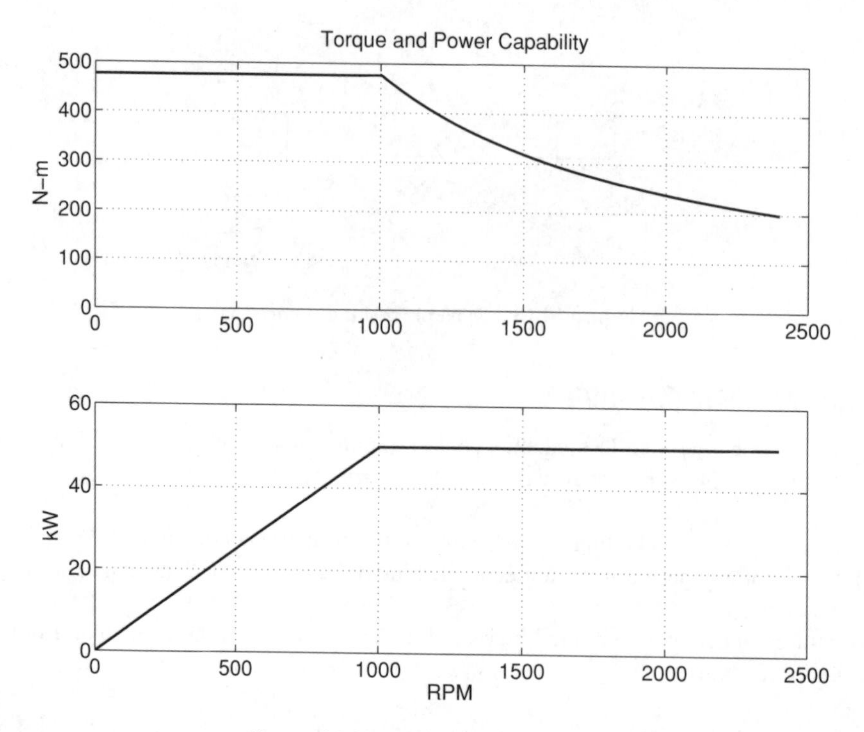

Figure 14.11 DC machine capability

14.5 Series Connection

A widely used connection for DC machines is the 'series connection' in which the field winding is connected in series with the armature, as shown in Figure 14.12. A machine to be connected this way is physically different from a shunt or separately excited machine in that the field winding is built for much higher current (consequently fewer turns). In a shunt connection the field current should be low to reduce the power drawn by the field winding, while in a series connected machine the field resistance should be low to reduce power dissipated by the field.

Current in the series connected machine, connected to a source is:

$$I_{\mathrm{a}} = I_{\mathrm{f}} = \frac{V}{R_{\mathrm{a}} + R_{\mathrm{f}} + G\Omega}$$

Figure 14.12 Series connection

And then torque is:

$$T_e = \frac{GV^2}{(R_a + R_f + G\Omega)^2}$$

It is important to note that this machine has no 'zero torque' speed, leading to the possibility that an unloaded machine might accelerate to dangerous speeds. This can be particularly problematic because the commutator, made of pieces of relatively heavy material tied together with non- conductors, is not very strong. To illustrate how such a machine would work, the machine used for the shunt and separately excited connections was modified with a field winding that has a resistance of about 0.6 Ω and a motor constant of 0.0476 N m/A^2. Figure 14.13 shows torque and power vs. speed for this machine connected to a 600 V source.

Speed control of series connected machines can be achieved with voltage control and many appliances using this type of machine use choppers or phase control. An older form of control that is still used in traction applications is the series dropping resistor. Adding resistance to the series circuit produces the set of torque/speed curves shown in Figure 14.14. This method of control is cheap but not without efficiency issues as shown in Figure 14.15.

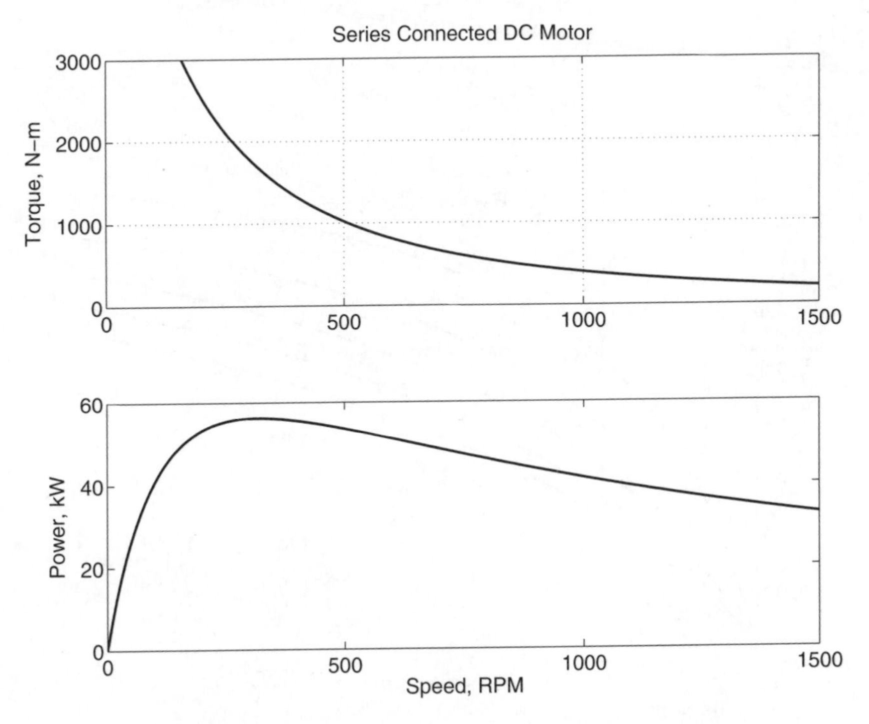

Figure 14.13 Torque and power for a series connected machine

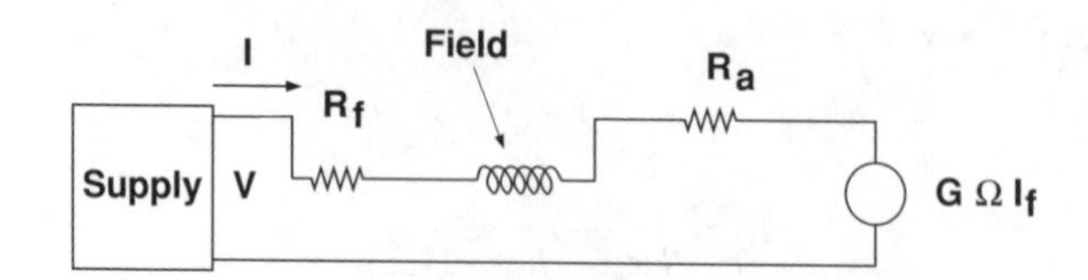

Figure 14.14 Resistance control of a series connected motor

14.6 Universal Motors

A variation on this class of machine is the very widely used 'universal motor', in which the stator and rotor (field and armature) of the machine are both constructed to operate with alternating current. An example of such a motor is shown in Figure 14.16. This one is taken from a handheld appliance, where such motors are very useful because they can be made to turn very fast and thus produce very high power density.

In a universal motor, both the field and armature are made of laminated steel. Note that such a machine will operate just as it would have with direct current, with the only addition being the reactive impedance of the two windings. Working with RMS quantities:

$$\mathbf{I} = \frac{\mathbf{V}}{R_a + R_f + G\Omega + j\omega(L_a + L_f)}$$

$$T_e = \frac{|\mathbf{V}|^2}{(R_a + R_f + G\Omega)^2 + (\omega L_a + \omega L_f)^2}$$

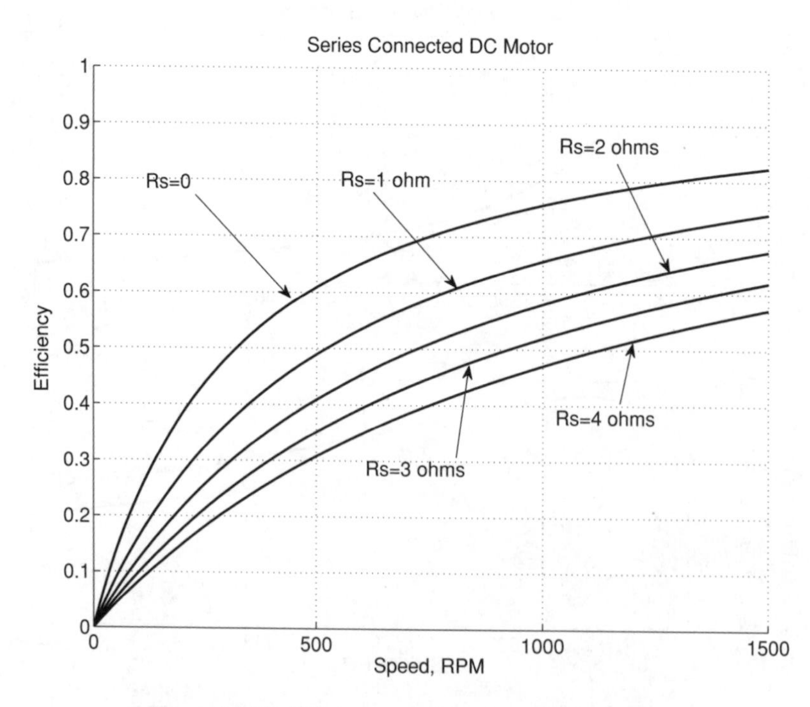

Figure 14.15 Efficiency as affected by resistive control

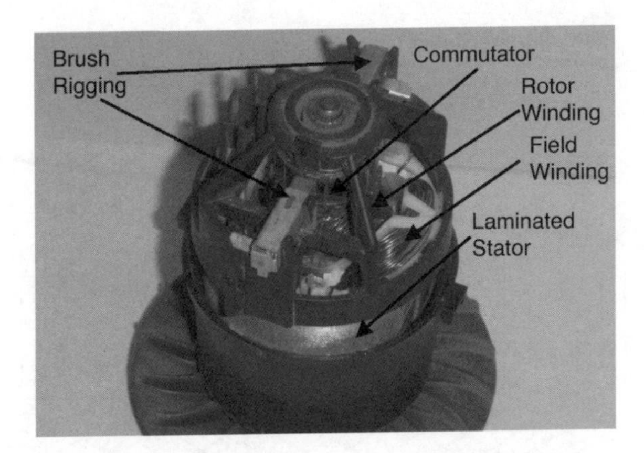

Figure 14.16 Universal motor from a leaf blower. Photo by author

where ω is the electrical supply frequency and Ω is rotational speed. Note that, unlike other AC machines, the universal motor is not limited in speed to the supply frequency. Universal motors typically turn substantially faster than the 3,600 r.p.m. limit of AC motors, and this is one reason why they are so widely used: with the high rotational speeds it is possible to produce more power per unit mass (and more power per unit cost). They are widely used in appliances and tools that operate from the AC power line, including vacuum cleaners, egg beaters, routers, and other machines that can be driven at high shaft speeds.

14.7 Commutator

The commutator is what makes this machine work. It is the interaction of the brushes and the commutator that keep currents underneath the field poles in a repetitive switching operation.

To start, take a look at the picture shown in Figure 14.17. Represented are a pair of poles (shaded) and a pair of brushes. Conductors make a group of closed paths. In this case, current from one of the brushes takes two parallel paths. It is possible to follow one of those paths around a closed loop, under each of the two poles (remember that the poles are of opposite polarity) to the opposite brush. Open commutator segments (most of them) do not carry current into or out of the machine.

A commutation interval occurs when the current in one coil must be reversed. (See Figure 14.18 In the simplest form this involves a brush bridging between two commutator segments, shorting out that coil. The resistance of the brush causes the current to decay. When the brush leaves the leading segment the current in the leading coil must reverse.

The commutation process, reversal of current in a coil is accomplished in different ways, sometimes in combination. *Resistive* commutation is the process relied upon in small machines. When the current in one coil must be reversed (because it has left one pole and is approaching the other), that coil is shorted by the brushes. The brush resistance causes the current in the coil to decay. Then the leading commutator segment leaves the brush the current *must* reverse (the trailing coil has current in it), and there is often sparking.

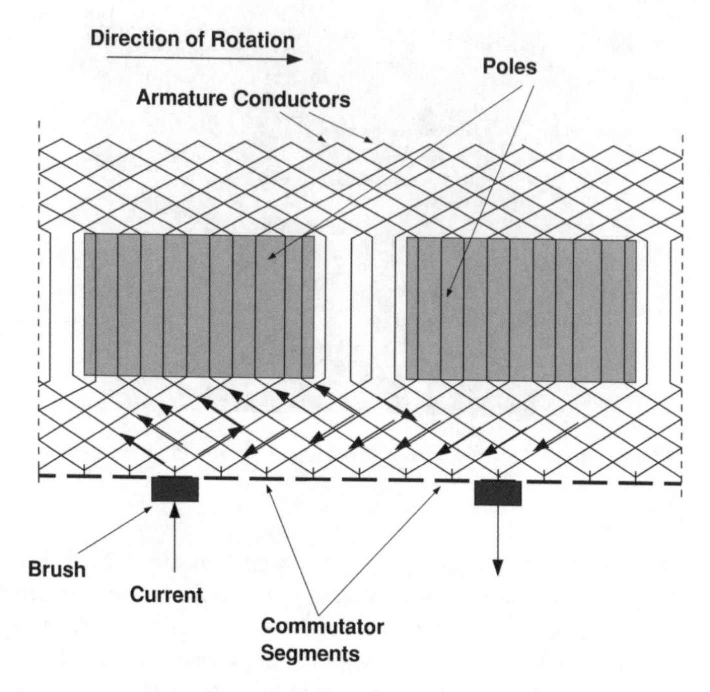

Figure 14.17 Commutator and current paths

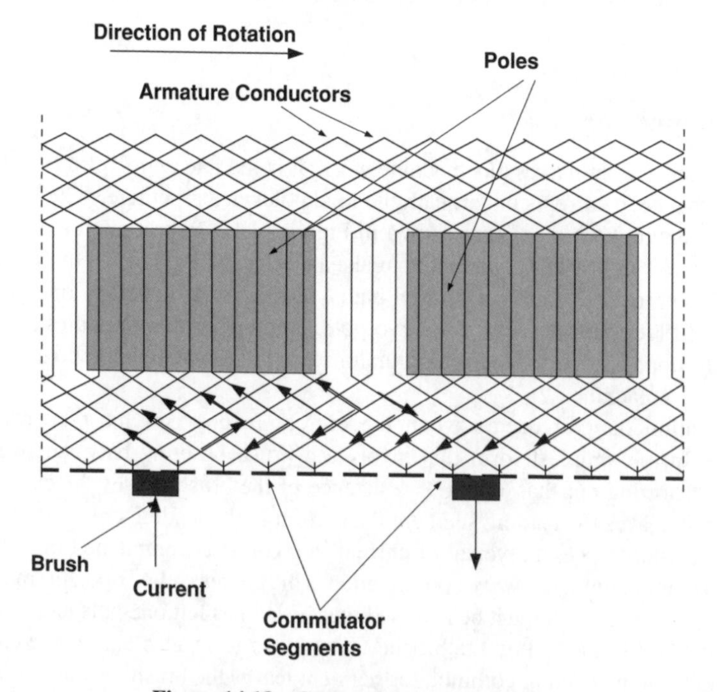

Figure 14.18 Commutator at commutation

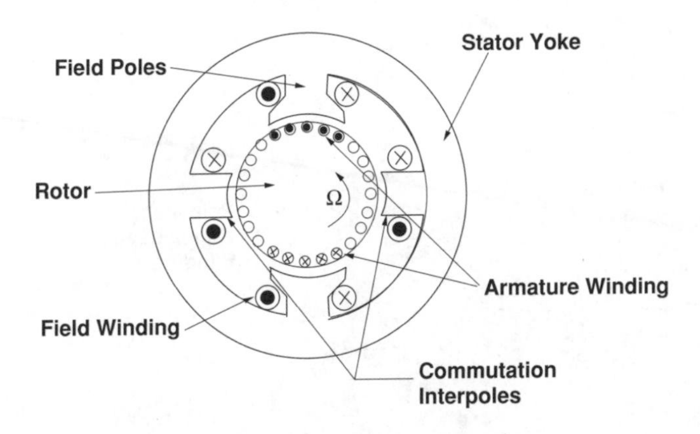

Figure 14.19 Commutation interpoles

14.7.1 Commutation Interpoles

In larger machines the commutation process would involve too much sparking, which causes brush wear, noxious gases (ozone) that promote corrosion, etc. In these cases it is common to use separate commutation interpoles. These are separate, usually narrow or seemingly vestigal pole pieces which carry armature current. They are arranged in such a way that the flux from the interpole drives current in the commutated coil in the proper direction. The coil being commutated is located physically between the active poles and the interpole is therefore in the right spot to influence commutation. The interpole is wound with armature current (it is in series with the main brushes). It is easy to see that the interpole must have a flux density proportional to the current to be commutated. Since the speed with which the coil must be commutated is proportional to rotational velocity and so is the voltage induced by the interpole, if the right number of turns are put around the interpole, commutation can be made to be quite accurate. An illustration of the location of commutation interpoles is shown in Figure 14.19.

14.7.2 Compensation

The analysis of commutator machines often ignores armature reaction flux. Obviously these machines *do* produce armature reaction flux, in quadrature with the main field. Normally, commutator machines are highly salient and the quadrature inductance is lower than direct-axis inductance, but there is still flux produced. This adds to the flux density on one side of the main poles (possibly leading to saturation). To make the flux distribution more uniform and therefore to avoid this saturation effect of quadrature axis flux, it is common in very highly rated machines to wind compensation coils: essentially mirror-images of the armature coils, but this time wound in slots in the surface of the field poles. Such coils will have the same number of ampere-turns as the armature. Normally they have the same number of turns and are connected directly in series with the armature brushes. What they do is to almost exactly cancel the flux produced by the armature coils, leaving only the main flux produced by the field winding. One might think of these coils as providing a reaction torque, produced in exactly

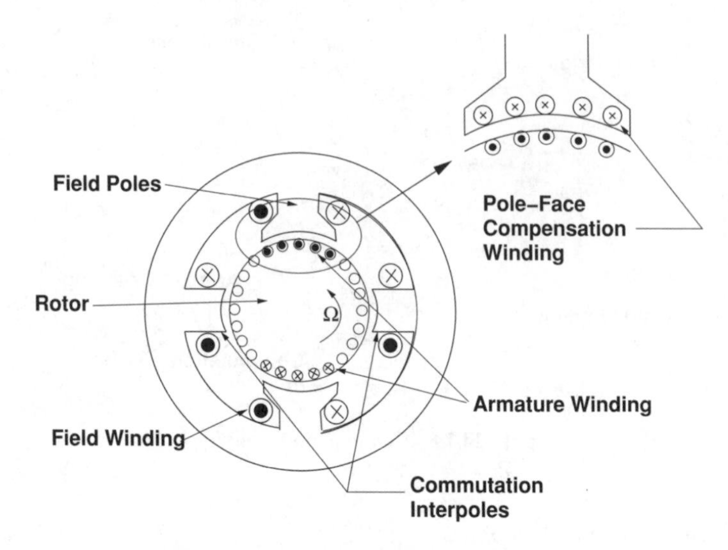

Figure 14.20 Pole face compensation winding

the same way as main torque is produced by the armature. A cartoon view of this is shown in Figure 14.20.

14.8 Compound Wound DC Machines

These machines are not very common any longer because the relatively more complex functions they performed are now done by induction machines and by permanent magnet brushless motors, but they are mentioned here for completeness and for historic significance. It is possible to tailor the output of a DC motor or generator by having *both* series and shunt field windings. For example, generators have internal voltage drop due to armature resistance and possibly prime mover 'droop'. These may be compensated for by strengthening the field from the main, 'shunt' field winding, using a second, series-connected field winding. In motors the torque–speed characteristic can be modified, usually in a way that strengthens the magnetic flux in the motor in response to torque demand. This gives the motor a more pronounced droop characteristic and reduces armature current drawn in response to torque demand. This kind of winding was typically used for driving intermittent or variable loads such as punch presses. In many applications it is necessary to have a 'zero torque' speed that is within reasonable bounds but a torque characteristic more like that of a series-connected machine, and the compound winding does this.

There are two possible methods of connecting the two field windings: so-called 'short shunt' and 'long shunt' connections, and these are shown in Figures 14.21 and 14.22.

In the compound machine, the internal voltage is:

$$E_{\mathrm{a}} = \Omega\Phi(F)$$

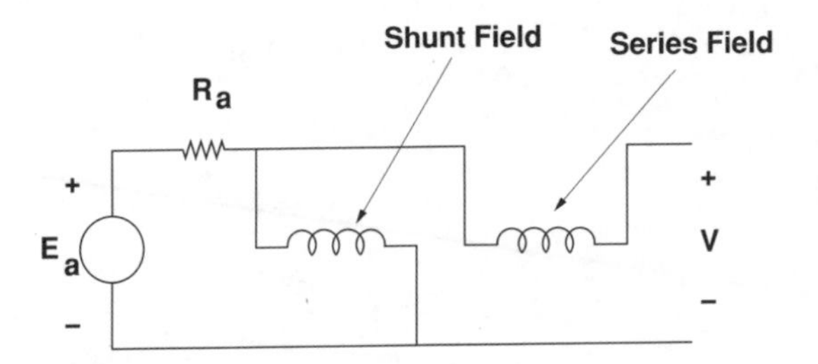

Figure 14.21 Short shunt compound machine connection

and torque is:

$$T_m = I_s \Phi(F)$$

where the MMF is:

$$F = N_f I_f + N_s I_s$$

Here, N_f is the number of turns of the 'shunt' field winding and N_s is the number of turns of the 'series' field winding. For the 'short shunt' connection, the voltage across the shunt field, and hence the shunt field current is affected by the resistance of the series field winding:

$$I_f = \frac{V_t - R_s I_t}{R_f}$$

where V_t is terminal voltage and R_s is the resistance of the series field. R_f is resistance of the shunt field. and for this connection the current in the series field winding is just the terminal current: $I_s = I_t$.

In the long-shunt connection, shunt field voltage is terminal voltage and the series field has current:

$$I_s = I_t - I_f$$

If the machine can be regarded as being magnetically linear,

$$\Phi(F) = G_f I_f + G_s I_s$$

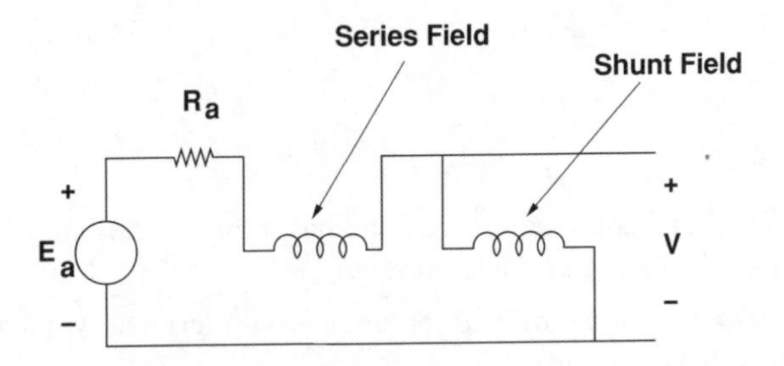

Figure 14.22 Long shunt compound machine connection

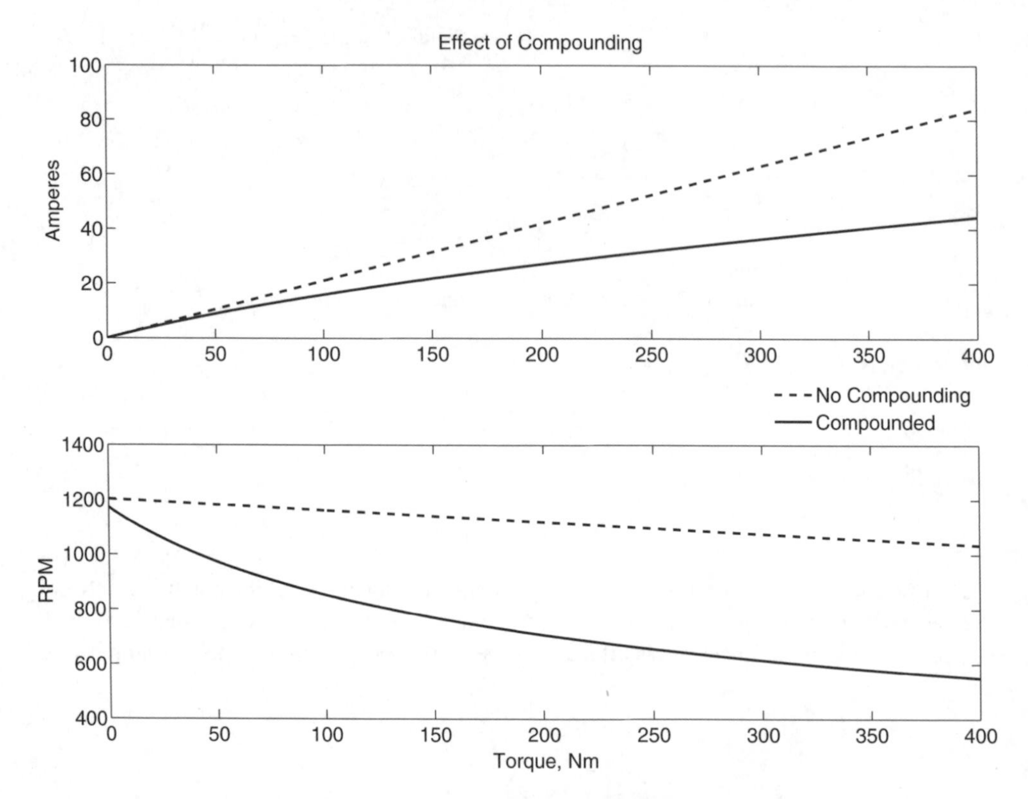

Figure 14.23 Comparison of compounded and uncompounded motors

so that:

$$E_a = \Omega(G_f I_f + G_s I_s)$$

And the two field coefficients are related by:

$$G_s = G_f \frac{N_s}{N_f}$$

Figure 14.23 shows how the armature current and speed are affected by a series field winding with a coefficient of 20% of the shunt field in the motor of Table 14.1.

14.9 Problems

1. A separately excited DC commutator machine has a motor constant $G = 1$ V$_s$/A and an armature resistance of $R_a = 1\ \Omega$. Field current is 1 A.
   (a) Running as a motor with zero load torque and with terminal voltage $V_a = 110V\,DC$, what is motor speed in r.p.m.?
   (b) Running as a motor with a load torque of 10 N m and with a the same terminal voltage $V_a = 110$ V DC, what is armature current, motor speed, power in and power out?

(c) Assuming field resistance is $R_f = 110\ \Omega$ and the motor is operated in shunt (field and armature connected to the same source), what is motor efficiency when torque is 10 N m?

(d) Estimate and plot power out, power in and efficiency of this motor for torque over the range of $1 < T_m < 20$ N m.

2. A shunt connected DC motor is known to have an armature resistance $R_a = \frac{1}{2}\ \Omega$ operating with a terminal voltage of $V_a = 100$ V DC and turning at a speed of $\Omega = 180$ radians/second, drawing 10 A from the source. Field current is 1 A.

(a) What is the motor constant G?

(b) How much torque is the motor producing?

3. A series-connected motor is operating from a 600 V DC source. It is drawing 100 A and producing 50 kW of mechanical output power at 200 Radians/second. What are:

(a) Armature plus field resistance?

(b) Motor constant G?

4. A series connected motor has a motor constant $G = 0.00625$ H and a combined armature plus field resistance $R_a + R_f = \frac{1}{8}\ \Omega$. This motor is operated from a voltage source of 600 V DC.

(a) How much output power is it producing when it is turning at $\Omega = 100$ radians/s?

(b) What is the input power?

5. A motor produces 400 kW at 1,000 r.p.m, connected to a 600 V DC source, drawing 800 A. This is a series connected motor, and its resistance (series field plus armature) is $\frac{1}{8}\ \Omega$. The load has a cubic relationship to speed, as one might see in a fan or windage load of, say, a vehicle:

$$P = P_0 \left(\frac{\Omega}{\Omega_0} \right)^3$$

With that load, and with a series connected motor, estimate and plot speed vs. terminal voltage.

6. A motor produces 400 kW at 1,000 r.p.m., connected to a 550 V DC source, drawing 800 A. This is a separately excited motor, and its armature resistance is $\frac{1}{16}\ \Omega$. Field current is not large enough to figure in the problem The load has a cubic relationship to speed, as one might see in a fan or windage load of, say, a vehicle:

$$P = P_0 \left(\frac{\Omega}{\Omega_0} \right)^3$$

With that load, and with this separately excited motor, estimate and plot speed vs. terminal voltage.

7. A DC generator has the following characteristics:

Field resistance	R_f	73 Ω
Armature resistance	R_a	2 Ω
Number of field turns	500	

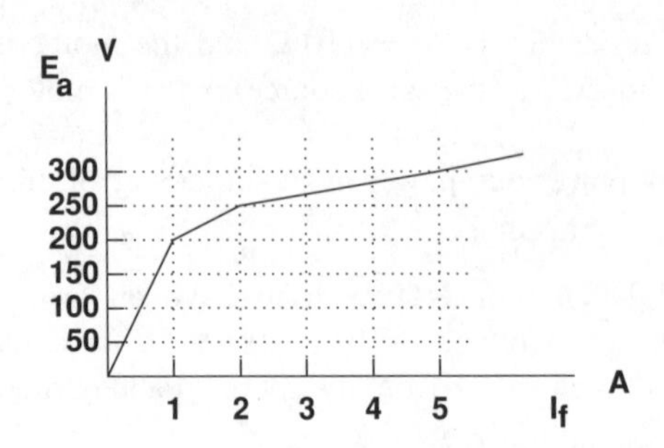

Figure 14.24 DC generator test curve at 1200 r.p.m.

Operating at a speed of 1200 r.p.m., the machine exhibits the saturation curve shown in Figure 14.24.

(a) This is to be a self-excited machine (a 'dynamo'), so the field winding is connected across the armature terminals. At what speed will this machine self-excite?
(b) Operating at 1,500 r.p.m., what is the steady state voltage if the machine is otherwise unloaded?
(c) Calculate the output voltage as a function of load current, with the machine turning at a steady 1,500 r.p.m..
(d) Now, the machine is to be compounded by use of a series field winding to make it a 'stiffer' voltage source. How many turns should there be to make it 'flat' compounded? (Zero apparent output impedance).

8. This problem concerns a compound-wound DC motor. The series winding is connected *cumulatively*, in such a way that motor current in the series field reinforces the flux from the shunt field winding. Here is some data:

Shunt field resistance	R_f	300 Ω
Armature resistance	R_a	0.25 Ω
Series field resistance	R_s	0.2 Ω
Number of turns in shunt field	N_f	500
Number of turns in series field	N_s	20

The machine has been tested, and with the armature winding open and 2.0 amperes in the shunt field, operating at 800 r.p.m., the armature voltage is measured at 600 V. Ignoring any mechanical losses (friction, windage, etc.), calculate and plot torque–speed and current–speed curves for this motor operating with a terminal voltage of 600 V:

(a) operating with the series field not connected,
(b) operating with the series field connected cumulatively in *long shunt*, and
(c) operating with the series field connected cumulatively in *short shunt*.

15

Permanent Magnets in Electric Machines

Of all changes in materials technology over the last few decades, advances in permanent magnets have had arguably the largest impact on electric machines. Permanent magnets are often suitable as replacements for the field windings in machines; that is, they can produce the fundamental interaction field. This does four things:

1. Since the permanent magnet is lossless it eliminates the energy required for excitation, usually improving the efficiency of the machine.
2. Since eliminating the excitation loss reduces the heat load it is often possible to make PM machines more compact.
3. Because the excitation is lossless, it is possible to build machines with high pole count. That reduces the amount of back iron required and so can be used to make the machine substantially lighter.
4. Less appreciated is the fact that modern permanent magnets have very large coercive force densities which permit vastly larger air gaps than conventional field windings, and this in turn permits design flexibility which can result in even better electric machines.

These advantages come not without cost. Permanent magnet materials have special characteristics which must be taken into account in machine design. The highest performance permanent magnets are brittle, some have chemical sensitivities, all are sensitive to high temperatures, most have sensitivity to demagnetizing fields, and proper machine design requires understanding the materials well. This chapter will not make the reader into a seasoned permanent magnet machine designer. It contains, however, a means to develop some of the mathematical skills required and to point to some of the important issues involved.

15.1 Permanent Magnets

Permanent magnet materials are just materials with very wide hysteresis loops. Figure 15.1 is an approximation to the hysteresis curve of one of the more popular ceramic magnet materials.

Electric Power Principles: Sources, Conversion, Distribution and Use James L. Kirtley

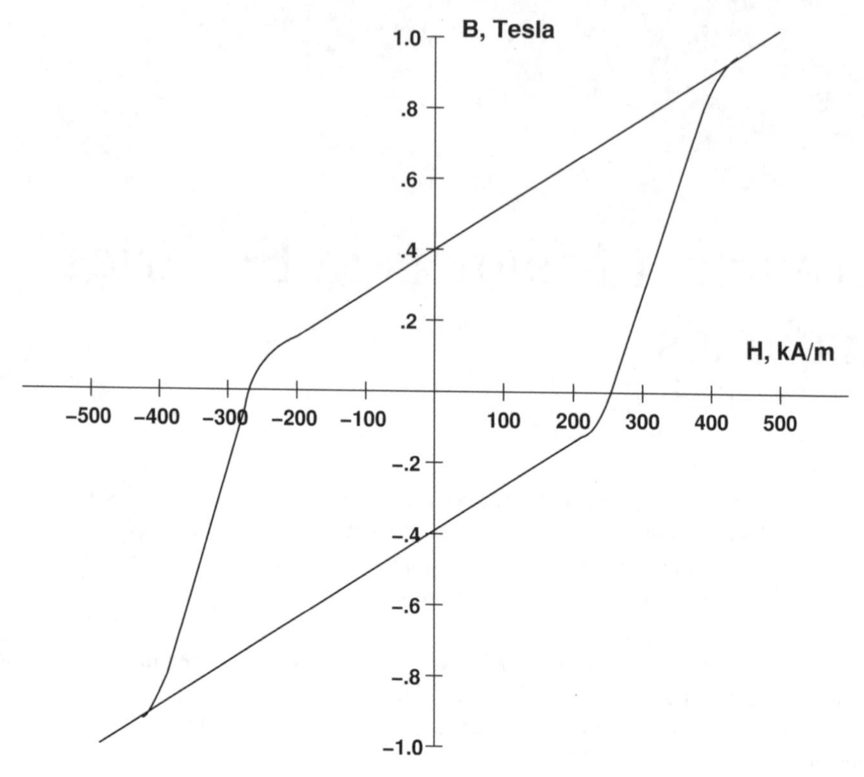

Figure 15.1 Hysteresis loop Of ceramic permanent magnet

Note that this hysteresis loop is so wide that you can see the effect of the permeability of free space.

It is usual to display only part of the magnetic characteristic of permanent magnet materials (see Figure 15.2), the upper left quadrant of this picture, because that is where the material is normally operated. Note a few important characteristics of what is called the 'demagnetization curve'. The remanent flux density B_r, is the value of flux density in the material with zero magnetic field H. The coercive field H_c is approximately the magnetic field at which permanent changes in magnetization will occur.

Shown also in Figure 15.2 are loci of constant 'energy product'. This quantity is unfortunately named, for although it has the same units as energy it represents real energy in only a fairly general sense. It is the product of flux density and field intensity. As you already know, there are three commonly used systems of units for magnetic field quantities, and these systems are often mixed up to form very confusing units. In the Imperial system of units, field intensity H is measured in *amperes per inch* and flux density B in *lines per square inch*. In CGS units flux density is measured in *Gauss* (or kilogauss) and magnetic field intensity in *Oersteds*. In SI the unit of flux density is the *Tesla*, which is *1 Weber per square meter*, and the unit of field intensity is the *ampere per meter*. Of these, only the last one, A/m is obvious: a *Weber* is a Volt-Second (Vs); a *Gauss* is 10^{-4} Tesla; and finally, 1 *Oersted* is that field intensity required to produce 1 Gauss in the permeability of free space. Since the permeability

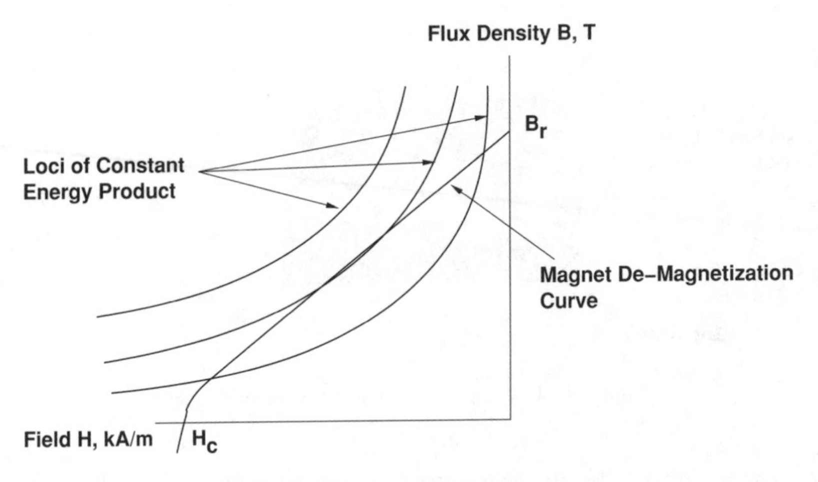

Figure 15.2 Demagnetization curve

of free space $\mu_0 = 4\pi \times 10^{-7} H/\text{m}$, this means that 1 Oe is about 79.58 A/m. Commonly, the energy product is cited in *MGOe* (mega-Gauss-Oersted)s; 1 MGOe is equal to 7.958 kJ/m^3. A commonly used measure for the performance of a permanent magnet material is the maximum energy product, the largest value of this product along the demagnetization curve.

15.1.1 Permanent Magnets in Magnetic Circuits

To start to understand how these materials might be useful, consider the situation shown in Figure 15.3: a piece of permanent magnet material is wrapped in a magnetic circuit with effectively infinite permeability. Assume this has some (finite) depth in the direction you cannot see. Now, if we take Ampere's Law around the path described by the dotted line,

$$\oint \vec{H} \cdot \mathrm{d}\vec{\ell} = 0$$

since there is no current anywhere in the problem. If magnetization is upwards, as indicated by the arrow, this would indicate that the flux density in the permanent magnet material is

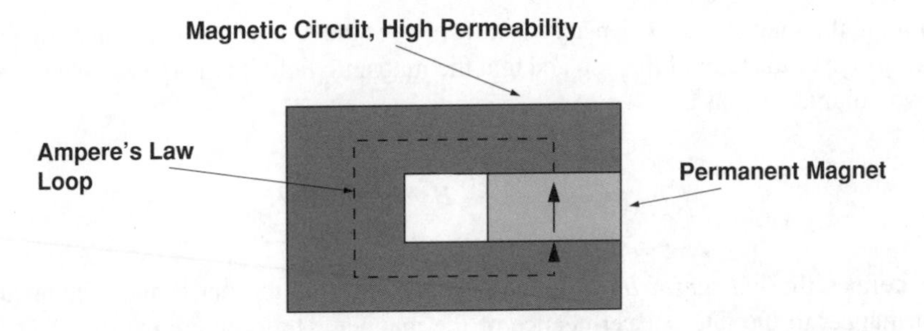

Figure 15.3 Permanent magnet in a 'keeper' magnetic circuit

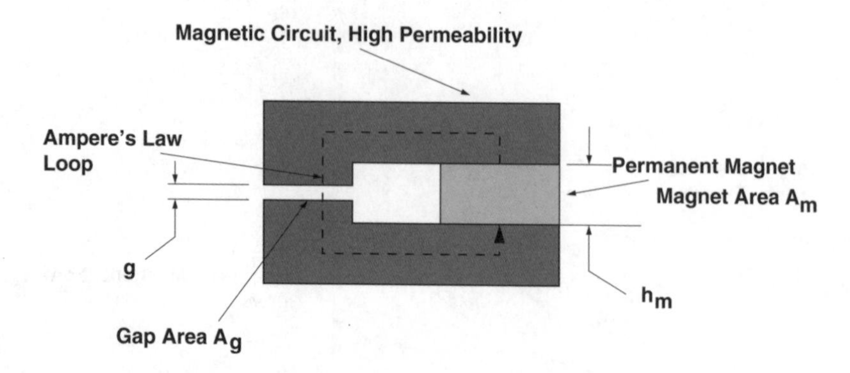

Figure 15.4 Permanent magnet driving an air-gap

equal to the remanent flux density (also upward). This kind of magnetic circuit is often called a 'keeper' because, by setting magnetic field intensity in the magnet to zero it preserves the magnetization of the magnet.

15.1.2 Load Line Analysis

A second problem is illustrated in Figure 15.4, in which the same magnet is embedded in a magnetic circuit with an air gap. Assume that the gap has width g and area A_g. The magnet has height h_m and area A_m. For convenience, take the positive reference direction to be up (as seen here) in the magnet and down in the air-gap. Ampere's Law becomes:

$$\oint \vec{H} \cdot \mathrm{d}\vec{\ell} = H_m h_m + H_g g$$

Gauss' Law could be written for either the upper or lower piece of the magnetic circuit. Assuming that the only substantive flux leaving or entering the magnetic circuit is either in the magnet or the gap:

$$\oiint \vec{B} \cdot \mathrm{d}\vec{A} = B_m A_m - \mu_0 H_g A_g$$

Assuming the magnetic flux density is uniform over the area of the permanent magnet and also uniform over the area of the gap and that the magnetic field intensity is also uniform over these two volumes of space:

$$B_m = -\mu_0 \frac{A_g}{A_m} \frac{h_m}{g} H_m = -\mu_0 \mathcal{P}_u H_m$$

This defines the *unit permeance*, essentially the ratio of the permeance facing the permanent magnet to the internal permeance of the magnet. The problem can be, if necessary, solved graphically in situations in which the relationship between B_m and H_m is inherently

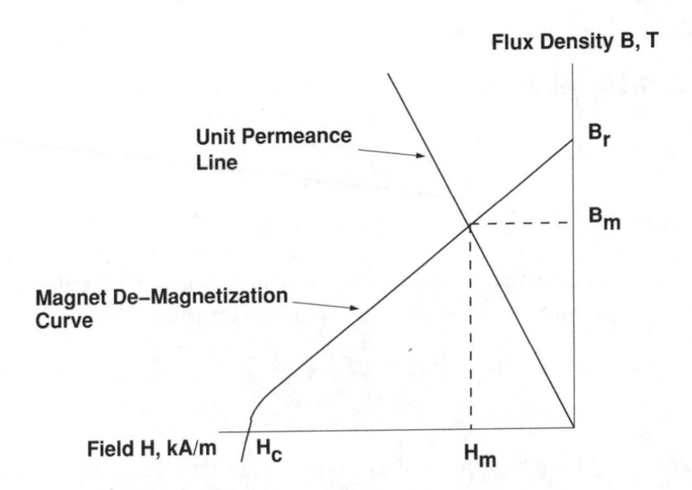

Figure 15.5 Load line, unit permeance analysis

non-linear, as shown in Figure 15.5. This is similar to a 'load line' analysis of a non-linear electronic circuit.

15.1.2.1 Very Hard Magnets

For large unit permeances the slope of the magnet characteristic is fairly constant. In fact, for most of the permanent magnets used in machines (the one important exception is the now rarely used ALNICO magnet), it is generally acceptable to approximate the demagnetization curve with:

$$\vec{B}_{\mathrm{m}} = \mu_{\mathrm{m}}(\vec{H}_{\mathrm{m}} + \vec{M}_0)$$

Here, the magnetization M_0 is fixed. Further, for almost all of the practical magnet materials the magnet permeability is nearly the same as that of free space ($\mu_{\mathrm{m}} \approx \mu_0$). With that in mind, consider the problem shown in Figure 15.6, in which the magnet fills only part of a gap in a magnetic circuit. However, here the magnet and gap areas are essentially the same. The magnet may be represented as simply a magnetization.

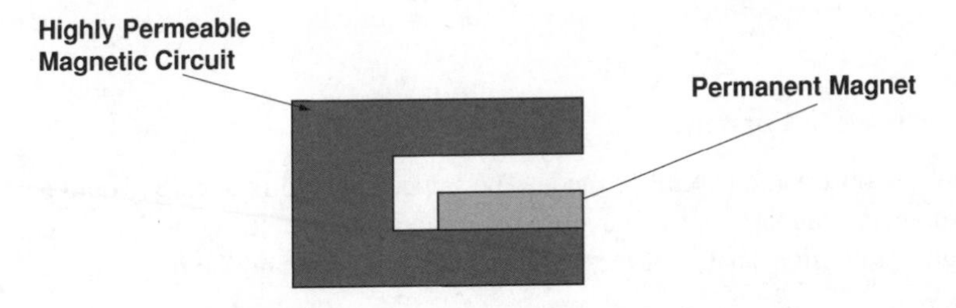

Figure 15.6 Surface magnet primitive problem

15.1.2.2 Surface Magnet Analysis

In the region of the magnet and the air-gap, Ampere's Law and Gauss' law can be written:

$$\nabla \times \vec{H} = 0$$
$$\nabla \cdot \mu_0(\vec{H}_\mathrm{m} + \vec{M}_0) = 0$$
$$\nabla \cdot \mu_0 \vec{H}_\mathrm{g} = 0$$

If in the magnet the magnetization is constant, the divergence of H in the magnet is zero. Because there is no current here, H is curl free, so that everywhere:

$$\vec{H} = -\nabla\psi$$
$$\nabla^2\psi = 0$$

That is, magnetic field can be expressed as the gradient of a scalar potential which satisfies Laplace's Equation. It is also pretty clear that, if the scalar potential is assigned to have a value of zero anywhere on the surface of the magnetic circuit it will be zero over all of the magnetic circuit (i.e. at both the top of the gap and the bottom of the magnet). Finally, note that the divergence of M is zero everywhere except at the top surface of the magnet where it is singular! In fact, we can note that the fields are the same as they would be in a situation in which there is a there is a magnetic charge density (ρ_m) at the top surface of the magnet:

$$\rho_\mathrm{m} = -\nabla \cdot \vec{M}$$

At the top of the magnet there is a discontinuous change in M and so this equivalent of a magnetic surface charge σ_m. Using H_g to note the magnetic field above the magnet and H_m to note the magnetic field in the magnet,

$$\mu_0 H_\mathrm{g} = \mu_0 (H_\mathrm{m} + M_0)$$
$$\sigma_\mathrm{m} = M_0 = H_\mathrm{g} - H_\mathrm{m}$$

and then to satisfy the potential condition, if h_m is the height of the magnet and g is the gap:

$$g H_\mathrm{g} = h_\mathrm{m} H_\mathrm{m}$$

Solving,

$$H_\mathrm{g} = M_0 \frac{h_\mathrm{m}}{h_\mathrm{m} + g}$$

15.1.2.3 Amperian Currents

Now, one more observation could be made. The same air-gap flux density would be produced if the permanent magnet were to have a surface current around its periphery equal to the magnetization intensity. That is, if the surface current runs around the magnet:

$$K_\phi = M_0$$

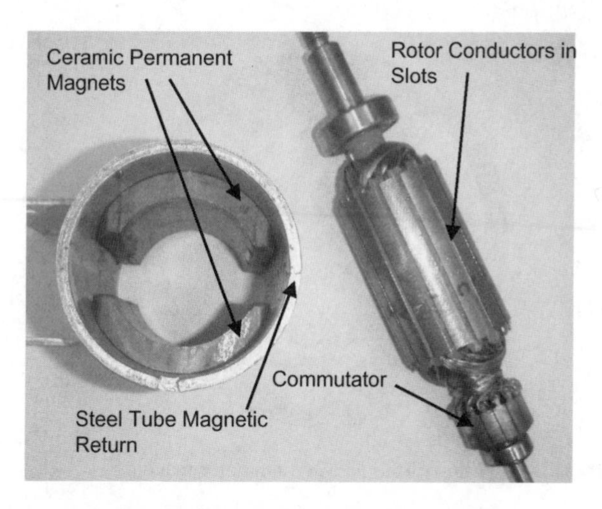

Figure 15.7 Rotor and stator of PM commutator motor; photo by author

This would produce an MMF in the gap of:

$$F = K_\phi h_m$$

and then since the magnetic field is just the MMF divided by the total gap:

$$H_g = \frac{F}{h_m + g} = M_0 \frac{h_m}{h_m + g}$$

The real utility of permanent magnets comes about from the relatively large magnetizations: numbers of a few to several thousand amperes per meter are common, and these would translate into enormous current densities in magnets of ordinary size.

15.2 Commutator Machines

Figure 15.7 is a photograph of a two pole commutator motor with permanent magnet excitation, and Figure 15.8 is a cartoon picture of a cross-section of the geometry of a two-pole commutator machine using permanent magnets. This is the most common geometry that is used for permanent magnet DC motors. The rotor (armature) of the machine is a conventional, windings-in-slots type, similar to those already seen for wound field commutator machines. The field magnets are fastened (often just bonded) to the inside of a steel tube that serves as the magnetic flux return path.

Assume for the purpose of first-order analysis of this machine that the magnet is describable by its remanent flux density B_r and has permeability of μ_0. Analysis of this motor is done in two steps:

1. First, estimate the useful magnetic flux density
2. Second, find voltage generated in the armature.

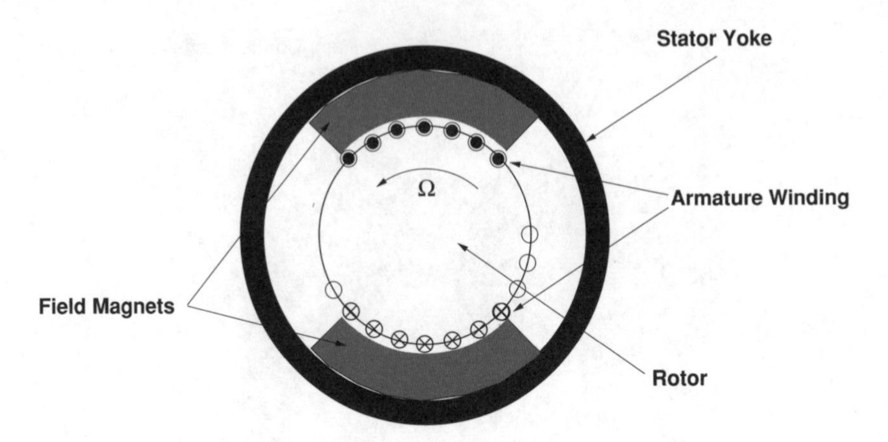

Figure 15.8 PM commutator machine

Using the basics of the analysis presented above, radial magnetic flux density at the air-gap may be estimated to be:

$$B_d = \frac{B_r}{1 + \frac{1}{\mathcal{P}_c}}$$

where the effective unit permeance is:

$$\mathcal{P}_c = \frac{f_l}{f_r} \frac{h_m}{g} \frac{A_g}{A_m}$$

A book on this topic by Ireland (1968) suggests values for the two 'fudge factors':

1. The 'leakage factor' f_l is cited as being about 1.1.
2. The 'reluctance factor' f_r is cited as being about 1.2.

The ratio of areas of the gap and magnet are estimated as being in the middle of their radial extent:

$$\frac{A_g}{A_m} = \frac{R + \frac{g}{2}}{R + g + \frac{h_m}{2}}$$

A second correction is required to correct the effective length for electrical interaction. The reason for this is that the magnets produce fringing fields, as if they were longer than the actual 'stack length' of the rotor (sometimes they actually are). This is purely empirical, and Ireland gives a value for effective length for voltage generation of:

$$\ell_{eff} = \frac{\ell^*}{f_l}$$

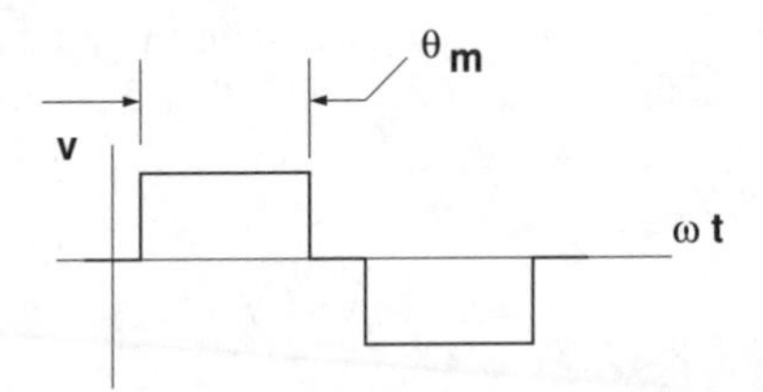

Figure 15.9 Voltage induced in one conductor

where $\ell^* = \ell + 2NR$, and the empirical coefficient

$$N \approx \frac{A}{B} \log \left(1 + B \frac{h_{\mathrm{m}}}{R} \right)$$

where

$$B = 7.4 - 9.0 \frac{h_{\mathrm{m}}}{R}$$
$$A = 0.9$$

15.2.1 Voltage

It is, in this case, simplest to consider voltage generated in a single wire first. If the machine is running at angular velocity Ω, speed voltage is, while the wire is under a magnet,

$$v_s = \Omega R \ell B_{\mathrm{r}}$$

If the magnets have angular extent θ_{m} the voltage induced in a wire will have a waveform as shown in Figure 15.9; it is pulse-like and has the same shape as the magnetic field of the magnets.

The voltage produced by a coil is actually made up of two waveforms of exactly this form, but separated in time by the 'coil throw' angle. If the coil throw angle is larger than the magnet angle, the two voltage waveforms add to look like Figure 15.10. There are actually two coil-side waveforms that add with a slight phase shift.

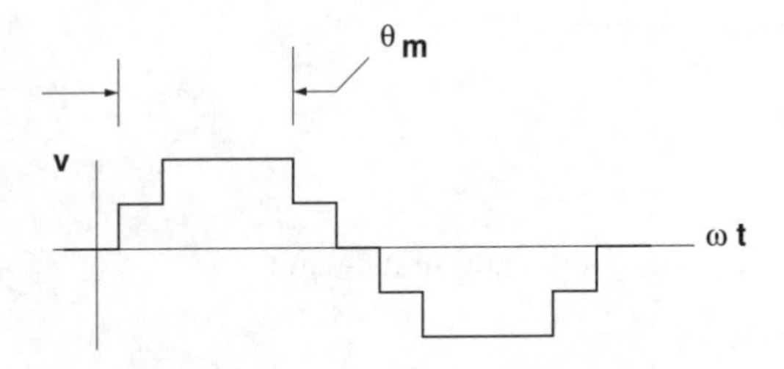

Figure 15.10 Voltage induced in a coil

If, on the other hand, the coil thrown is smaller than the magnet angle, the picture is the same, only the width of the pulses is that of the coil rather than the magnet. In either case the average voltage generated by a coil is:

$$v = \Omega R \ell N_{\mathrm{s}} \frac{\theta^*}{\pi} B_{\mathrm{d}}$$

where θ^* is the lesser of the coil throw or magnet angles and N_{s} is the number of series turns in the coil. This gives the number of 'active' turns:

$$\frac{C_{\mathrm{a}}}{m} = N_{\mathrm{s}} \frac{\theta^*}{\pi} = \frac{C_{\mathrm{tot}}}{m} \frac{\theta^*}{\pi}$$

Here, C_{a} is the number of *active* conductors, C_{tot} is the total number of conductors and m is the number of parallel paths. The motor coefficient is then:

$$K = \frac{R \ell_{\mathrm{eff}} C_{\mathrm{tot}} B_{\mathrm{d}}}{m} \frac{\theta^*}{\pi}$$

This type of machine is very much like a separately excited DC machine with fixed field current, so that induced back voltage and torque are:

$$E_{\mathrm{b}} = K\Omega$$
$$T = KI$$

15.2.2 Armature Resistance

The last element needed for first-order prediction of performance of the motor is the value of armature resistance. The armature resistance is simply determined by the length and area of the wire and by the number of parallel paths (usually equal to 2 for small commutator motors). If we note N_{c} as the number of coils and N_{a} as the number of turns per coil,

$$N_{\mathrm{s}} = \frac{N_{\mathrm{c}} N_{\mathrm{a}}}{m}$$

Total armature resistance is given by:

$$R_{\mathrm{a}} = 2 \rho_{\mathrm{w}} \ell_{\mathrm{t}} \frac{N_{\mathrm{s}}}{m}$$

where ρ_{w} is the resistivity (per unit length) of the wire:

$$\rho_{\mathrm{w}} = \frac{1}{\frac{\pi}{4} d_{\mathrm{w}}^2 \sigma_{\mathrm{w}}}$$

(d_w is wire diameter, σ_w is wire conductivity and ℓ_t is length of one half-turn). This length depends on how the machine is wound, but a good first-order guess might be something like this:

$$\ell_t \approx \ell + \pi R$$

15.3 Brushless PM Machines

Permanent magnets can be used as substitutes for the field winding of synchronous machines. Coupled with power semiconductors operating under position control permanent magnet synchronous machines are often called 'brushless DC motors'. Because of the size and efficiency advantages of electric machines with high performance permanent magnets, such machines perform well in servo applications and for traction motors. The drive motors of many hybrid electric automobiles are of this class of machines.

15.4 Motor Morphologies

There are, of course, many ways of building permanent magnet motors, but only a few are described here. Once these are understood, evaluations of most other geometrical arrangements will be fairly straightforward. It should be understood that the 'rotor inside' vs. 'rotor outside' distinction is in fact trivial, with very few exceptions.

15.4.1 Surface Magnet Machines

Figure 15.11 shows the basic *magnetic* morphology of the motor with magnets mounted on the surface of the rotor and an otherwise conventional stator winding. This sketch does not show some of the important mechanical aspects of the machine, such as the means for fastening the

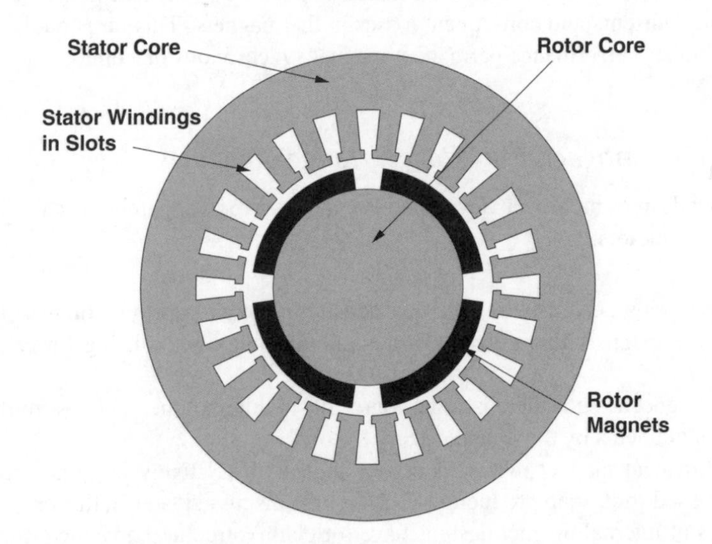

Figure 15.11 Axial view of a surface mount motor

permanent magnets to the rotor. In addition, this sketch and the other sketches to follow are not necessarily to a scale that would result in workable machines.

This figure shows an axial section of a four-pole ($p = 2$) machine. The four magnets are mounted on a cylindrical rotor 'core', or shaft, made of ferromagnetic material. Typically this would simply be a steel shaft. In some applications the magnets may be simply bonded to the steel. For applications in which a glue joint is not satisfactory (e.g., for high-speed machines) some sort of rotor banding or retaining ring structure is required.

The stator winding of this machine is very much like that of an induction motor, consisting of wires located in slots in the surface of the stator core. The stator core itself is made of laminated ferromagnetic material (usually silicon iron sheets), the character and thickness of the sheets determined by operating frequency and efficiency requirements. They are required to carry alternating magnetic fields, so must be laminated to reduce eddy current losses.

This sort of machine is simple in construction. Note that the operating magnetic flux density in the air-gap is nearly the same as in the magnets, so that this sort of machine cannot have air-gap flux densities higher than that of the remanent flux density of the magnets. If low cost ferrite magnets are used, this means relatively low induction and consequently relatively low efficiency and power density. (Note the qualifier 'relatively' here!) Note, however, that with modern, high performance permanent magnet materials in which remanent flux densities can be on the order of 1.2 T, air-gap working flux densities can be on the order of 1 T (peak). With the requirement for slots to carry the armature current, this may be a practical limit for air-gap flux density anyway. Note that flux in the stator teeth will be higher than air-gap flux density because the area of the teeth is smaller than the total area because of the need for slots to carry current.

It is also important to note that the magnets in this design are really in the 'air gap' of the machine (that is, in the magnetic gap), and therefore are exposed to all of the time- and space-harmonics of the stator winding MMF. Because some permanent magnets have electrical conductivity (particularly the higher performance magnets), any asynchronous fields will tend to produce eddy currents and consequent losses in the magnets. This turns out to be a major limitation on higher performance permanent magnet synchronous machines.

15.4.2 Interior Magnet, Flux Concentrating Machines

Interior magnet designs have been developed to counter several apparent or real shortcomings of surface mount motors:

- Flux-concentrating designs allow the flux density in the air-gap to be higher than the flux density in the magnets themselves. This is an advantage when using lower performing magnets.
- In interior magnet designs there is some degree of shielding of the magnets from high order space harmonic fields by the pole pieces.
- Interior permanent magnet machines can be built with relatively large negative saliency that can be used for torque production. This will be discussed later in this chapter.
- Some types of internal magnet designs have (or claim) structural advantages over surface mount magnet designs.

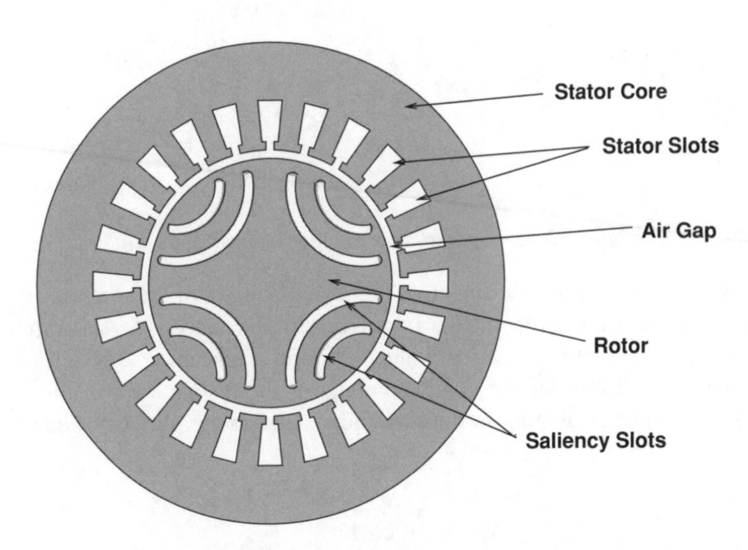

Figure 15.12 Axial view of internal magnet motor

A morphology for an internal magnet motor is shown in cartoon form in Figure 15.12. This geometry can also be used for highly salient synchronous machines without permanent magnets: such machines would run on the saliency torque alone and are called *synchronous reluctance* motors. However, the saliency slots may be filled with permanent magnet material, giving them some internally generated flux as well. The rotor iron tends to short out the magnets, so that the 'bridges' around the ends of the permanent magnets must be relatively thin. They are normally saturated. One can see that flux lines that cross the permanent magnet slots have a much higher magnetic reluctance than flux lines that run parallel to the magnet slots.

At first sight, these machines appear to be quite complicated to analyze, and that judgment seems to hold up.

15.4.3 Operation

In determining the rating of a machine, one must consider two separate sets of parameters. The first set, the elementary rating parameters, consist of the machine inductances, internal flux linkage and stator resistance. From these and a few assumptions about base and maximum speed it is possible to get a first estimate of the rating and performance of the motor. More detailed performance estimates, including efficiency in sustained operation, require estimation of other parameters.

15.4.3.1 Voltage and Current: Round Rotor

Consider the equivalent circuit shown in Figure 15.13. This is actually the equivalent circuit which describes all *round rotor* synchronous machines. It is directly equivalent to only some of the machines dealt with here, but it will serve to illustrate one or two important points. Note that stator winding resistance is not shown in Figure 15.13. This is because the winding resistance is typically small, as it must be if the machine is to have good efficiency.

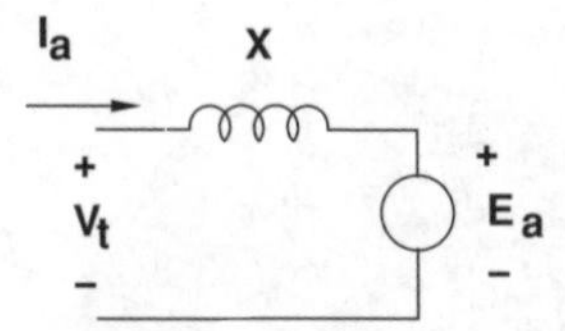

Figure 15.13 Synchronous machine equivalent circuit

What is shown here is the equivalent circuit of a single phase of the machine. Most motors are three-phase, but it is not difficult to carry out most of the analysis for an arbitrary number of phases. The circuit shows an internal voltage E_a and a reactance X which together with the terminal current I determine the terminal voltage V. In this picture armature resistance is ignored. If the machine is running in the sinusoidal steady state, the major quantities are of the form:

$$E_a = \omega\sqrt{2}\lambda_a \cos(\omega t + \delta)$$
$$V_t = \sqrt{2}V \cos \omega t$$
$$I_a = \sqrt{2}I \cos(\omega t - \psi)$$

The machine is in synchronous operation if the internal and external voltages are at the same frequency and have a constant (or slowly changing) phase relationship (δ). The relationship between the major variables may be visualized by the phasor diagram shown in Figure 15.14. The internal voltage is just the time derivative of the internal flux from the permanent magnets, and the voltage drop in the machine reactance is also the time derivative of flux produced by armature current in the air-gap and in the 'leakage' inductances of the machine. By convention, the angle ψ is positive when current I lags voltage V and the angle δ is positive then internal voltage E_a leads terminal voltage V. So both of these angles have negative sign in the situation shown in Figure 15.14.

If there are q phases, the *time average* power produced by this machine is simply:

$$P = qVI\cos\psi$$

if V and I are both taken as RMS quantities, as suggested in the expressions above.

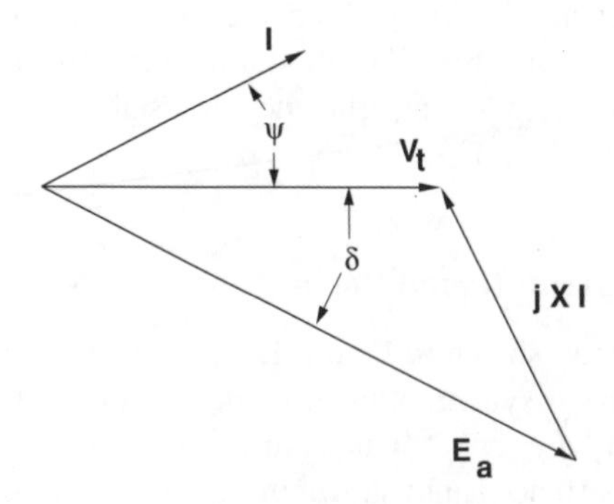

Figure 15.14 Phasor diagram for A synchronous machine

For most polyphase machines operating in what is called 'balanced' operation (all phases doing the same thing with uniform phase differences between phases), torque (and consequently power) are approximately constant. If power dissipated in the machine armature is negligible, it must be true that power absorbed by the internal voltage source is the same as terminal power, or:

$$P = qE_aI\cos(\psi + \delta)$$

Since in the steady state:

$$P = \frac{\omega}{p}T$$

where T is torque and ω/p is mechanical rotational speed, torque can be derived from the terminal quantities by simply:

$$T = p\frac{q}{2}\lambda_aI\cos(\psi + \delta)$$

In principle, then, to determine the torque and hence power rating of a machine it is only necessary to determine the internal flux, the terminal current capability, and the speed capability of the rotor. In fact it is *almost* that simple. Unfortunately, the model shown in Figure 15.13 is not quite complete for some of the motors of interest, so it is necessary go one more level into machine theory.

15.4.4 A Little Two-Reaction Theory

The material in this subsection is framed in terms of three-phase ($q = 3$) machine theory, but it is actually generalizable to an arbitrary number of phases. This is the same theoretical framework as was encountered in Chapter 9.

Suppose the machine at hand whose three-phase armature can be characterized by *internal* fluxes and inductance which may, in general, not be constant but are a function of rotor position. Note that the simple model presented in the previous subsection does not conform to this picture, because it assumes a constant terminal inductance. In that case:

$$\boldsymbol{\lambda}_{ph} = \mathbf{L}_{ph}\mathbf{I}_{ph} + \boldsymbol{\lambda}_R \qquad (15.1)$$

where $\boldsymbol{\lambda}_R$ is the set of internally produced fluxes (from the permanent magnets) and the stator winding may have both self- and mutual-inductances.

A transformation on these stator fluxes is carried out in the following way: each armature quantity, including flux, current and voltage, is projected into a coordinate system that is fixed to the rotor. This is the *Park's Transform*.

It is straightforward to show that balanced polyphase quantities in the stationary, or phase variable frame, translate into *constant* quantities in the so-called '*d*-*q*' frame. For example:

$$
\begin{aligned}
I_{\mathrm{a}} &= I\cos\omega t \\
I_{\mathrm{b}} &= I\cos\left(\omega t - \frac{2\pi}{3}\right) \\
I_{\mathrm{c}} &= I\cos\left(\omega t + \frac{2\pi}{3}\right) \\
\theta &= \omega t + \theta_0
\end{aligned}
$$

maps to:

$$
\begin{aligned}
I_d &= I\cos\theta_0 \\
I_q &= -I\sin\theta_0
\end{aligned}
$$

If $\theta = \omega t + \theta_0$, the transformation coordinate system is chosen correctly and the d-axis will correspond with the axis on which the rotor magnets are making positive flux. That happens if, when $\theta = 0$, phase A is linking maximum positive flux from the permanent magnets. If this is the case, the *internal* fluxes are:

$$
\begin{aligned}
\lambda_{\mathrm{af}} &= \lambda_{\mathrm{f}}\cos\theta \\
\lambda_{\mathrm{bf}} &= \lambda_{\mathrm{f}}\cos\left(\theta - \frac{2\pi}{3}\right) \\
\lambda_{\mathrm{cf}} &= \lambda_{\mathrm{f}}\cos\left(\theta + \frac{2\pi}{3}\right)
\end{aligned}
$$

If the fluxes are expressed in the d-q frame:

$$
\boldsymbol{\lambda}_{dq} = \mathbf{L}_{dq}\mathbf{I}_{dq} + \boldsymbol{\lambda}_{\mathrm{R}} = \mathbf{T}\mathbf{L}_{\mathrm{ph}}\mathbf{T}^{-1}\mathbf{I}_{dq} + \boldsymbol{\lambda}_{\mathrm{R}} \tag{15.2}
$$

Two things should be noted here. The first is that, if the coordinate system has been chosen as described above, the flux induced by the rotor is, in the d-q frame, simply:

$$
\boldsymbol{\lambda}_{\mathrm{R}} = \begin{bmatrix} \lambda_{\mathrm{f}} \\ 0 \\ 0 \end{bmatrix} \tag{15.3}
$$

That is, the magnets produce flux *only* on the d-axis.

The second thing to note is that the inductances in the d-q frame are *independent of rotor position* and have no mutual terms. That is:

$$
\mathbf{L}_{dq} = \mathbf{T}\mathbf{L}_{\mathrm{ph}}\mathbf{T}^{-1} = \begin{bmatrix} L_d & 0 & 0 \\ 0 & L_q & 0 \\ 0 & 0 & L_0 \end{bmatrix} \tag{15.4}
$$

The assertion that inductances in the d-q frame are constant requires some assumptions about the relationships between stator winding self and mutual inductances, but it is close enough to being true and analyses that use it have proven to be close enough to being correct that it (the assertion) has held up to the test of time. In fact the deviations from independence on rotor position are small. Independence of axes (that is, absence of mutual inductances in the d-q frame) is correct because the two axes are physically orthogonal. The third, or 'zero' axis in this analysis has very little effect. It doesn't couple to anything else and has neither flux nor current anyway. Note that the direct- and quadrature-axis inductances are in, principle, straightforward to compute. They are:

- **direct axis** the ratio of flux to current of one of the armature phases (corrected for the fact of multiple phases) with the rotor aligned with the axis of the phase, and
- **quadrature axis** the ratio of flux to current of one of the phases with the rotor aligned 90 electrical degrees away from the axis of that phase.

Armature voltage is, ignoring resistance, given by:

$$\mathbf{V}_{\text{ph}} = \frac{\text{d}}{\text{d}t}\boldsymbol{\lambda}_{\text{ph}} = \frac{\text{d}}{\text{d}t}\mathbf{T}^{-1}\boldsymbol{\lambda}_{dq}$$

and that the *transformed* armature voltage must be:

$$\begin{aligned}\mathbf{V}_{dq} &= \mathbf{T}\mathbf{V}_{\text{ph}} \\ &= \mathbf{T}\frac{\text{d}}{\text{d}t}(\mathbf{T}^{-1}\boldsymbol{\lambda}_{dq}) \\ &= \frac{\text{d}}{\text{d}t}\boldsymbol{\lambda}_{dq} + \left(\mathbf{T}\frac{\text{d}}{\text{d}t}\mathbf{T}^{-1}\right)\boldsymbol{\lambda}_{dq}\end{aligned}$$

The second term in this expresses 'speed voltage'. A good deal of straightforward but tedious manipulation yields:

$$\mathbf{T}\frac{\text{d}}{\text{d}t}\mathbf{T}^{-1} = \begin{bmatrix} 0 & -\frac{\text{d}\theta}{\text{d}t} & 0 \\ \frac{\text{d}\theta}{\text{d}t} & 0 & 0 \\ 0 & 0 & 0 \end{bmatrix}$$

The direct- and quadrature-axis voltage expressions are then:

$$V_d = \frac{\text{d}\lambda_d}{\text{d}t} - \omega\lambda_q \qquad (15.5)$$

$$V_q = \frac{\text{d}\lambda_q}{\text{d}t} + \omega\lambda_d \qquad (15.6)$$

where

$$\omega = \frac{\text{d}\theta}{\text{d}t}$$

Instantaneous *power* is given by:

$$P = v_a i_a + v_b i_b + v_c i_c$$

Using the transformations given above, this can be shown to be:

$$P = \frac{3}{2} V_d I_d + \frac{3}{2} V_q I_q + 3 V_0 I_0$$

which, in turn, is:

$$P = \omega \frac{3}{2} (\lambda_d I_q - \lambda_q I_d) + \frac{3}{2} \left(\frac{d\lambda_d}{dt} I_d + \frac{d\lambda_q}{dt} I_q \right) + 3 \frac{d\lambda_0}{dt} I_0 \tag{15.7}$$

Then, noting that $\omega = p\Omega$ and that Equation 15.7 describes electrical terminal power as the sum of shaft power and rate of change of stored energy, one may deduce that torque is given by:

$$T = \frac{3}{2} p (\lambda_d I_q - \lambda_q I_d) \tag{15.8}$$

Noting that, in general, L_d and L_q are not necessarily equal,

$$\lambda_d = L_d I_d + \lambda_f \tag{15.9}$$

$$\lambda_q = L_q I_q \tag{15.10}$$

then torque is given by:

$$T = p \frac{3}{2} \left(\lambda_f + (L_d - L_q) I_d \right) I_q \tag{15.11}$$

15.4.5 Finding Torque Capability

For high performance drives, one may generally assume that the power supply, generally an inverter, can supply currents in the correct spatial relationship to the rotor to produce torque in some reasonably effective fashion. Given a required torque, the required values of I_d and I_q can be determined. Then the power supply, given some means of determining where the rotor is (the instantaneous value of θ), will use the inverse Park's transformation to determine the instantaneous valued required for phase currents. This is the essence of what is known as 'field oriented control', or putting stator currents in the correct location *in space* to produce the required torque.

There are three things to consider here in determining torque capability of a motor:

- armature current is limited, generally by heating,
- a second limit is the voltage capability of the supply, particularly at high speed, and
- if the machine is operating within these two limits, there is an optimal placement of currents to get the most torque per unit of current and thus minimize losses.

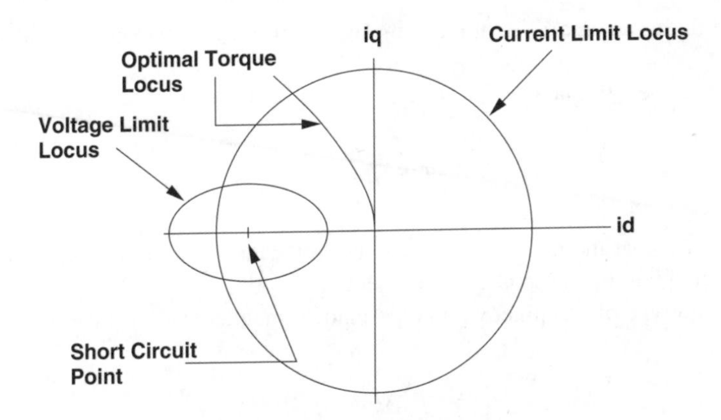

Figure 15.15 Limits to operation

The discussion of current placement is carried out using, as a tool to visualize what is going on, the I_d, I_q plane. Operation in the steady state implies a single point on this plane. A simple illustration is shown in Figure 15.15. The thermally limited armature current capability is represented as a circle around the origin, since the magnitude of armature current is just the length of a vector from the origin in this space. Note that since in general, for permanent magnet machines with buried magnets, $L_d < L_q$, so the optimal operation of the machine will be with negative I_d. Optimum operation will in general follow a curve in the I_d, I_q plane as shown.

Finally, an ellipse describes the *voltage* limit. To start, consider what would happen if the terminals of the machine were to be short-circuited so that $V = 0$. If the machine is operating at sufficiently high speed so that armature resistance is negligible, armature current would be simply:

$$I_d = -\frac{\lambda_f}{L_d}$$

$$I_q = 0$$

Loci of constant flux turn out to be ellipses around this point on the plane. Since terminal flux is proportional to voltage and inversely proportional to frequency, if the machine is operating with a given terminal voltage, the ability of that voltage to command current in the I_d, I_q plane is an ellipse whose size 'shrinks' as speed increases.

15.4.5.1 Optimal Currents

There are three cases to consider here:

1. Round rotor machines, that is, machines with no saliency but with permanent magnets.
2. Synchronous reluctance machines, with saliency but no permanent magnets.
3. Salient machines with permanent magnets.

The first of these, with permanent magnets and round rotors, were considered in an earlier subsection.

For these machines, torque is:

$$T = p\frac{3}{2}\lambda_f I_q$$

It is pretty clear that the maximum torque per unit armature current is achieved when the current is injected into the $q-$axis, so that $I_d = 0$.

For the second type of machine, with no permanent magnet at all, torque is given by:

$$T = -p\frac{3}{2}(L_q - L_d)I_d I_q$$

For this type of machine, it is straightforward to show that torque is proportional to the square of armature current and that the maximum torque per unit of current will be achieved when the direct and quadrature axis currents have the same magnitude:

$$I_d = -\frac{I_a}{\sqrt{2}}$$

$$I_q = \frac{I_a}{\sqrt{2}}$$

This type of machine is called a synchronous reluctance machine.

It is the third case, with both saliency and permanent magnets that is interesting.

To simplify the mathematics involved in this estimation, normalize reactances, fluxes, currents and torques. First, define the *base* flux to be simply $\lambda_B = \lambda_f$ and the *base* current I_B to be the armature capability. Then it is possible to define two *per-unit* reactances:

$$x_d = \frac{L_d I_B}{\lambda_B} \tag{15.12}$$

$$x_q = \frac{L_q I_B}{\lambda_B} \tag{15.13}$$

Next, define the *base torque* to be:

$$T_B = p\frac{3}{2}\lambda_B I_B$$

and then, given *per-unit* currents i_d and i_q, the *per-unit* torque is simply:

$$t_e = \left(1 - (x_q - x_d)i_d\right) i_q \tag{15.14}$$

It is fairly straightforward (but a bit tedious) to show that the locus of current-optimal operation (that is, the largest torque for a given current magnitude or the smallest current magnitude for a given torque) is along the curve:

$$i_d = -\sqrt{\frac{i_a^2}{2} + 2\left(\frac{1}{4\left(x_q - x_d\right)}\right)^2 - \frac{1}{2(x_q - x_d)}\sqrt{\left(\frac{1}{4(x_q - x_d)}\right)^2 + \frac{i_a^2}{2}}} \quad (15.15)$$

$$i_q = \sqrt{\frac{i_a^2}{2} - 2\left(\frac{1}{4\left(x_q - x_d\right)}\right)^2 + \frac{1}{2(x_q - x_d)}\sqrt{\left(\frac{1}{4(x_q - x_d)}\right)^2 + \frac{i_a^2}{2}}} \quad (15.16)$$

Looking at Figure 15.15, one can see that for small armature current, the optimal current curve is near the quadrature axis, as it would be for a non-salient machine, while for higher currents the optimal current curve tends toward the direction of the synchronous reluctance case. This is because for small currents almost all flux is produced by the permanent magnets. For larger currents the d-axis current has a stronger interaction with the quadrature axis current, through machine saliency.

15.4.5.2 Rating

The 'rating point' will be the point along the optimal current curve when $i_a = 1$, or where this curve crosses the armature capability circle in the i_d, i_q plane. In general, the 'per-unit' torque will *not* be unity at the rating, so that the rated, or 'base speed' torque is not the base torque, but:

$$T_r = T_B \times t_e \quad (15.17)$$

where t_c is calculated at the rating point (that is, $i_a = 1$ and i_d and i_q as per Equations 15.15 and 15.16).

For sufficiently low speeds, the power electronic drive can command the optimal current to produce torque up to rated. However, for speeds higher than the 'base speed', this is no longer true. Define a per-unit terminal flux:

$$\psi = \frac{V}{\omega \lambda_B}$$

Operation at a given flux magnitude implies:

$$\psi^2 = (1 + x_d i_d)^2 + (x_q i_q)^2$$

which is an ellipse in the i_d, i_q plane. The *base speed* is that speed at which this ellipse crosses the point where the optimal current curve crosses the armature capability. Operation at the highest attainable torque (for a given speed) generally implies d-axis currents that are higher than those on the optimal current locus. What is happening here is the (negative) d-axis current serves to reduce effective machine flux and hence voltage which is limiting q-axis current. Thus operation above the base speed is often referred to as 'flux weakening'.

The strategy for picking the correct trajectory for current in the i_d, i_q plane depends on the value of the per-unit reactance x_d. For values of $x_d > 1$, it is possible to produce *some* torque at *any* speed. For values of $x_d < 1$, there is a speed for which no point in the armature current capability is within the voltage limiting ellipse, so that useful torque has gone to zero. Generally, the maximum torque operating point is the intersection of the armature current limit and the voltage limiting ellipse:

$$i_d = \frac{x_d}{x_q^2 - x_d^2} - \sqrt{\left(\frac{x_d}{x_q^2 - x_d^2}\right)^2 + \frac{x_q^2 - \psi^2 + 1}{x_q^2 - x_d^2}} \tag{15.18}$$

$$i_q = \sqrt{1 - i_d^2} \tag{15.19}$$

It may be that there is no intersection between the armature capability and the voltage limiting ellipse. If this is the case and if $x_d < 1$, torque capability at the given speed is zero.

If, on the other hand, $x_d > 1$, it may be that the intersection between the voltage limiting ellipse and the armature current limit is *not* the maximum torque point. To find out, we calculate the maximum torque point on the voltage limiting ellipse. This is done in the usual way by differentiating torque with respect to i_d while holding the relationship between i_d and i_q to be on the ellipse. The algebra is a bit messy, and results in:

$$i_d = -\frac{3x_d(x_q - x_d) - x_d^2}{4x_d^2(x_q - x_d)} - \sqrt{\left(\frac{3x_d(x_q - x_d) - x_d^2}{4x_d^2(x_q - x_d)}\right)^2 + \frac{(x_q - x_d)\left(\psi^2 - 1\right) + x_d}{2(x_q - x_d)x_d^2}} \tag{15.20}$$

(15.21)

$$i_q = \frac{1}{x_q}\sqrt{\psi^2 - (1 + x_d i_d)^2} \tag{15.22}$$

Ordinarily, it is probably easiest to compute Equations 15.21 and 15.22 first, then test to see if the currents are outside the armature capability, and if they are, use Equations 15.18 and 15.19.

These expressions provide the capability to estimate the torque–speed curve for a salient machine. As an example, the machine described by the parameters cited in Table 15.1 is a (nominal) 3 HP, 4-pole, 3,000 r.p.m. machine.

The rated operating point turns out to have the attributes listed in Table 15.2.

The loci of operation in the I_d, I_q plane is shown in Figure 15.16. The armature current limit is shown only in the second and third quadrants, so shows up as a semicircle. The two ellipses correspond with the rated point (the larger ellipse) and with a speed that is 12 times rated (36,000 r.p.m.). Figure 15.17 shows the torque–speed and power–speed curves. Note that

Table 15.1 Example machine

d-axis inductance	2.53 mH
q-axis inductance	6.38 mH
Internal flux	58.1 mWb
Armature current	30 A

Table 15.2 Operating characteristics of example machine

Per-unit d-axis current at rating point	i_d	−0.5924
Per-unit q-axis current at rating point	i_q	0.8056
Per-unit d-axis reactance	x_d	1.306
Per-unit q-axis reactance	x_q	3.294
Rated torque (Nm)	T_r	13.76
Terminal voltage at base point (V)		146

this sort of machine only approximates 'constant power' operation at speeds above the 'base' or rating point speed.

A second set of curves shows what happens if the internal flux is a bit bigger (in this case, 90 mT). This puts the short circuit current (marked as a '+' symbol) outside of the armature capability. The machine torque becomes zero at finite speed. Figures 15.18 and 15.19 illustrate what this machine will do.

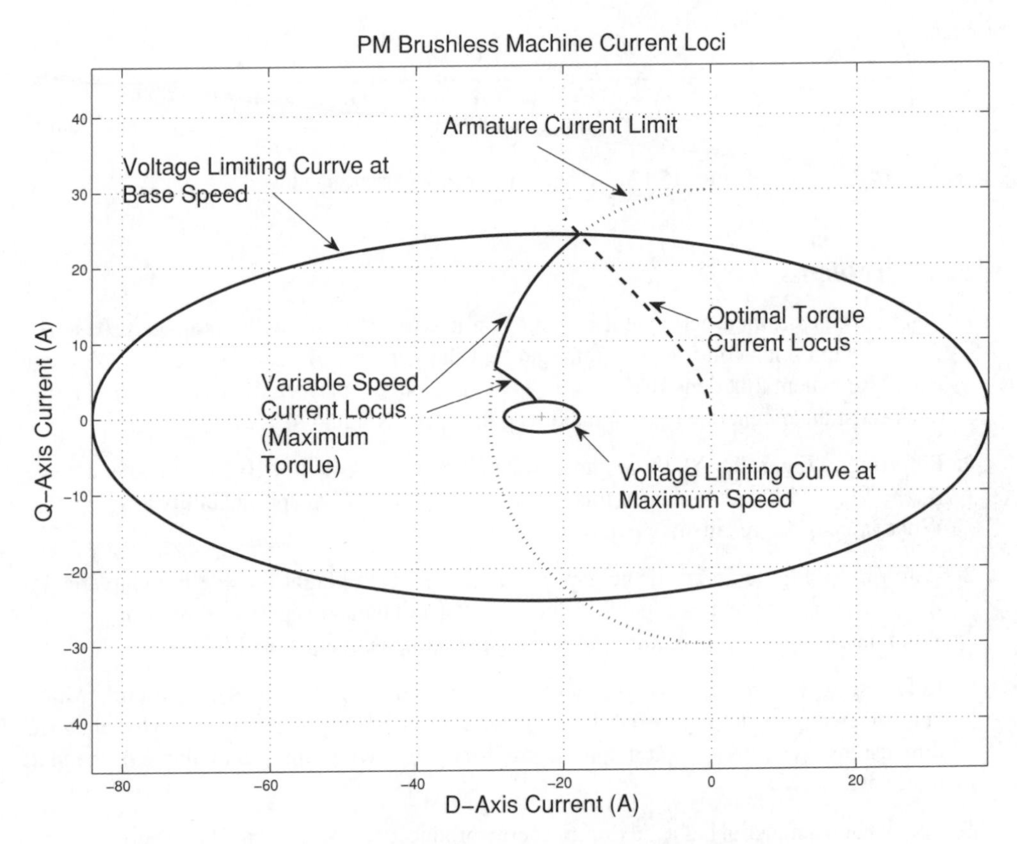

Figure 15.16 Operating current loci of example machine

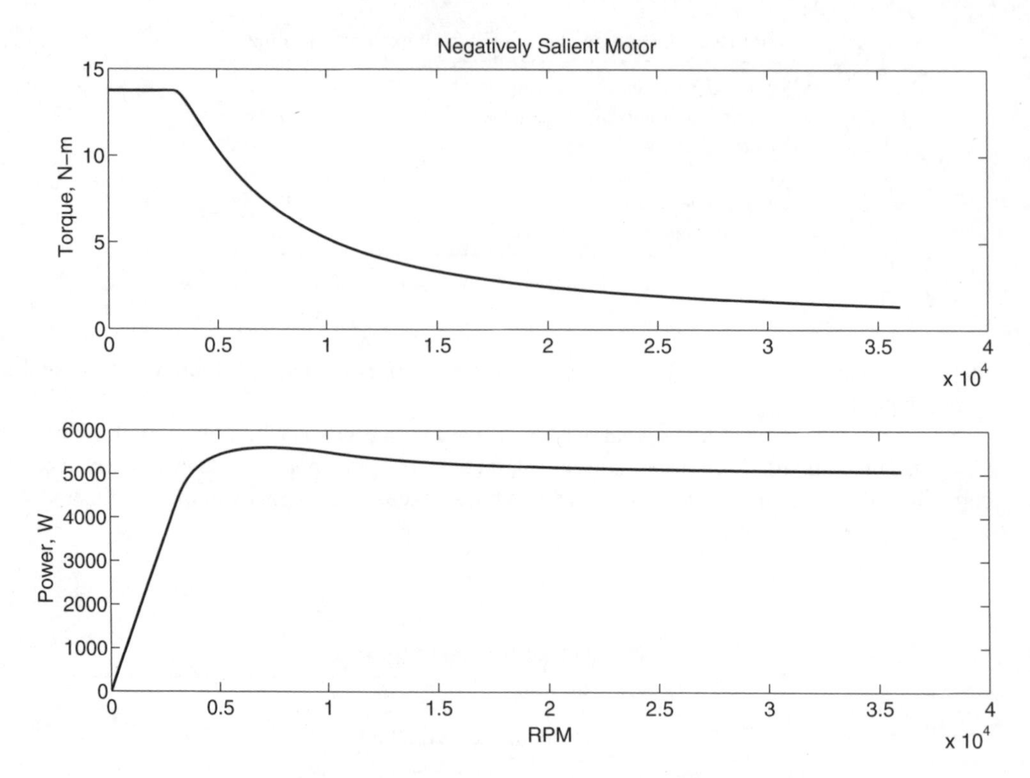

Figure 15.17 Torque– and power–speed capability

15.5 Problems

1. Figure 15.20 shows a permanent magnet in a magnetic circuit with an air-gap. Assume that the gap dimension is g = 2 mm and that the gap area A_g and magnet area A_m are equal. Remanent flux density is $B_{rem} = 1.4$ T and magnet permeability is $\mu_m = \mu_0$. How thick must the magnet be (h_m) to make gap flux density equal to 1 T?

2. Referring to Figure 15.20, if the gap g = 2 mm, gap area is $A_g = 10\ \text{cm}^2$, magnet area is $A_m = 40\ \text{cm}^2$. Magnet remanent flux density is $B_{rem} = 0.4$ T and permeability is $\mu_m = \mu_0$. What value of h_m results in a gap flux density of 1 T?

3. Referring to Figure 15.20, if the gap $g = 2$ mm, magnet height $h_m = 1$ cm Gap area is $A_g = 10\ \text{cm}^2$. Magnet remanent flux density is 0.4 T and magnet permeability is $\mu_m = \mu_0$. What magnet area A_m is required to make gap flux density equal to 1 T?

4. Referring to Figure 15.20, if the gap $g = 2$ mm and gap area is $A_g = 10\ \text{cm}^2$. Magnet remanent flux density is 0.4 T and magnet permeability is $\mu_m = \mu_0$. What magnet dimensions h_m and A_m result in gap flux density of 1 T with minimum volume of magnet material?

5. Some magnet material has a maximum energy product of 50 MGOe, and has a permeability $\mu_m = \mu_0$. What is the remanent flux density?

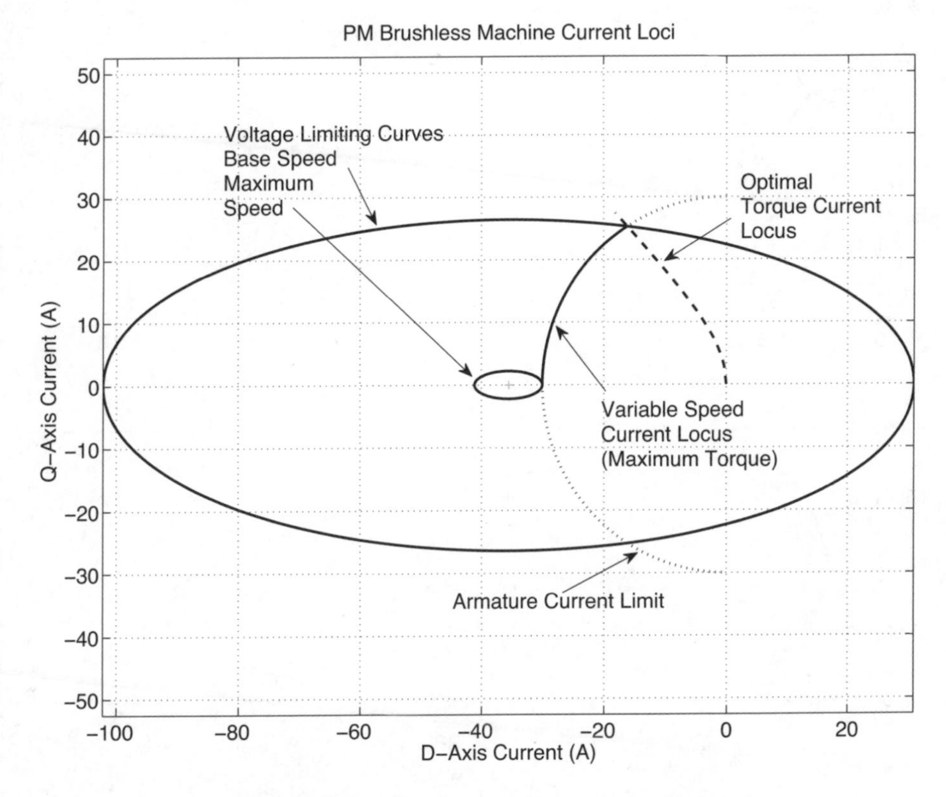

Figure 15.18 Operating current loci of example machine

6. A magnet material has a remanent flux density of 1.4 T and a permeability $\mu_m = 1.05\,\mu_0$. What is its maximum energy product?

7. A small permanent magnet commutator (DC) motor has a motor constant of $K = 1\text{mWb}$. If run unloaded (ignoring friction and windage), how fast does it turn with a supply of 12 V?

8. A small permanent magnet commutator (DC) motor turns at 6,000 r.p.m. when connected to a 12 V source. Ignoring friction and windage load, what is its motor constant K?

9. A commutator (DC) motor has a motor constant $K = 0.02$ Wb, and armature (plus brush) resistance $R_a = 2\Omega$. It is connected to a 12 V source.
   (a) What is it's no-load speed (ignoring friction and windage)?
   (b) How fast is it turning when producing mechanical power $P_m = 12$ W?
   (c) What torque is it producing when mechanical power is 10 W?
   (d) How much current is it drawing when producing mechanical power of 10 W?
   (e) Ignoring friction and windage, what is the efficiency of the motor when it is producing 10 W?

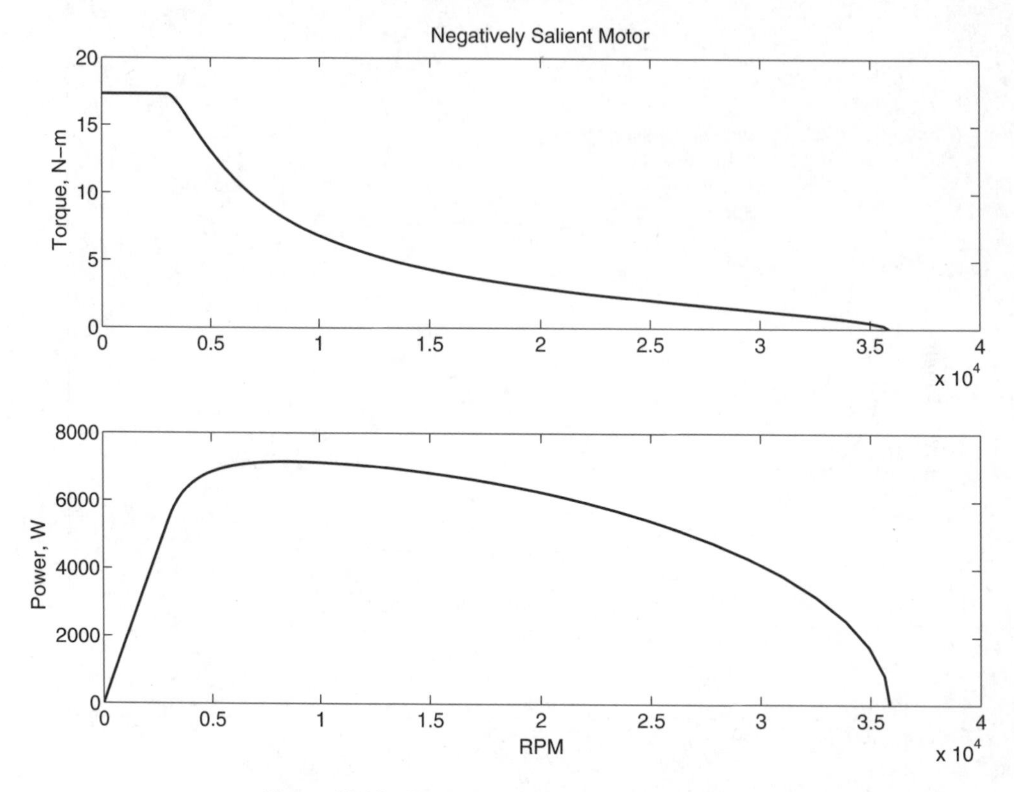

Figure 15.19 Torque– and power–speed capability

10. A commutator (DC) motor turns at 3,000 R.P.M. with a 12 V supply and no mechanical load.

(a) How much torque does it produce at stall with $I_a = 10$ A?

(b) If armature resistance $R_a = 1\Omega$ and maximum current is 10 A, how much power can this motor produce with a 12 V supply? How efficient is it?

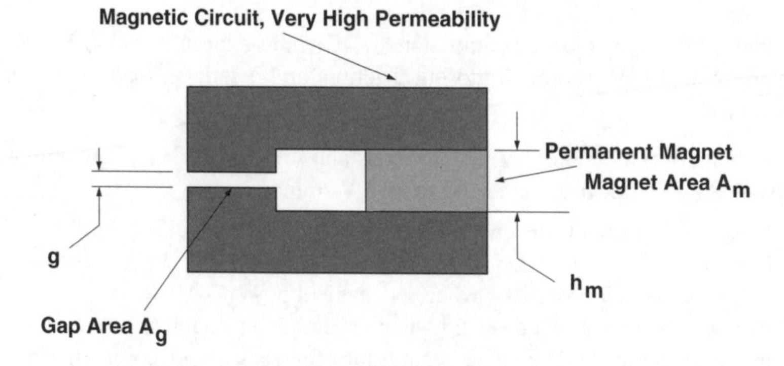

Figure 15.20 Permanent magnet driving an air gap

11. A round rotor, three-phase PM machine produces a 120 V (RMS), 200 Hz sine wave when turning at 4,000 r.p.m.

(a) How many pole pairs has this machine?
(b) What is it's internal flux λ_0?

12. A round rotor, three-phase permanent magnet synchronous machine has internal flux $\lambda_0 = 0.4$ Wb and a synchronous inductance $L_d = 50$ mH. It is a four-pole machine. The armature can carry 4 A (peak).

(a) How much torque can this machine produce as a motor?
(b) If the machine is to have a base speed of 4,000 r.p.m., what is the motor rating?
(c) For a base speed of 4,000 r.p.m., what is the peak line-to-line voltage that must be produced by the power supply?
(d) With that power supply voltage, what is the zero torque speed, above which the machine can produce no torque at all?

13. A three-phase, six-pole (p = 3) machine has direct axis inductance $L_d = 2.5$ mH and a quadrature axis inductance of $L_q = 7.5$ mH. The phase resistance is $R_a = 100$ mΩ. It has a rated armature current of $I_a = 30$ A peak and a 'base' speed of 3,000 r.p.m. The issue here is choice of permanent magnet material. Available to the designer are magnets that can produce internal fluxes of between 10 and 100 mWb.

(a) If $\lambda_0 = 10$ mWb, what is 'rated' torque (at the intersection of optimal torque locus with rated current?
(b) If $\lambda_0 = 100$ mWb, what is 'rated' torque (at the intersection of optimal torque locus with rated current?
(c) What is peak, line–line voltage at 3,000 r.p.m. and rated torque with $\lambda_0 = 10$ mWb?
(d) What is peak, line-line voltage at 3,000 r.p.m. and rated torque with $\lambda_0 = 100$ mWb?
(e) What are efficiency and power factor at this base point with $\lambda_0 = 10$ mWb?
(f) What are efficiency and power factor at this base point with $\lambda_0 = 100$ mWb?

14. This problem concerns a buried magnet permanent magnet motor, perhaps suitable for use as a starter-generator for an engine. Basic data on this machine is:

Peak internal flux (phase)	0.009	Wb
Direct axis inductance	60	μH
Quadrature axis inductance	240	μH
Number of poles	12	

The machine is to be operated with a power supply that can produce a phase voltage of 42 V, peak, line–line, and a terminal current limit of 180 A, peak. The machine can operate over a speed range of 0 to 6,000 r.p.m.

(a) Find the locus of current that optimally produces torque for this machine. Plot I_d vs. I_q.
(b) What is the *base* speed for this machine? This is the speed at which the power supply voltage just matches the required back voltage when the machine is producing torque optimally.
(c) What is the maximum torque it can make?

(d) What is the power factor at this operating point?
(e) Compute and plot the torque vs. speed and power vs. speed capability of this machine from zero to maximum speed.

Reference

Ireland, J.R. (1968) Ceramic Permanent Magnet Motors; electrical and magnetic design and applications. New York: McGraw-Hill.

Index

Electric Power Principles: Sources, Conversion, Distribution and Use James L. Kirtley